Desert Olive Oil Cultivation:
Advanced Biotechnologies

Desert Olive Oil Cultivation: Advanced Biotechnologies

Zeev Wiesman

Plant Lipid Biotechnology Lab (PLBL), Department of
Biotechnology Engineering, Ben Gurion University
of the Negev, Beer Sheva, 84105, Israel

Amsterdam • Boston • Heidelberg • London • New York • Oxford
Paris • San Diego • San Francisco • Singapore • Sydney • Tokyo

Academic Press is an imprint of Elsevier

Academic Press is an imprint of Elsevier
30 Corporate Drive, Suite 400, Burlington, MA 01803, USA
525 B Street, Suite 1900, San Diego, CA 92101-4495, USA
32 Jamestown Road, Londan, NW1 7BY, UK
360 Park Avenue South, New York, NY 10010-1710, USA

First edition 2009

Library of Congress Cataloging-in-Publication Data
A catalog record for this book is available from the Library of Congress

British Library Cataloguing in Publication Data
A catalogue record for this book is available from the British Library

ISBN: 978-0-12-374257-5

For information on all Academic Press publications
visit our website at www.elsevierdirect.com

Working together to grow
libraries in developing countries

www.elsevier.com | www.bookaid.org | www.sabre.org

ELSEVIER   BOOK AID International   Sabre Foundation

Contents

Foreword

For more than 40 years we have been traveling to the south of Israel, inspired by David Ben Gurion's vision of making the desert bloom and the commitment of Ben Gurion University of the Negev to fulfil this dream. Zeev Wiesman's decade long research, synthesized in *Desert Olive Oil Cultivation*, represents the best of BGU and of Israel. Zeev's pioneering work benefits not only farmers in the Negev, but also provides knowledge and tools to assist the 100s of millions of people around the world who live in a semi-arid climate zone to be more economically productive and to live healthier and more hopeful lives.

In the interest of full disclosure we should mention that we are benefiting personally from Zeev's work. Having learned about desert olive oil and the advantages of growing organically in a desert environment we are now importing into the US a superior tasting organic olive oil grown from orchards in the Negev.

Marvin Israelow and Dorian Goldman

Preface

The present annual economic value of worldwide olive oil production is estimated to exceed €10 billion. The olive oil industry is traditionally concentrated in the semi-arid Mediterranean region, where high-value olive oil is extracted from olive fruit by crushing in an oil mill. Increasing awareness of the nutritional benefits and superior flavor of high-quality olive oil has led to a rapidly increasing international demand for this oil. Thus, in recent decades a significant wave of establishment of new olive plantations has taken place world-wide, including Australia, South and North America, South Africa, and Southeast Asia. For a variety of economic, environmental, and sociological reasons, this new trend of olive cultivation is mainly concentrated in desert-marginal environments.

The major problems preventing an increase in olive oil supply are the dependence of the industry on traditional production methods and inefficient olive oil quality controlling bio-technologies.

Aiming to contribute to the current need for technological modernization of the international olive oil industry with emphasis on the available and economically feasible desert environments, we decided to summarize in the present book our research and development experience in the successful Negev Desert olive oil initiative during the past 25 years. We tried to close the gaps of desert-specific cultivation know-how with high-quality and cost-effective olive oil agro-production technologies. We also intensively surveyed the reported literature and introduce in this book many additional relevant aspects to cover most of the available information. All of this was done to aid the understanding of interested readers regarding the intensive cultivation needs of olive trees, the desert environment's available resources (e.g., water, soil, and climate), and their efficient utilization for successful olive production. We also discuss advanced novel bio-technologies specifically developed for Negev Desert high olive oil quality extraction and other high added-value product-processing techniques including solid and liquid waste material recycling. In particular, these technologies were directed to characterize the Negev effect on metabolomic fingerprinting of a major olive oil component (triglycerides) and minor components (e.g., polyphenol, tocopherols, and phytosterols). The fact that many common and new varieties of olives cultivated with saline water irrigation in Negev Desert conditions meet the international olive oil quality standards is presented. Furthermore, in some cases the desert stress-limiting conditions were found to increase the olive oil quality, so that these environmental disadvantages may be converted to unique nutritional branding advantages. Novel prediction models regarding olive varieties' oil potential and their optimal date of harvesting are also presented.

Social aspects as well as basic economics and marketing of conventional and organic olive oil produced in the Negev Desert environment are demonstrated in two specific test cases. It is suggested that all these aspects covered may be adopted with some specific needed modifications for many other desert environments on all continents.

The mission of completing this book could not have been carried out without the great help, support, and dedication of Dr. Bishnu Chapagain, who was deeply involved in recent years in the R&D efforts in the olive oil metabolomic field and later also in putting together all the collected data. Well-deserved acknowledgments are offered to all the staff of the Phyto-Lipid Biotechnology Lab (PLBL) with special thanks to Shirley and Shahar Nizri, and collaborating BGU units (e.g., Industrial Engineering and Management Department, Analytical Equipment Unit, the Zuckerberg Institute for Water Research, and the offices of the Rector and the President of Ben-Gurion University of the Negev).

This Negev Desert initiative was also supported by a long list of external groups and companies including: Prof. G. Bianchi's Oil Technology group at Pescara, Italy, Technical Section of IOC, Spain, Peres Center for Peace, Israeli Ministries of Science, Commerce and Industry, Environment Protection, and Health, Re'em Farm, Nahal Boker Farm, Halperin Farm, Neot Smadar, Beit Nir, Yavene group, Hader Darawsh Olive Oil Mills, Zabarga Olive Oil Mill, Talil Nursery, and many small olive oil and by-products producers. Last but not least, this Negev Desert olive oil R&D mission could not have been finalized without the financial and great moral support of the David Dibner Fund, Israelow-Goldman Fund and AABGU.

> "Environmentally Intelligent Balanced Bio-Technologies can Produce Desert Golden Olive Oil for the Benefit of all Mankind in the 21st Century."

Part 1
Introduction to Desert Olive Oil

CHAPTER 1

The current status of and major trends in the world olive oil industry

1.1 The current status of the olive oil industry

Olive oil is produced solely from the fruit of the olive tree (*Olea europaea* L) and differs from most of the other vegetable oils in the method of extraction, allowing it to be consumed in crude form, hence conserving its vitamins and other natural healthy high-value compounds. Olive oil is the most widespread dietary fat in the ancient Mediterranean world. Olive culture has been closely connected with the rise and fall of Mediterranean empires and other advanced civilizations throughout the ages. Traditionally, the olive oil industry has played a significant economic role only in the Mediterranean basin countries, including Spain, Portugal, Italy, Greece, Turkey, Tunisia, Morocco and Syria. However, in recent decades olive cultivation has become more important in other countries, such as Australia, New Zealand, the United States, countries in South America, South Africa and India. Another traditionally important area is the large, dry region of Central Asia, but its comparative remoteness and isolation from Western civilization means that only meager information in regard to the olive oil cultivation in this huge and historically important area is available today.

Most olive-growing areas lie between the latitudes 30° and 45° north and south of the equator (Figure 1.1), although in Australia some of the recently established commercial olive orchards

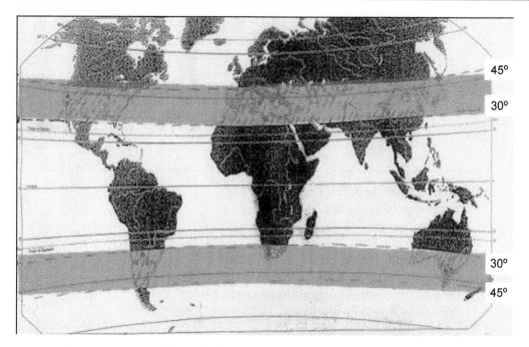

Figure 1.1: Geographical distribution of areas suitable for olive cultivation.

are nearer to the equator than to the 30° latitude and are producing a good yield; this may be because of their altitude or for other geographic reasons.

1.1.1 World olive oil production

According to the report of the International Olive Oil Council (IOOC), Mediterranean countries account for around 97 percent of the world's olive cultivation, estimated at about 10,000,000 hectares. There are more than 800 million olive trees currently grown throughout the world, of which greater than 90 percent are grown for oil production and the rest for table olives. It is estimated that more than 2,500,000 tons of olive oil are produced annually throughout the world. Since the mid-1990s, Spain has consistently been the largest producer; in the year 2004/05 it produced 826,300 tons of olive oil (Table 1.1).

The pattern of production of olive oil during these years shows big fluctuations from one year to the next; however, Spain, Italy and Greece remain the three largest olive oil producing countries, dominating the world annual olive oil production. This signifies a high level of uncertainty regarding production levels. In the year 2004/05, Spain, Italy and Greece produced 32, 28 and 13.5 percent of the world's olive oil, respectively (Table 1.2). However, the recent expansion of the olive oil industry and significant contribution to the global olive oil market by several other countries, such as Australia and the United States, may lead to stabilization of the market in the near future.

Table 1.1: Olive oil production by country (1000 tons; bi-yearly average; 1990/91 to 2004/05 on alternate year basis)

Country	Year							
	90/91	92/93	94/95	96/97	98/99	00/01	02/03	04/05
Spain	616.2	586.1	431.5	1037.1	748.5	1187.3	1142.3	826.3
Italy	474.7	500.0	558.4	521.1	592.4	540.5	587.7	732.9
Greece	296.5	330.5	329.8	390.6	397.5	355.3	437.4	346.9
Tunisia	215.0	165.0	65.0	200.0	202.5	72.5	176.0	134.8
Syria	62.5	81.6	92.4	101.8	112.5	130.4	157.3	170.0
Turkey	70.0	53.0	102.5	120.0	117.5	125.0	120.0	122.5
Morocco	65.0	39.0	40.0	65.0	62.5	37.5	64.4	85.8
Others Mediterranean	37.9	59.2	44.0	98.4	85.2	86.5	78.4	85.6
Others Europe	51.5	35.6	49.9	49.5	51.2	41.3	47.5	56.2
Rest of the world	19.0	12.5	11.7	11.6	11.8	10.6	14.4	15.9

Source: FAOSTAT (www.fao.org).

Table 1.2: Main olive oil producing countries in the world in 2004/05

Country	Oil production (%)
Spain	32.1
Italy	28.4
Greece	13.5
Tunisia	5.2
Syria	6.6
Turkey	4.8
Morocco	3.3

Source: FAOSTAT (www.fao.org).

The changes in levels of production of olive oil over the past decade have shown a heterogeneous pattern. If we compare the 1990–93 and 2002–05 four-year averages, among the traditional olive oil producing countries the greatest production increases were found in Syria (+127 percent), Turkey (+97 percent) and Spain (+64 percent). Among the Mediterranean countries, the largest production increase was in Israel (+1900 percent) over the same period of time. Cyprus, Croatia, France, Jordan and Slovenia also increased their olive oil production, while Chile, the USA and Australia produced a significantly high increment as well (Table 1.3).

1.1.2 Olive oil consumption

Olive oil has a long history of medicinal, magical, and even fiduciary uses. Medical properties of olive oil were reported by many ancient Greek writers and philosophers, with the Greek

Table 1.3: Production of olive oil (1000 tons) and percent changes by country (4-year averages; 1990–93 to 2002–05)

Country	1990–1993	2002–2003	% Change
Spain	601.1	984.28	63.7
Italy	487.3	660.3	35.5
Greece	313.5	392.2	25.1
Syria	72.0	163.7	127.2
Morocco	52.0	75.1	44.1
Turkey	61.5	121.3	97.2
Tunisia	190.0	155.4	−18.2
Other Mediterranean			
Jordan	9.6	20.3	111.7
Palestinian territories	–	15.9	–
Lebanon	5.1	6.0	17.7
Israel	0.23	4.53	1900.0
Algeria	25.9	26.7	3.1
Libya	7.8	8.5	8.8
Other European			
Portugal	34.7	39.2	13.1
Cyprus	1.6	3.2	96.7
France	2.0	3.8	87.7
Macedonia	–	2.13	–
Croatia	1.0	2.0	100.0
Albania	2.6	1.1	−57.1
Slovenia	0.04	0.23	411.2
Serbia & Montenegro	0.11	0.13	22.4
Rest of the world			
Argentina	11.7	11.0	−6.0
Chile	1.04	1.42	36.7
USA	0.61	0.88	44.0
Iran	1.3	0.79	−39
El Salvador	0.52	0.53	1.4
Mexico	0.4	0.2	−50
Australia	0.07	0.16	143.9
Azerbaijan	–	0.1	–
Afghanistan	0.08	0.08	0.0

Source: FAOSTAT (www.fao.org).

Table 1.4: Main olive oil consuming countries of the world in 2003

Country	Oil consumption (%)
Spain	19.3
Italy	30.3
Greece	6.9
France	3.9
USA	8.2
Syria	4.8
Morocco	3.0
Others EU	8.7
Others Mediterranean	7.4
Rest of the world	7.6

Source: FAOSTAT (www.fao.org).

poet Homer referring to it as "liquid gold." Today, olive oil still retains some of this historical mystique, with clinically proven health benefits, exotic flavor infusions, and a variety of applications. It is believed that the renewed interest in olive oils will follow a parallel path to those forged by wine, coffee and, most recently, chocolate.

In terms of olive oil consumption, the main producing countries are also the main consuming countries. The European Union accounts for 71 percent of global olive oil consumption, while Mediterranean basin countries account for 77 percent of world consumption. Italy has the highest consumption of olive oil (30 percent of overall world consumption in 2003), followed by Spain (19 percent) and the United States (8 percent); other main consumers of olive oil are Greece (7 percent), Syria (5 percent), France (4 percent) and Morocco (3 percent). The non-Mediterranean European countries account for 9 percent of world consumption in total (Table 1.4). Canada, Australia, Japan and New Zealand are non-Mediterranean and non-European countries with high olive oil consumption rates.

The consumption of a country can be calculated as the product of *per capita* consumption in that country, and its demographic size. Greece, Italy and Spain, in that order, were the countries with the highest annual *per capita* consumption in 2003 (Table 1.5). Variation in *per capita* olive oil consumption can be seen across the countries, even among the traditional consumers (Grigg, 2001).

The world consumption of olive oil has been growing steadily over the years (Table 1.6). The evolution of production and consumption shows a slight growth from the 1980s to the early 1990s. In the early 2000s, there was a strong increase in both the production and consumption patterns of olive oil. These figures show that the prospects for global demand for olive oil definitely appear promising. However, future developments are most likely to be linked to

Table 1.5: *Per capita* olive oil consumption of top ten and selected countries of the world in 1990 and 2003

Country	Consumption (kg)	
	1990	2003
Top ten		
Greece	17.5	15.6
Italy	12.3	13.1
Spain	10.8	11.7
Syria	5.2	6.7
Palestinian territories		5.1
Portugal	4.0	5.1
Tunisia	3.8	4.7
Jordan	4.6	3.7
Cyprus	2.4	2.6
Morocco	1.9	2.5
Selected		
France	0.6	1.6
Australia	0.6	1.4
Canada	0.3	0.8
United Kingdom	0.1	0.8
United States	0.4	0.7
Netherlands	0.1	0.7
Turkey	1.1	0.6
Switzerland	0.1	0.6
Germany	0.1	0.4

Source: FAOSTAT (www.fao.org).

developments in consumers' attitudes towards the "Mediterranean diet" and health concerns, changes in income, and the effectiveness of further strategic marketing activities (Mili, 2006).

1.1.3 Olive oil trade

Though a large portion of olive oil is for domestic consumption, a significant percentage enters the world oil trade. However, olive oil production has not yet reached the stage of competing with other sources of edible vegetable oils, such as soybean, palm and canola. Olive oil constitutes just 2 percent of all edible oils, and is considered to be a niche market directed at discerning oil consumers. Intensive promotion efforts emphasizing the various health and industrially important properties of olive oil, together with significant progress in the understanding of human nutrition and diet needs, have yielded a continuing increase in demand for olive oil among consumers throughout the world.

Table 1.6: Olive oil consumption by country (1000 tons; 1991 to 2003 on alternate year basis)

Country	Year						
	1991	1993	1995	1997	1999	2001	2003
Spain	457.5	455.3	453.8	464.9	459.6	475.2	481.3
Italy	698.0	672.0	681.7	698.0	724.0	751.1	754.4
Greece	181.7	187.7	183.6	187.7	186.0	187.0	171.5
USA	92.0	116.2	114.2	153.7	155.8	208.5	205.4
Syria	69.2	74.2	79.7	95.0	110.8	145.0	118.9
France	30.5	40.4	44.4	66.6	77.9	97.6	97.4
Morocco	54.1	51.5	40.8	37.6	54.0	44.7	75.5
Others Mediterranean	79.4	100.5	118.6	134.3	161.9	179.7	216.0
Others Europe	79.4	100.5	118.6	134.3	161.9	179.7	216.0
Rest of the world	82.1	79.9	96.3	141.9	152.0	182.1	188.9

Source: FAOSTAT (www.fao.org).

Table 1.7: Export of olive oil, 1990–2004 on alternate year basis (1000 tons)

Country	Year						
	1990	1992	1994	1996	1998	2000	2002	2004
Spain	200	170	220	200	350	350	500	600
Italy	100	120	150	150	200	280	300	380
Tunisia	50	100	160	30	120	100	20	150
Greece	90	140	100	120	120	100	90	50

Source: UNCTAD, based on data from the Food and Agriculture Organization (FAO) of the United Nations.

Olive oil export data (Table 1.7) show that the main producing countries are also the main exporting counties. Once again, the Mediterranean basin countries provide more than 95 percent of total exports. Spain and Italy are not only the main world producers of olive oil, but also the largest exporters; the third largest exporter is Tunisia, which is the fifth largest producer. Greece is the fourth largest exporter (Table 1.7).

In terms of value of exports, Spain is first and Italy is second. In 2004, Spain exported oil worth US$ 2033.6 million and Italy exported oil worth US$ 1656.5 million (Table 1.8). Among the main exporting countries, olive oil exports declined only in Greece.

The data in Table 1.9 provide confirmative conclusions drawn from the analysis of world consumption of olive oil. They show that the olive oil trade is mainly intra-European, although there are also trade flows with industrialized countries outside the Mediterranean area that have

Table 1.8: Export of olive oil, 1990–2004, on alternate year basis (US$ millions)

Country	Year							
	1990	1992	1994	1996	1998	2000	2002	2004
Spain	673.5	437.6	741.0	1029.5	894.1	803.9	1140.9	2033.6
Italy	332.8	422.2	397.2	859.5	588.2	826.0	779.7	1656.5
Tunisia	121.9	156.9	301.9	120.3	186.5	193.0	39.3	568.6
Greece	255.3	449.5	293.6	613.3	272.8	195.2	176.5	151.6
Turkey	4.7	18.9	21.0	74.4	74.2	29.1	43.3	133.0
Others Europe	90.9	78.8	85.8	133.5	115.0	114.3	102.4	150.4
Others Mediterranean	47.3	16.4	8.0	50.7	15.5	19.4	8.5	57.5
Rest of the world	15.1	24.2	14.6	30.8	36.7	34.3	29.6	40.1

Source: FAOSTAT (www.fao.org).

Table 1.9: Import of olive oil in 2004 (US$ millions) by major countries and their world share

Country	Import (US$ millions)	World share (%)
Italy	2003	40.2
USA	747	15.0
France	317	6.4
Spain	273	5.5
United Kingdom	185	3.7
Germany	176	3.5
Portugal	173	3.5
Japan	140	2.8
Australia	109	2.2
Others Europe	346	6.9
Others Mediterranean	16	0.3
Rest of the world	502	10.1

Source: FAOSTAT (www.fao.org).

not traditionally been consumers, such as the US, Germany, the UK, Japan and Australia. The US, the third largest consumer of olive oil, imported US$747 million worth of oil in 2004, representing 15 percent of global imports. However, unit sales have grown at a lower rate, and the market has been impacted by the rising price of olive oil. The US market is heavily dependent on imports from Europe, and the strengthening of the euro against the dollar has contributed significantly to the increase in prices.

This current status indicates that the market for olive oil will continue to grow. The aggregate demand will continue to expand in *per capita* consumption levels in both developed and developing countries. However, as availability of land is the limiting factor in the traditional olive oil producing countries, the boost has to come from new areas. Likely areas might be the dry, arid lands in different parts of the world, where cultivation of trees could be valuable not only because of their marketable product but also from the environmental point of view.

1.2 Market-differentiating trends in the olive oil industry

Following an intensive and continuing marketing campaign by the IOC, the increased demand for and consumption of olive oil has led to a consequent increase in its production in recent years, as shown above. However, now that many producers are players in the world olive oil market, there is strong competition between them. This scenario has led to an increase in the quality of olive oil available in the world markets, which is a good result from the consumers' point of view. From the producers' point of view, though, there is now a clear need to differentiate their oils from others, and to demonstrate their unique qualities. Today, therefore, several differentiating trends are common in the world olive oil market.

1.2.1 Premium olive oil quality
1.2.1.1 Production of virgin olive oil

Olive oil qualifies as a natural product, and virgin olive oil can have a designation of origin when it meets specific criteria, associated with particular health traits, that are consistent on chemical analysis. These analytical parameters include triglyceride components and other minor antioxidant components delivered in the oil, and organoleptic characteristics affecting taste, aroma and flavors, including various phenolic components; the oils can be differentiated mainly by their genetic origin, geographic regions of cultivation, and organic methods of cultivation. Premium-quality olive oil is categorized as extra virgin olive oil, followed by (in decreasing order of quality) virgin olive oil, ordinary virgin olive oil, lampante virgin olive oil, and olive pomace oils. For further details, see Chapter 11.

1.2.1.2 Common olive oil types

Cold-pressed olive oil

"Cold pressed" is an outdated and irrelevant label nowadays. Current extraction technologies no longer use the traditional pressing procedure ("first press") followed by mixing the olive paste with steam or hot water, and a subsequent second pressing cycle to optimize oil extraction. The relatively high temperatures used in this process meant that the second pressing produced lower-quality oil owing to evaporation of some of the volatiles, which affects the flavor.

Today, oil extraction is based on a tri- or di-phase system, and the paste is almost always warmed only to room temperature during the malaxation process. Since the olives ripen and are harvested in the winter, limited heating is necessary and the temperature used is below the evaporation temperature of most olive oil volatiles. IOC Standard Regulation 1019 of 2002 still considers this procedure to be "cold pressing" if the paste is not heated above 27 °C (80 °F). As overheating can lead to degradation of the flavor of the oil to the point where it will not qualify as more profitable extra-virgin olive oil, producers would lose money if they were to use higher temperatures in an attempt to extract a little more oil.

First press olive oil

When oil was produced by hydraulic presses, the "first press" oil was of higher quality than that produced by subsequent pressings using higher pressure and temperatures. The term is no longer officially recognized, as advanced technologies mean that extraction is continuous and therefore there is no second pressing.

Early harvest olive oil

Olive oil produced at the beginning of the season ("early harvest oil") comes from younger, greener olives. These contain less oil than black olives, and have a more bitter flavor due to their higher polyphenol content. In recent years, many studies have reported on metabolic changes between the oil produced from young olives and that from more mature olives. The two types can be easily differentiated; oil from green olives is pungent, and is described as having astringent, grassy and green leaf flavors.

Main harvest olive oil

Olives are usually harvested when they are purple-black, fully ripe, and contain the optimum level of oil. However, delaying the harvest increases the risk that the fruit will be damaged by frost. Usually, purple-black olives have a light, mellow taste with little bitterness and more floral flavors. Flavor notes of peach, melon, apple and banana are recognized, and terms such as perfumy, buttery, fruity, rotund, soave and sweet are often used to describe the oil.

Blended olive oil

Most commercial brands of olive oil are blends of oil from many different varieties, regions, and even countries. Because olive oil from exactly the same grove tastes different from year to year, depending on the weather, blenders must use oil from a number of different sources in order to achieve consistency of flavor.

Blending an oil that is high in polyphenols with one that is not may upgrade the final product and increase its shelf life. Sometimes olive oil is blended with canola or other vegetable oils; if this is the case, it must be stated on the label. Illegal blending with cheaper vegetable oils can be profitable for the unscrupulous, and is not easily detected, even using the most advanced chromatographic and spectral technologies, if the mix contains less than 10–20 percent of the lower-quality oil.

Flavored olive oil

Many consumers like olive oil with added flavorings. Technically, oils that have had herbs or fruits infused in them cannot be called olive oil and must, according to IOC regulations, be termed "fruit juice". In reality, few producers comply with this ruling and labels such as "lemon-infused olive oil" and "basil olive oil" can be seen. As these flavored oils are so popular, some national councils are currently trying to establish clearer labeling standards.

1.2.2 Olive oil, health trends and the Mediterranean diet

Factors such as its healthy fat content, the influence of the Mediterranean diet, olive oil's premium image, flavor innovations and a variety of other applications have established olive oil as a natural choice in Western consumers' minds and diets. Major marketers generate consumer interest in olive oil through programs and events, through olive oil tasting bars, and also by introducing new products with innovative flavors.

The Mediterranean diet pyramid that has been widely brought to the attention of the majority of Western countries' populations in recent years has been well integrated by the majority of these consumers. The Mediterranean diet, which represents the optimal, traditional diet of the Mediterranean population, is based on the dietary traditions of Crete and southern Italy. It is structured in the light of nutrition research carried out in 1993 and presented by Professor Walter Willet during the 1993 International Conference on the Diets of the Mediterranean, held in Cambridge, Massachusetts (Kafakos and Riboli, 2003).

The Mediterranean diet pyramid underlines the importance of the foods making up the principal food groups. Each of these individual food groups offers some (but not all) of the nutrients required by humans. Food from one group cannot replace that of another group, and all the groups are necessary for a healthy diet.

In brief, the basic food groups of the Mediterranean diet, in descending order of the quantity and frequency advised, are as follows:

- *Grains*. These form the basis of the majority of meals in Mediterranean countries, and include bread (wholemeal or otherwise), pasta, couscous and rice.

- *Fruit and vegetables*. Meals are more flavorsome when in-season products are selected and they are cooked very simply. In most Mediterranean countries, dessert is generally fruit.

- *Legumes and nuts*. A wide variety of legumes and nuts, such as chickpeas, lentils, haricot beans, pine kernels, almonds, hazelnuts, walnuts, are used in cooking.

- *Olive oil and olives*. Olive oils (regular and extra virgin) and table olives are common in the diet throughout the Mediterranean region. Regular olive oil is normally used for cooking, and virgin olive oil, which is appropriate for all uses, is consumed raw to best appreciate its aroma and flavor and to benefit fully from all its natural components.

The proportion of fats delivered mainly by olive oil in the traditional Mediterranean diet is more than 40 percent kcal/day, of which 8 percent are saturated fats, 3 percent are polyunsaturated and 29 percent are monounsaturated fats.

- *Dairy products*. Cheese, yoghurt and other dairy products are regularly consumed, with no special mention of milk.

- *Fish*. This is offered as a primary protein, before eggs and poultry.

- *Wine*. Wine is consumed in moderation, primarily with meals (one or two glasses per day), especially in European parts of the Mediterranean. It is taken optionally, and avoided when it puts individuals or others at risk.

1.2.3 Olive oil as a preventive medicine

Additional direct medicinal findings recently reported regarding the benefits of olive oil in a wide spectrum of diseases that greatly affect human life and its quality have further stimulated the consumption of the oil among people in developed countries. Studies have shown a direct correlation between the consumption of olive oil and the prevention or reduction of many diseases; some of these are described briefly below.

Cardiovascular disease

Cardiovascular disease is well accepted as being the top cause of death in the industrialized world today. A wide variety of studies have documented that arteriosclerosis (where cholesterol-rich patches are deposited on the walls of the arteries, thus stopping blood from reaching the tissues and obstructing the functioning of vital organs such as the heart and brain) is closely linked to dietary habits, lifestyle, and some aspects of economic development. The progression of arteriosclerosis depends on many factors, the most important of which are high blood cholesterol, high blood pressure, diabetes, and cigarette smoking.

Interestingly, it has been reported that the lowest rates of death from coronary heart disease currently occur in countries where olive oil is virtually the only fat consumed (Ferro-Luzzi and Branca, 1995).

Cancer

Cancer is one of the chief causes of death in developed countries, and its incidence is increasing. It is now conceded that there is a relationship between diet and the development of a large number of malignant tumors. Cell oxidation is one of the major risks in the formation of cancer: the more susceptible the cell is to oxygen, the greater the risk of cancer. The types of cancer most closely associated with diet are colon-rectal, prostate and breast. Recent research has revealed that the type of fat seems to have more implications for cancer incidence than the quantity of fat.

Epidemiological studies suggest that olive oil exerts a protective effect against certain malignant tumors (breast, prostate, endometrial, digestive tract, and others).

Much has still to be discovered about how olive oil affects cancer, and concrete data are still lacking regarding the mechanism behind the beneficial role it plays in the prevention or inhibition of different types of cancer. However, according to the information available at present, olive oil may act simultaneously during the different stages involved in the process of the formation of cancerous cells.

High blood pressure

Various research studies have reported a close relationship between diet and blood pressure. Certain foods can raise blood pressure, as well as having an effect on body weight. High blood pressure (arterial hypertension) is diagnosed when blood pressure readings are constantly above 140/90 mmHg. Hypertension is one of the chief coronary risk factors in the development of arteriosclerosis. Along with high blood cholesterol, cigarette smoking, obesity and diabetes, it is one of the main health problems of the developed world, with one in every four adults being hypertensive. Like other risk factors, lifestyle can contribute to high blood pressure. Hypertension increases the risk of early death because of the damage to the body's arteries, especially those that supply blood to the heart, kidneys, brain and eyes.

There is recent evidence that when olive oil is consumed, the daily dose of drugs needed to control blood pressure in hypertensive patients can be decreased – possibly because of a reduction in nitric acid caused by polyphenols (Psaltopoulou *et al.*, 2004).

Diabetes

Diabetes mellitus is a leading health problem in developed countries, and the sixth highest cause of death. It is a major metabolic disease, and is potentially very serious because it can cause many complications that seriously damage health, such as cardiovascular disease, kidney failure, blindness, peripheral circulation disorders.

An olive-oil-rich diet is not only a good alternative in the treatment of diabetes; it may also help to prevent or delay the onset of the disease. It does this by preventing insulin resistance and its possible pernicious implications by raising HDL cholesterol, lowering triglycerides, and ensuring better blood-sugar level control and lower blood pressure. It has been demonstrated that a diet that is rich in olive oil, low in saturated fats, and moderately rich in carbohydrates and soluble fiber from fruit, vegetables, pulses and grains is the most effective approach for diabetics. Besides lowering the "bad" low-density lipoproteins, this type of diet improves blood-sugar control and enhances insulin sensitivity. These benefits have been documented in both childhood and adult diabetes (ADA, 1994).

Obesity

Obesity is a major health issue in the West because many people eat large amounts and get little physical exercise. Nowadays, in cities especially, people are adopting a sedentary, stressful lifestyle. Over half the population of some industrialized countries is overweight, leading to increased risk of high blood pressure, diabetes, high cholesterol and triglycerides – all factors that increase the risk of cardiovascular disease.

Olive oil is a nutrient of great biological value. Like all other fats and oils it is high in calories (9 kcal per gram), which could lead to the assumption that it might contribute to obesity. However, experience shows that there is less obesity amongst the Mediterranean peoples, who consume the most olive oil.

It has been demonstrated that an olive-oil-rich diet leads to greater and longer-lasting weight loss than a low-fat diet. It is also accepted more easily, because it tastes good and it stimulates the consumption of vegetables (Bes-Rastrollo *et al.*, 2007).

The immune system

It has been demonstrated that olive oil plays an important role in the immune system (Pablo *et al.*, 2004). The immune system defends the body against invasion by foreign substances (toxins, micro-organisms, parasites, tumor processes, etc.) by coordinating specific and non-specific mechanisms.

The non-specific or innate defenses are the front-line protection against micro-organisms. They consist of the skin, mucous membranes, the complement system (complement, a group of some 20 proteins manufactured in the liver, helps to destroy micro-organisms), hormonal factors, etc., and their action is not affected by prior contact with a foreign substance.

Specific mechanisms occur following exposure to a foreign substance, and require the involvement of B-lymphocytes (the humoral system) and T-lymphocytes (cell system). Innate immunity responds in a similar way to the majority of microbes, whereas the specific immune response varies according to the type of micro-organism in order to eliminate it as effectively as possible.

Rheumatoid arthritis

Rheumatoid arthritis is a chronic inflammatory immune disease of unknown cause that affects the joints. Genes, infective factors, hormones and diet have been suggested as possible factors in its onset. Studies have suggested that olive oil may help to alleviate its symptoms, and results of a recently published study suggest that regular consumption of olive oil may reduce the risk of developing rheumatoid arthritis (Linos *et al.*, 1999). The authors found that people whose diets contained high levels of olive oil had a reduced risk of suffering the disease, and those who consumed less olive oil had a 2.5-fold greater likelihood of developing rheumatoid arthritis than those who consumed it more frequently. Although the mechanism involved is not yet clear, it is suspected that antioxidants exert a beneficial effect.

The digestive system

Olive oil has a number of immediate effects throughout the digestive system. In ancient times it was recommended for assorted digestive disorders, and its beneficial properties are now being corroborated by epidemiological studies and a wealth of scientific data (Alarcón de la Lastra *et al.*, 2001).

The stomach

When olive oil reaches the stomach, it does not reduce the tonus of the muscular ring or sphincter at the base of the esophagus. Because of this, it reduces the risk of the flow or reflux of food and gastric juices from the stomach to the esophagus.

Olive oil also partially inhibits gastric motility. As a result, the gastric contents of the stomach are released more slowly and gradually into the duodenum, giving a greater sensation of "fullness," and favoring the digestion and absorption of nutrients in the intestine (Romero *et al.*, 2007).

The hepato-biliary system

Olive oil is a cholagogue, ensuring optimal bile drainage and complete emptying of the gall bladder. It is also cholecystokinetic – i.e., it stimulates the contraction of the gall bladder – which is extremely helpful in the treatment and prevention of disorders of the bile ducts. It stimulates the synthesis of bile salts in the liver, and increases the amount of cholesterol excreted by the liver.

In short, owing to its beneficial effect on the muscle tone and activity of the gall bladder, olive oil stimulates the digestion of lipids, because they are emulsified by the bile, and prevents the onset of gallstones.

The pancreas

When olive oil is consumed, it causes a small amount of secretion by the pancreas. The organ is therefore stimulated to work a little, but efficiently, and enough to carry out all its digestive functions. Olive oil is therefore recommended in diseases where pancreatic function has to be maintained, such as pancreatic failure, chronic pancreatitis, cystic fibrosis, malabsorption syndromes.

The intestines

Owing to the sitosterol it contains, olive oil limits cholesterol absorption by the small intestine. It also stimulates the absorption of various nutrients (calcium, iron, magnesium). Olive oil is therefore a fat that is digested and absorbed really well. It has choice properties, and a mild laxative effect that helps to combat constipation and bad breath.

Ageing

Olive oil is rich in various antioxidants (vitamin E, polyphenols, etc.), which play a positive, biological role in eliminating free radicals (the molecules involved in some chronic diseases and ageing) and in extending life expectancy. This has been demonstrated in several epidemiological studies. Many age-related conditions are influenced by diet – in particular, osteoporosis and deteriorating cognitive function. Osteoporosis is a reduction in bone tissue mass that increases the risk of fractures. There are two types; type I occurs in middle-aged post-menopausal women, and type II in the elderly (Quiles *et al.*, 2004).

Olive oil appears to have a favorable effect on bone calcification, and bone mineralization improves when more olive oil is consumed. It helps calcium absorption, thereby playing an important part during the period of growth and in the prevention of osteoporosis.

Cognitive function

Olive-oil-rich diets may reduce memory loss in healthy elderly people. A lower possibility of suffering age-related cognitive decline has been observed in a study conducted on elderly people administered diets containing a large amount of monounsaturated fats, and especially olive oil (International Olive Council, www.internationaloliveoil.org).

Exactly how large quantities of these fats prevent cognitive decline is not known; however, the effect is believed to occur because the monounsaturated fatty acids may help to maintain the structure of the brain-cell membranes, since the demand for these acids appears to grow during ageing. The same study observed that the quantity of olive oil consumed was inversely proportional to age-related cognitive decline and memory loss, dementia, and Alzheimer disease.

The skin

In human beings, ageing leads to gradual structural and functional skin damage. Skin tissue goes through a number of changes. Chief among these are that the inner and outer layers of the skin (dermis and epidermis) grow thinner, elasticity is lost, the area joining the dermis to the epidermis becomes less cushioned, fibrosis occurs with the accumulation of collagen, and the tissue becomes less able to fight against and repair damage.

External factors, such as solar radiation, speed up ageing by generating free radicals. Though cells are equipped with mechanisms that neutralize their action, it is possible to reduce cell damage by using inhibitors that lower the risk. One such natural inhibitor is olive oil, the lipid profile of which is very similar to that of human skin.

In addition to polyphenols, olive oil has a large proportion of vitamins A, D and K, as well as vitamin E – the main source of protection against the free radicals that produce cell oxidation. This makes it beneficial in specific therapies to treat skin disorders such as acne, psoriasis and seborrheic eczemas.

1.2.4 Organic olive oil production

Due to an international urgent need to feed millions of hungry people following World War II, the strong industrial capacity that had been built up during the years of war shifted to food production. To intensify and speed up the growth of crops, the production of large amounts of agricultural products by the chemical industry was increased, and the field of pesticides in particular was developed significantly. Indeed, the capacity of food production in the Western world reached such a stage that, after a relatively short period of time, there was no hunger in this part of the world. However, as a result of this intensive wave of food production, followed by the continuous imbalanced initiatives of the giant chemical companies and the lack of control and regulation by the governments, significant environmental risks to human health have developed and have only recently been recognized by the majority of the population in the Western world.

Today, the continuously increasing fraction of the consumers' expenditure on food is focused on safer cultivation of fresh and final products. In recent years, this trend has led to more and more olive oil consumers, mainly in the EU and the United States. Along with this general trend, organic olive oil has become a high-demand product in Western countries as consumers are willing to pay higher prices for olive oil cultivated by methods avoiding toxic substances or invasive techniques. Such methods involve respecting the needs and qualities of the environment in which the olives are cultivated, and taking into consideration all the parameters necessary to ensure that resources such as soil, water, native fauna and flora come to no harm and are incorporated into the equilibrium of the cultivation.

The production of organic olive oil implies maximum respect for the environment, because of the sole use of natural products in olive farming. Consequently, organic crops remain free of all those chemical products that cause enormous damage to the ecosystem and to human health. Awareness of the environment has increased the trend for organic production of olive oil drastically. In Italy, Spain, Greece, and various other European countries, organic olive growing has been substituted for regular cultivation mainly in traditional extensive plantations. In the USA, Japan and Australia this trend is also starting to gain many new consumers, and the trend is expected to grow further in the coming years.

1.2.5 Varietals and geographical trends

Today, various firms that have been traditionally active in olive oil production and marketing, in particular from Italy, Spain and Greece, are promoting olive oil with brand names based on specific geographically originated environments. Olive oils produced in Toscana, Crete, Cyprus, Negev and elsewhere can be found in the markets of Europe, the USA and Australia. The concept behind this trend is based on the effect of the environmental conditions of the area of olive cultivation on the quality of the oil produced. Many reports have demonstrated the influence of climate, soil and other environmental conditions on the composition of olive oil, and the interaction of the genetic material with environmental conditions may also be a factor. These trends support differentiation of various olive oils, and have encouraged new and unique branding that assists in their sales.

1.2.6 Processing-related trends

Most olive oil today is produced by using advanced vertical centrifugation technology that consists mainly of tri-phase systems. Such olive oil production systems are highly efficient in terms of oil yield; however, owing to the use of water for washing, the oil obtained is relatively low in water-soluble compounds such as polyphenols. There is a trend towards differentiating olive oil produced via traditional methodologies consisting of pneumatic or hydraulic pressing of the olives via the di-phase system. The di-phase system avoids using water for oil separation from the non-lipid fraction of the olive extraction, and therefore produces higher-quality oil. As well as the increased quality, the environmental benefits of reducing the volume of olive oil mill wastewater are also promoted in marketing the olive oil.

1.2.7 Large-scale expansion into desert environments

In recent years, mainly for cost reasons, establishment of new large-scale olive plots in marginal lands has become attractive in the olive oil industry. Cultivation of large farms significantly increases the efficacy of production costs, such as labor and agricultural machinery. Large enough units justify scaling up the entire production chain, including olive production and oil processing facilities, and thus optimization of net income. This approach enables the establishment of a unique olive oil brand name that may ease competition in the world market. The main available land today that is suitable for olive oil cultivation is found in the desert environments widely distributed throughout the world. Indeed, this trend is now operating in a number of regions, including Argentina, Arizona and California in the USA, South Africa, North Africa, the Middle East and Australia.

In addition to purely economic reasons, environmental factors are encouraging this trend. The current scenario has shown that combating desertification can effectively be achieved by olive oil cultivation in the desert environment. Moreover, political and social factors include the aim of improving the lives of impoverished farmers, who are widespread in many dry and underdeveloped regions. These reasons may also increase the chance of attracting international investment in desert olive oil cultivation. This approach is demonstrated in the Western Desert of Egypt, the La Riocha, Catamarca and San Louis regions of Argentina, and many other desert areas in the Middle East.

It is well recognized by many olive oil experts that the traditional and extensive olive cultivation methods used in the leading olive oil producing countries, such as Italy, in the twentieth century are no longer suitable for twenty-first century needs.

Intensive cultivation methodologies have recently been developed and engineered specifically for large-scale olive oil cultivation in desert areas. These technologies include the selection and engineering of desert-suitable plant material, and stimulation of growth and development with an emphasis on proper irrigation technologies based on saline and recycled water, which are available in most desert environments. Various technologies can stimulate the productivity of olive trees, and mechanical biotechnologies are increasing the automation of desert olive farming, with an emphasis on the harvesting and processing of oil.

References

ADA (American Diabetics Association) (1994). Position Statement: nutritional recommendations and principles for people with diabetes mellitus. *Diabetes Care*, 17, 519–522.

Alarcón de la Lastra, C. A., Barranco, M. D., Motilva, V., & Herrerias, J. M. (2001). Mediterrranean diet and health: biological importance of olive oil. *Current Pharm Des*, 7, 933–950.

Bes-Rastrollo, M., Sánchez-Villegas, A., de la Fuente, C., de Irala, J., Martínez, J. A., & Martínez-González, M. A. (2007). Olive oil consumption and weight change: the SUN prospective cohort study. *Lipids*, 41, 249–256.

de Pablo, M. A., Puertollano, M. A., & Álvarez de Cienfuegos, G. (2004). Olive oil and immune system functions: potential involvement in immunonutrition. *Grasas y Aceites (Sevilla)*, 55, 42–51.

Ferro-Luzzi, A., & Branca, F. (1995). Mediterranean diet, Italian type: prototype of a healthy diet. *Am J Clin Nutr*, 61, 1338S–1345S.

Grigg, D. (2001). Olive oil, the Mediterranean and the World. *GeoJournal*, 53, 163–172.

Kafakos A, Riboli E. (2003). *Proceedings of the International Conference on Health Benefits of Mediterranean Diets*, 1st edition. Utrecht: European Geography Association.

Linos, A., Kaklamani, V. G., Kaklamani, E., Koumantaki, Y., Giziaki, E., Papzoglou, S., & Mantzoros, C. (1999). Dietary factors in relation to rheumatoid arthritis: a role for olive oil and cooked vegetables? *Am J Clin Nutr*, 70, 1077–1082.

Mili, S. (2006). Olive oil marketing on non-traditional markets: prospects and strategies. *New Medit*, 5(1), 27–37.

Psaltopoulou, T., Naska, A., Orfanos, P., Trichopoulos, D., Mountokalakis, T., & Trichopoulou, A. (2004). Olive oil, the Mediterranean diet, and arterial blood pressure: the Greek European Prospective Investigation into Cancer and Nutrition (EPIC) study. *Am J Clin Nutr*, 80, 1012–1018.

Quiles, J. L., Ochoa, J. J., Ramirez-Tortosa, C., Battino, M., Huertas, J. R., Martin, Y., & Mataix, J. (2004). Dietary fat type (virgin olive vs sunflower oils) affects age-related changes in DNA double-strand-breaks, antioxidant capacity and blood lipids in rats. *Exp Gerontol*, 39, 1189–1198.

Romero, C., Medina, E., Vargas, J., Brenes, M., & de Castro, A. (2007). In vitro activity of olive oil polyphenols against *Helicobacter pylori*. *J Agr Food Chem*, 55, 680–686.

Further reading

Amirante, P., Clodoveo, M. L., Dugo, G., Leone, A., & Tamborrino, A. (2006). Advance technology in virgin olive oil production from traditional and de-stoned pastes: influence of the introduction of a heat exchanger on oil quality. *Food Chemistry*, 19, 797–805.

Anania G, Pupo D'Andrea MR, (2007). The global market for olive oil: actors, trends, policies, prospects and research needs. Paper presented at the 103rd EASE Seminar on Adding Value to the Agro-Food Supply Chain in the Future EuroMediterranean Space, Barcelona, Spain, 23–25 April 2007.

Beauchamp, G. K., Keast, R. S. J., Morel, D., Lin, J., Pika, J., Han, Q., Lee, C. H., Smith, III A. B., & Breslin, P. A. S. (2005). Ibuprofen-like activity in extra-virgin olive oil. *Nature*, 437, 45–46.

Beaufoy G. (2000). *The Environmental Impact of Olive Oil Production in the European Union: Practical options for improving the environmental impact*. Report for European Forum for Nature Conservation and Pastoralism, UK.

Berry, E. M., Eisenberg, S., Haratz, D., Frielander, Y., Norman, Y., Kaufmann, N. A., & Stein, Y. (1991). Effects of diets rich in monounsaturated fatty acids on plasma lipoproteins – The Jerusalem Nutrition Study: high MUFAs vs high PUFAs. *Am J Clin Nutr*, 53, 899–907.

Berry, E. M., Eisenberg, S., Friedlander, Y., Harats, D., Kaufmann, N. A., Norman, Y., & Stein, Y. (1995). Effects of diets rich in monounsaturated fatty acids on plasma lipoproteins. The Jerusalem Nutrition Study: monoun- saturated vs saturated fatty acids. *Nutr Metab Cardiovasc Dis*, 5, 55–62.

Cohen, L. A., Epstein, M., Pittman, B., & Rivenson, A. (2000). The influence of different varieties of olive oil on N-methylnitrosourea (NMU)-induced mammary tumorigenesis. *Anticancer Res*, 20, 2307–2312.

Olives Australia. http://www.oliveaustralia.com.au (accessed 25 September 2007).

Solfrizzi, V., Panza, F., & Torres, F. (1999). High monounsaturated fatty acids intake protects against age-related cognitive decline. *Neurology*, 52, 1563–1569.

The Olive Oil Source. (www.oliveoilsource.com/definitions.htm).

USDA. (United States Department of Agriculture Data and Statistics), http://www.usda.gov (accessed 14 September 2007).

CHAPTER 2

Why olives and deserts?

2.1 Availability of land due to global climate changes and desertification

Deserts are unique, naturally occurring environments characterized by low and irregular rainfall of less than 100 mm a year on average, and extreme temperatures. The desert experiences extreme temperature variations, ranging from 58 °C to below freezing. Desert landscapes across the world vary greatly, ranging from sand dunes to gravel plains, rocky cliffs and salt flats. The dry conditions found in deserts have resulted in a limited number of human inhabitants (about 13 percent of the world's population is estimated to live in deserts) (Figure 2.1). Despite this, deserts sustain a diverse range of plants and animals, as many species, including reptiles and mammals, have developed ingenious ways to cope with these extreme conditions. It is reported that deserts support a wide diversity of life, with many species of flora and fauna evolving in the dry conditions – for example, there are an estimated 1200 plant species in the Sahara Desert alone. Because of their delicate ecosystems, deserts are particularly vulnerable to habitat disturbance.

One-fourth of the Earth's land is threatened by desertification, according to estimates by the United Nations Environment Program (UNEP; http://www.unep.org). The livelihoods of over a billion people in more than 100 countries are also jeopardized by desertification, as farming and grazing land become less productive.

Desertification is the term used to describe land degradation in its most extreme form. It can be further defined as land degradation that results in barren, desert-like conditions. It occurs in arid,

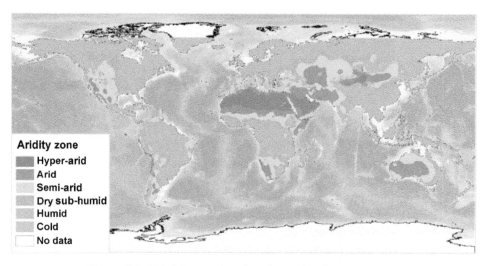

Figure 2.1: World map showing the major desert areas.
Source: http://earthtrends.wri.org.

semi-arid and dry sub-humid areas known as drylands, which can be found on all continents except Antarctica, and affects the livelihoods of millions of people. The causes of desertification are many and complex, and can vary greatly from one location to another. Causes may include deforestation, overgrazing, and poorly managed irrigation. At a higher level, factors such as population increases, economic instability and government policies can increase the potential for these unsustainable land management practices to occur. Both natural and human-induced climate change can compound desertification, and desertification is in turn thought to contribute to climate change.

The end result of desertification is barren and unproductive land that cannot be used for food production or other agricultural purposes, and has little biodiversity. Desertification is estimated to make 12 million hectares of land useless for cultivation every year. In some countries up to 70 percent of the land is vulnerable to desertification, and thousands of people are being forced to leave their homes in search of better living conditions. This can affect people living in non-dryland areas, making desertification an issue relevant to people everywhere. However, there is a huge amount of desert or desert-like land created by the global climate change and other desertification factors that is now available for potential agriculture.

2.2 Genetic and physio-morphological adaptation of the olive tree to desert conditions

Due to a long evolutionary period of natural selection of olive plant material to desert conditions, the majority of olive varieties are more or less adapted to these environments. Both genetic and environmental effects are expressed in typical physio-morphological systems that characterize many of the desert-adapted plant species.

The metabolism of olive trees, as in all terrestrial plants, operates in an aqueous phase, placing them in the hostile interface between a transiently wet soil and a relatively dry atmosphere. In this sense, plant growth can be considered as resulting from an interchange of internal water for carbon dioxide from the atmosphere, required for photosynthesis. The loss of water from plant leaves via transpiration establishes internal flows that eventually draw replacement water from the soil via the roots. Rates of water flow into and within the tree depend upon the gradients of water potential and hydraulic conductance, with the xylem providing a high-conductance, direct internal pathway between the roots and canopy. The internal water status of plants thus varies dynamically in response to the balance between loss and uptake. The important short-term dynamic is diurnal (Connor and Fereres, 2005).

Evaporative demand increases as the day advances, and plant water content falls to a minimum around midday, provided soil water content is high. It recovers in the evening, so the plant may then approach equilibrium with the water potential of the soil. As the root zone dries, however, the leaf water potential falls further each day and, despite gradual control of water loss by stomatal closure, recovery slows until the soil is re-wetted by rainfall or irrigation. After a prolonged dry period, plant water potential is much lower than that of the environment – even by dawn the following day. If a serious internal water deficit persists, the plant's metabolism is disrupted and it may eventually die from desiccation (Fernández *et al.*, 1997).

Growth and survival therefore require adaptation of the plant to the uptake of water, and conservation of the internal water status that is appropriate to the environmental patterns of water supply and demand. The special features by which the evergreen olive is able to maintain an adequate internal water status during severe summer drought derive from its ability to restrict loss of water to the atmosphere and withstand the substantial internal water deficit that is required to maximize extraction of water from the soil. In practice, orchard management greatly assists this balance between uptake and loss by adjusting the size of the transpiring canopy that intercepts radiation, by controlling the ground cover that minimizes or prevents non-tree transpiration and, in some situations, by full or deficit irrigation. Canopy volume and cover are managed through planting density and pruning (Pastor Munoz Cobo and Humanes Guillen, 1996; Gucci and Cantini, 2000), while ground cover is controlled either by tillage or by herbicides (Pastor *et al.*, 1998).

The following are some of the important mechanisms that have been developed by olive trees in order for them to survive under desert conditions.

2.2.1 Ability to collect water from the soil

It is suggested that newly formed roots of the olive trees provide the water uptake capacity, while older roots (which have survived harsh conditions and predation to undergo secondary thickening) provide the framework for exploration, the conduit for transport to foliage via the trunk, and anchorage to the soil (Connor and Fereres, 2005). The total length and branching of the roots varies significantly between olive varieties, and is closely related to cultural practices and especially on irrigation methodology. The success of olive cultivation in marginal soils is claimed to be attributable (at least in part) to its root system, regarding not only its extent but also its plasticity and capacity to react quickly to changes in soil water content.

As in many desert-cultivated plant species, colonization of the roots of olives by mychorrizae is known to affect root morphology and assist the uptake of water and nutrients, especially under conditions of low fertility and limited water supply. Arbuscular mycorrhizae have been recorded in olives (Fernández *et al.*, 1997).

2.2.2 Leaf anatomy protects against water loss

Olive leaves are well adapted to conditions of water shortage (Connor and Fereres, 2005). Olive leaves are small (5–6 cm long and 1–1.5 cm at their widest point), sclerophyllous, and have stomata on the lower (abaxial) surface only. The specific leaf mass (SLM) of olive leaves ranges from 190 to 220 g/m^2 for cv Picual in field-grown conditions, and this value will be much smaller (130 g/m^2) when plants are grown under controlled conditions (Gucci *et al.*, 1997). In olive leaves the leaf surfaces, especially the abaxial ones, are covered with wax sheets and peltate trichomes, which are the characteristic scales supported above the epidermis on single cells.

High reflectivity, combined with a small leaf size, sensed with dissipation of sensed heat, thus minimizing differences between leaf and air temperatures – a feature particularly important when stomata close under conditions of water shortage.

The internal structure of the leaf is comprised of two layers of elongated palisade cells, one associated with each epidermis, that enclose the mesophyll with characteristically thick cell walls, dispersed vascular bundles, and lignified strengthening tissues. The upper and lower palisade layers are usually three cells and one cell deep, respectively. The stomatal characteristics of the olive leaves, combined with the waxy cuticle and trichomes, afford good control over water loss by transpiration. The conductance of the waxy cuticle is negligible, so leaf conductance to water vapor transfer from sub-stomatal cavities to the boundary layer is essentially equal to the stomatal conductance (Fernández *et al.*, 1997).

Though the leaf size and structure of olive trees vary among cultivars, it can be seen that, in general, under water stress and mainly in dry areas, leaves are smaller and thinner, and are composed of more, smaller, and more densely packed cells in each tissue type. Trichomes and stomata are more numerous under water stress conditions, and the net result is higher reflectivity, less conductive cuticles, improved stomatal control, and a smaller cell area exposed to evaporation within the mesophyll tissue (Bongi *et al.*, 1987a). In olive leaves, an established difference in cultivar response to water shortage is found; however, there are no differences between the cultivars regarding cell wall elasticity or osmotic adjustment, greater response in stomata, and trichome densities, which are consistent with its perceived greater drought tolerance and survival in water stress conditions (Bosabalidis and Kofidis, 2002).

Many plants have, however, developed the ability to further decrease osmotic potential during water shortages, and maintain turgor, metabolism and water uptake, by the accumulation of osmotically active ions and metabolites. This is known as osmotic adjustment, and in olives the accumulation of mannitol plays a major role under water stress conditions (Flora and Matore, 1993; Dichio *et al.*, 2003).

2.2.3 *Water hydraulic system of the olive tree*

The olive tree can be represented hydraulically as a conductor-capacitor model in which the canopy is connected in series to the root system by the xylem, and each of the three components is in turn connected in parallel to internal storage tissues (Connor and Fereres, 2005). On a diurnal basis, the active storage tissues are the sapwood, with associated cambium and phloem, and the canopy. The flows in the xylem are determined by the gradients of water potential and hydraulic conductance, while movement to and from the storage tissue is explained by storage volume and capacitance – i.e., the change in water content per unit change in water potential (Wullschleger *et al.*, 1998; Meinzer *et al.*, 2001).

These characteristic phenomena have evolved in olive plants to produce an efficient hydraulic system, and can assist olive plants in surviving under different water stress conditions. However, most studies in this field have been carried out mainly in olive plants cultivated in controlled environments, and further studies are required in olives cultivated under field conditions (Connor and Fereres, 2005).

2.2.4 *Ability to control transpiration*

Many studies have demonstrated the stomatal response in olive trees to leaf water status and the environment. These studies reveal that stomata respond in ways consistent with their role in controlling transpiration and maintaining leaf water status (Moriana *et al.*, 2002). Leaf conductance is small, and decreases as the water potential falls and the vapor pressure deficit increases.

It is suggested that the stomatal response of olive plants to water stress cannot be interpreted using simplified physical models of the continuum of water status in trees, because it appears that endogenous factors modulate responses in the long term. Hormonal regulation, dominated by abscisic acid (ABA), is well known to be closely involved in stomata operation during the day and night, in coordination with the water status of the tree (Araus *et al.*, 2002; Chaves and Oliveira, 2004).

The small, narrow leaf shape and the whole olive tree canopy design also have an impact on transpiration. These olive tree characteristics combine with the irrigation methodologies used to control the rate of transpiration in desert conditions, and in general ABA plays the central role in this as well (Li *et al.*, 2006).

2.3 Limited insect pest problems due to isolation and low atmospheric humidity

The large regions of desert available for agricultural use throughout the world enable selection of isolated areas. Choosing such areas that are also characterized by conditions of low atmospheric humidity provides a significant advantage for olive oil cultivation. Insect pests in general, and in particular the olive fly (*Batrocera olea*), are among the most serious limiting factors for high-quality olive oil production. It is well known that environments with low humidity, such as are common in desert conditions, reduce the risk of olive fly (Mazomenos *et al.*, 2002). In addition, the isolated areas in desert environments increase olive yield and oil quality, with minimum investment in pest control using toxic pesticides that are not always effective in coping with massive pest infestation. These environmental conditions also increase the feasibility of the cultivation of organic olive oil, which is in high demand today in the consumer-sensitive markets of the developed countries of Europe, North America, Australia and Japan.

2.4 Development of new biotechnologies for intensive cultivation

During recent decades, significant progress has been made in agricultural biotechnologies specifically related to desert cultivation. The most important field of development is concerned with controlled saline drip irrigation technologies. Since desert areas have very limited fresh water, and in most cases saline or recycled water is the only source for agriculture, drip irrigation technology is certainly an aid to growing olives in desert environments. This new and approved technology not only provides the source of water, but is also an essential tool for production of greater yields of both olives and olive oil using available water resources (Wiesman *et al.*, 2004). In most cases, saline or recycled water from sewage provides low-cost irrigation for cultivation. Harvesting technology has also been developed specifically for large-scale desert olive production. Following the increasing labor costs in most olive-growing regions, the recent development of automated technology that includes an efficient system for prediction of the optimized olive maturation date, mechanical harvesting, and oil processing will certainly help to reduce the cost of cultivation and ensure high-quality production.

2.5 Economic sustainability

Generally, desert or drylands are considered to be marginal lands, and are therefore probably the least expensive areas for olive oil cultivation these days. Governments all over the world are encouraging the use of such lands, focusing in particular on agriculture and especially tree cultivation. International organizations are also strongly supporting the use of these lands for agriculture rather than other uses, for eco-economic and political reasons. There is therefore a high likelihood of being able to establish large olive plantations in such areas, and the establishment of large production units rather than small ones is always justifiable economically. Studies have proved that it is possible, under desert conditions, to produce both the quantity and quality of olive oil required for a commercial business (Wiesman *et al.*, 2004). There is a strong opportunity for the new desert olive oil brands, produced in different world desert environments, to compete with conventionally produced olive oil in the day-to-day expanding international olive oil market.

2.6 Ecological sustainability

The world population is increasing, and is projected to expand from 6.5 billion now to 9 billion by 2042. The global climate is also changing, due to deforestation and other human-induced phenomena. The world focus on forestation to combat global warming, the greenhouse effect and desertification has therefore become much stronger than previously. One of the primary programs of the United Nations Environmental Program (UNEP) is the "Plant for the Planet: Billion Tree Campaign." Planting fruit trees such as olives would ultimately help to fulfill this campaign. The initiative for planting olives in marginal drylands and/or deserts, which already has strong public support, would ultimately help this program succeed. Furthermore, olive plantations would not only encourage ecological balance due to forestation, but also provide food and protection for numerous species of microorganisms, animals, birds and other plant species, thus contributing decisively to biodiversity levels in their area.

References

Araus, J. L., Slafer, G. A., Reynolds, M. P., & Royo, C. (2002). Plant breeding and drought in C3 cereals: What should we breed for? *Ann Bot*, 89, 925–940.

Bongi, G., Mencuccini, M., & Fontanazza, G. (1987a). Photosynthesis of olive leaves: effect of light flux density, leaf age, temperature, peltates, and H_2O vapor pressure on gas exchange. *J Am Soc Hort Sci*, 112, 143–148.

Bosabalidis, A. M., & Kofidis, G. (2002). Comparative effects of drought stress on leaf anatomy of two olive cultivars. *Plant Sci*, 163, 375–379.

Chaves, M. M., & Oliveira, M. M. (2004). Mechanisms underlying plant resilience to water deficits: prospects for water-saving agriculture. *J Exp Bot*, 55, 2365–2384.

Connor, D. J., & Fereres, E. (2005). The physiology of adaptation and yield expression in olive. *Hort Rev*, 31, 155–229.

Dichio, B., Xiloyannis, C., Angelopoulos, K., Nuzzo, V., Bufo, S. A., & Celano, G. (2003). Drought-induced variations of water relations parameters in *Oleo europaea* L. *Plant Soil*, 257, 381–389.

Fernández, J. E., Moreno, F., Girón, I. F., & Blázquez, O. M. (1997). Stomatal control of water use in olive tree leaves. *Plant Soil*, 190, 179–192.

Flora, L. J., & Matore, M. A. (1993). Stachyose and mannitol transport in olive (*Olea europaea* L.). *Planta*, 189, 484–490.

Gucci, R., & Cantini, C. (2000). *Pruning and Training Systems for Modern Olive Growing*. Landlinks Press, CSIRO, Victoria.

Gucci, R., Lombardini, L., & Tattini, M. (1997). Analysis of leaf water relations in leaves of two olive (*Olea europaea*) cultivars differing in tolerance to salinity. *Tree Physiol*, 17, 13–21.

Li, S., Assmann, S. M., & Albert, R. (2006). Predicting essential components of signal transduction networks: a dynamic model of guard cell abscisic acid signaling. *PLoS Biol*, 4, e312.

Mazomenos, B. E., Pantazi-Mazomenou, A., & Stefanou, D. (2002). Attracting and killing the olive fruit fly *Bactrocera oleae* in Greece as a part of an integrated control system. Use of pheromones and other semiochemicals in integrated production. *IOBC wprs Bull*, 25, 137–146.

Meinzer, F. C., Clearwater, M. J., & Goldstein, G. (2001). Water transport in trees: current perspectives, new insights and some controversies. *Environ. Exp Bot*, 45, 239–262.

Moriana, A., Villalobos, F. J., & Fereres, E. (2002). Stomatal and photosynthetic responses of olive (*Olea europaea* L.). *leaves to water deficits. Plant Cell Environ*, 25, 395–405.

Pastor, M., Castro, J., Vega, V., & Humanes, M. D. (1998). Sistemas de manejo del suelo. In: D. Barranco, R. Fernández-Escobar, L. Rallo (Eds.), *El Cultivo del Olivo* (pp. 197–236). Junta de Andalucía y Mundi-Prensa, Madrid.

Pastor Munoz Cobo, M., & Humanes Guillen, J. M. (1996). *Poda del olivo. Moderna olivicultura*. Española, S.A, Madrid.

Wiesman, Z., Itzhak, D., & Dom, N. B. (2004). Optimization of saline water level for sustainable Barnea olive and oil production in desert conditions. *Scientia Horticulturae*, 100, 257–266.

Wullschleger, S. D., Meinzer, F. C., & Vertessy, R. A. (1998). A review of whole-plant water use studies in trees. *Tree Physiol*, 18, 499–512.

Further reading

Bongi, G., Soldatini, G. F., & Hubick, K. T. (1987b). Mechanism of photosynthesis in olive tree (*Olea europaea* L.). *Photosynthetica*, 21, 572–578.

Chartzoulakis, K., Patakas, A., & Bosabalidis, A. M. (1999). Changes in water relations, photosynthesis and leaf anatomy induced by intermittent drought in two olive cultivars. *Environ Exp Bot*, 42, 113–120.

Citernesi, A. S., Vitagliano, C., & Giovannetti, M. (1998). Plant growth and root system morphology of *Olea europaea* L. rooted cuttings as influenced by arbuscular mycorrhizas. *J Hort Sci Biotechnol*, 73, 647–654.

Fahn, A. (1986). Structural and functional properties of trichomes of xeromorphic leaves. *Ann Bot*, 57, 631–637.

Karabourniotis, G., Papadopoulos, K., Papamarkou, M., & Manetas, Y. (1992). Ultraviolet-B radiation absorbing capacity of leaf hairs. *Physiol Plant*, 86, 414–418.

Mariscal, M. J., Orgaz, F., & Villalobos, F. J. (2000a). Modelling and measurement of radiation interception by olive canopies. *Agr Forest Meteorol*, 100, 183–197.

Mariscal, M. J., Orgaz, F., & Villalobos, F. J. (2000b). Radiation-use efficiency and drymatter partitioning of a young olive (*Olea europaea*) orchard. *Tree Physiol*, 20, 65–72.

CHAPTER 3

Key characteristics of the desert environment

3.1 What is a desert?

As described in Chapter 2, a desert is a biome that is characterized by low moisture levels and infrequent and unpredictable precipitation. Deserts, also known as arid lands, are regions that gain less precipitation (rain, sleet, or snow) than their potential evapotranspiration (evaporation from the soil and plants plus transpiration from plants), causing a severely limiting environment for living things. This is because most rain evaporates before it ever hits the ground. A desert, then, is a region with very little vegetation cover and large surfaces of exposed bare soil, where

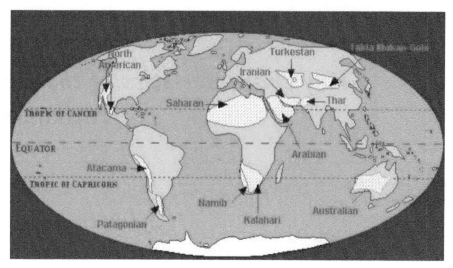

Figure 3.1: Major deserts on Earth.
Adapted from http://www.enchantedlearning.com/biomes/desert/desert.shtml.

the average annual rainfall is less than the amount needed to support optimum plant growth, and where plants and animals show clear adaptations for survival during long droughts (UNEP, 2007).

Deserts occur in six of the seven continents of the world, with Europe being the exception. Antarctica is entirely desert, and, apart from some fertile land along the east coast, most of Australia is desert. With the exception of the frozen deserts in Greenland, Argentina and Antarctica, most of the world's deserts are in two belts that lie within 25–35° of the equator. Deserts cover almost one-third of the Earth's land surface, and 13 percent of the world's population live in these areas (Figure 3.1). Some of them choose to live near oases, while others are nomads who wander throughout the desert. Deserts can be further characterized by their appearance and plant life. They may be flat, mountainous, divided by gorges and ravines, or covered by a sea of sand.

Due to natural phenomena and human interference, more and more land throughout the world is tending towards desert via the process of desertification, which limits conventional agriculture. The details of desertification and its causes will be discussed later in this chapter.

3.2 Desert climates

Although different deserts have different climates, the common denominator is that they are always extreme. Desert surfaces receive a little more than twice the solar radiation received by humid regions, and lose almost twice the heat at night. Hot deserts experience some of the highest temperatures and some of the greatest temperature contrasts to be found on the planet. In hot deserts, days are usually sunny and cloudless. During the summer, daytime air temperatures of between 44 °C and 47 °C are not unusual. Because there is little vegetation, rocks and soil are exposed to the sun. This may cause ground temperature in the hottest deserts to exceed 80 °C. Nights, however, are much cooler. The lack of cloud cover allows heat to escape, and the temperature may drop very quickly by 25 °C or more after the sun sets. At night, temperatures of 10 °C or less are not uncommon, and may even drop below freezing. The evaporation rate in hot deserts can be 20 times the annual precipitation rate, as most rain evaporates before it reaches the ground. The cold night temperatures provide favorable conditions for some crops to complete their reproductive cycle, which is one of the reasons why agricultural cultivation around desert regions is possible.

The cold deserts, on the other hand, are so cold that the air can only hold a small amount of moisture. There is very little precipitation, and all the surface water is locked in unusable blocks of ice. The winter in cold deserts at latitudes midway between the polar and equatorial regions can be harsh, with temperatures below freezing being common. Blizzards and violent winds often accompany the icy temperatures.

Although both extreme hot and cold deserts are generally not suitable for agricultural practices, the peripheral areas around such regions can be and people have been growing crops in these areas for thousands of years. With the application of modern technologies, additional desert or dryland marginal regions that are not currently in use can also be intensively cultivated.

Deserts are often classified by their geographical location and weather patterns, which are described below.

3.2.1 Trade-wind deserts

Trade-wind deserts include some of the largest deserts on Earth, such as the Sahara, the Kalahari, and the Australian deserts. They lie along the Tropics of Cancer and Capricorn, between 15° and 30° north and south of the equator. Hot, moist air rises near the equator, and as it rises it cools and drops heavy rain in tropical areas. Now cooler and drier, this air sinks and warms in the areas of the Tropics of Cancer and Capricorn.

The main feature of these deserts is that they are areas of very little precipitation and very high temperatures that occur where trade winds warm up and blow away cloud cover, causing the land to heat up vigorously.

3.2.2 Mid-latitude deserts

Mid-latitude deserts occur in the interiors of continents, between 30° and 50° north- and south-poleward of the subtropical high-pressure zones. These deserts are in interior drainage basins far from oceans or seas, and may undergo a wide range of annual temperatures. Air reaching these areas has long since dropped its moisture, creating dry conditions. The Sonoran or Gobi Deserts are typical mid-latitude deserts.

3.2.3 Rain-shadow deserts

Rain-shadow deserts are formed because of the "rain-shadow" effect, where nearby mountains prevent moisture-rich clouds from reaching areas on the lee (or protected) side of the range. Moist air approaching a mountain is forced upwards, where it cools, condenses, and often falls as rain. Once this air has passed over the top of the mountain and down the other side it can again expand, but by now it has unfortunately lost most of its moisture. Mountains often absorb far more precipitation than the areas around them, thus creating deserts in the leeside shadow of the range. Patagonia, in Argentina, is an example.

3.2.4 Coastal deserts

Coastal deserts are generally found in moderately cool to warm areas on the western edges of continents, particularly in the southern hemisphere. They are affected by cold ocean currents, which inhibit the formation of rain clouds generally created by warm oceans. Winter fogs frequently blanket coastal deserts and block solar radiation. Persistent high-pressure systems are also a factor, tending to block incoming storms. The Namib Desert in Africa and the Atacama Desert of Chile are good examples of this type of desert. The Atacama Desert is the driest desert on Earth.

3.2.5 Polar deserts

Polar deserts are frozen deserts, located in polar regions, which have an annual precipitation of less than 250 mm and temperatures that do not exceed 10 °C even during the warmest months. This is a consequence of the weak energy supplied by the sun, which remains below the horizon for the winter and never rises very high in the sky, even during the summer when it does not set. Polar deserts cover nearly 5 million square kilometers of the Earth's surface area, and consist mainly of bedrock or gravel plains.

3.3 Water in deserts

Since deserts are arid areas, water is coveted and preserved by all living things populating these regions. In all arid regions, a major challenge is the appropriate management of water – to conserve it, to use it efficiently and to avoid damage to the soil.

The little precipitation that falls in deserts is usually erratic, and varies from year to year. While a desert might have an annual average of 150 mm of precipitation, it may receive 200 mm one year, none in the second year, 300 mm in the third year, and 100 mm in the fourth. Thus, in arid environments, the annual average tells little about the actual rainfall. Some deserts do not receive any rainfall for years. As mentioned earlier, the availability of water is the most limiting factor in drylands or deserts. According to its availability, desert water is classified as surface or subsurface water.

3.3.1 Surface water

Surface water in the desert includes ephemeral streams, arroyos, lakes, ponds and puddles produced by intense rain. Since the ground is usually impermeable, the water quickly runs into streams that only exist during the rainfall. The swift water of these temporary streams is responsible for most of the erosion that takes place in the desert. Desert rain often never reaches the ocean; the streams usually end in lakes or streams that dry up when the rain stops. Desert lakes are generally shallow, temporary, and salty. Because these lakes are shallow and have a low bottom gradient, wind stress may cause the waters to move over many square kilometers. When small lakes dry up, they leave a salt crust or hardpan. Permanent streams in the desert are usually the result of water coming from outside the desert, and are often referred to as "exotic" water. The River Nile, for example, flows through a desert, but its source is high in the mountains of Central Africa.

3.3.2 Subsurface water

Subsurface water or groundwater is water that remains beneath the surface of the Earth and saturates the pores and fractures of sand, gravel and rock formations. This is the main source of water in desert regions. It is a major source of water for agricultural and industrial purposes, and an important source of drinking water. For desert habitants, this source of water is sometimes the only one they can rely on. It is not always accessible, or fresh enough to use without treatment, and sometimes it is difficult to locate or measure. It can be utilized, however, by drilling wells and pumping water. In arid areas it usually lies hundreds of meters below the surface. It comes from seepage which has collected, sometimes over many centuries, underneath the land, and is replenished by precipitation. Different climates and geologies lead to uneven distribution in both quantity and quality. Subsurface water is stored in and moves slowly through moderately to highly permeable rocks called aquifers. An aquifer may be a layer of gravel, sand, sandstone or cavernous limestone; the rubbly top or base of lava flows; or even a large body of massive rock, such as fractured granite, that has sizable cracks and fissures. The ability of the aquifer to store and transfer groundwater depends on the porosity (the amount of water a rock formation can hold) and permeability of the ground.

When it rains, the first few drops that hit the ground only penetrate the surface and serve to replace water resources that have evaporated or been used by plants during the preceding dry period. Any excess water then seeps to a water table, which is the area just above the aquifer, where soil or rocks are permanently saturated with water. The water table is an intermediate region that separates the groundwater zone that lies below it from the capillary fringe that lies above it. The water table may rise and fall from season to season and year to year, usually varying around an average depth. If the water table reaches the ground surface, water tends to seep out as a natural spring. In the end, every drop of ground water leaves the aquifer by outflow as a natural spring, or as seepage into a lake, a river or the sea, or by being pumped out via a well, but by that time it has been replaced by other water. Natural refilling of an aquifer at depth is a slow process, since groundwater moves slowly before reaching it.

Subsurface water contains many dissolved mineral and organic compounds that may, in high concentrations, affect the water's quality, sometimes to the extent of being hazardous to animals and plants. Since a high evaporation rate of precipitation exists in arid regions, subsurface water in these areas is often quite saline. This salinity is introduced from the accumulations of salts in the soil and on the surface, through the intermediary of floods and recharge waters. Dissolution of minerals by the slowly moving groundwater forming the aquifer contributes to the salinity build-up. Parent materials high in sodium (Na^+), magnesium (Mg^{2+}), calcium (Ca^{2+}), potassium (K^+), sulfate (SO_4^{2-}), bicarbonate (HCO_3^-), carbonate (CO_3^{2-}), nitrate (NO_3^-), and chloride (Cl^-) release these ions to form salts directly or influence soil chemistry in ways that result in salt accumulation. Under extreme conditions, brine may result from such a process – for example, on contact of groundwater with salt domes (Gat, 1979).

Demand for groundwater is influenced by the quality of the resource, with some groundwater being of higher quality than surface water while other such resources cannot be used without treatment. Most of the groundwater extracted is used for irrigation, occasionally mixed with other water supplies to meet quantity or quality requirements. Normally, moderate- to high-saline water is of limited use, while slightly saline water can sometimes be used in place of fresh water. Some settlements in the Israeli Negev Desert have utilized underground water for various agricultural purposes, including olive cultivation (Figure 3.2).

Global consumption of water is doubling every 20 years – more than twice the rate of human population growth – while pollution and over-extraction in many regions of the world have reduced the ability of supplies to meet demand. Over the past 70 years the global population has tripled, but water withdrawals have increased over six-fold.

3.4 Soil in deserts

The unique properties of desert soils are due to the environmental attributes mentioned above. The absence of water in deserts means that little or no chemical weathering can take place. Instead, erosion, frost, sedimentation, and big temperature fluctuations between day and night break down the rocky surface into sand or gravel. The low moisture also means that silt and sand-sized particles are easily blown away, while the remaining particles eventually form a tightly packed layer known as desert pavement. Finally, because of the negligible rainfall, much

Figure 3.2: Successful commercial agricultural farming, using underground saline water, in the southern Israeli Negev Desert.

of the soil in arid and semi-arid areas is almost totally devoid of decaying organic matter and micro-organisms. The result is a dry place, full of sand and desert pavement.

Most arid soils are high in salt content. Salinization is the accumulation of water-soluble salts in the soil to a level that impacts on agricultural production, environmental health and economic welfare. A soil is considered to be saline if the electrical conductivity of its saturation extract (ECe) is above 4 dS/m; however, the threshold value above which deleterious effects occur can vary depending on several factors, including plant type, soil water regime and climatic condition (Maas, 1986; Rengasamy, 2002). Excess salt has the same lethal effect on plants as drought: the high concentration of salt reduces the ability of plants to take up water, which interferes with their growth and reduces their vitality. Some salts, such as those containing sodium, can change the physical condition of the soil, reducing infiltration, increasing runoff and erosion, and impairing biological activity.

3.5 Solar irradiance in deserts

The amount of solar energy that reaches the Earth from the sun depends on the angle at which the rays enter the atmosphere. The Earth is positioned at the center of a gaseous atmosphere, which is bounded gravitationally to it. The circulation of our atmosphere is a complex process because of the Earth's rotation and the tilt of its axis. The heating rays coming from the sun to the Earth are filtered through the atmosphere prior to meeting the land, thus allowing the warmth of the rays to reach the ground while preventing a great deal of solar radiation; the extent to which this occurs depends on how far a particular area is from the equator.

The Earth's axis is inclined 23.5° from the ecliptic, and orbits around the sun. Due to this inclination, vertical rays from the sun strike at 23.5° north – the Tropic of Cancer – at the summer solstice in late June. At the winter solstice, the vertical rays strike at 23.5° south – the Tropic of Capricorn. This tilt causes some parts of the Earth to be closer to the warmth of the sun than others. The further an area tilts away from the sun, the greater the angle at which the sun's rays have to pass through the atmosphere, and thus the longer their path through the atmosphere before reaching the Earth's surface; thus, more radiation and heat are filtered out. The further a region is from the equator (either north or south), the lower the sun appears in the sky, and the surface area over which the same amount of solar energy is spread is larger. This is why poleward areas have fewer hours of sunlight and are significantly colder than more equatorial regions. If an area of the planet is tilted more directly toward the sun, the sun appears high in the sky and its rays pass directly through the atmosphere at a lesser angle, so a smaller amount of atmosphere filters out the heat radiation and there are more hours of daylight. The tilt of the axis allows differential heating of the Earth's surface, which causes seasonal changes in the global circulation. Areas that are closer to the equator (as is the case with most deserts) are tilted more directly toward the sun and deliver more intense energy per unit area, for more hours of sunshine in winter, and for greater portions of the year.

As the source of energy to the Earth is the sun, irradiance and/or radiance in a particular place plays a crucial role for the growth and development of plants. *Irradiance* is the power of electromagnetic radiation at a surface per unit area, and is commonly used when electromagnetic radiation is incident on the surface, whereas *radiation* is used interchangeably for radiation emerging from the surface. Plants in desert areas are exposed not only to a high irradiance but also to a combination of environmental stress conditions, including low water availability, extreme temperature fluctuations, high irradiance, and nutrient deprivation, so the effects are cumulative. Plants that have desert environments as their natural habitat may therefore reveal novel mechanisms and strategies that enable them to resist stressful conditions. Since olive trees have grown in most dryland conditions with ample irradiance, they have developed such mechanisms (as will be discussed in detail below). Although the olive is grown in regions of high sunlight, it has a low saturation irradiance compared to other fruit trees (Bongi and Palliotti, 1994).

3.6 Desertification

One of the biggest ecological concerns nowadays involves a process referred to as desertification. This has existed for thousand of years, due to climate and environmental fluctuations, but more recently abuse by humans has degraded the land to a hazardous extent. Large amounts of funds are now being transferred by different groups in order to raise awareness and fight desertification.

It involves multiple causes, and it proceeds at varying rates in different climates. Desertification may intensify a general climatic trend toward greater aridity, or it may initiate a change in local climate.

The process involves the degradation, in arid, semi-arid and dry sub-humid areas, of fertile lands capable of sustaining vegetation and various organisms, to "dead" lands. The new deserts are not necessarily hot, dry, sandy places; instead, they are any areas where the soil has been so mistreated by humans that it is now useless for growing crops. It usually occurs near arid regions, in a transitional zone between desert and the surrounding or peripheral areas. These transition zones have very fragile, delicately balanced ecosystems that support vegetation. This vegetation absorbs some of the heat carried by the hot winds, and protects the land from the prevailing winds. After rainfall, the vegetated areas are distinctly cooler than the surroundings. In these marginal areas, the demands and activities of the increasing populations that settle on the land stress the ecosystem beyond its tolerance limit. Degradation is caused by pounding the soil; livestock compact the substrate, increase the proportion of fine material and reduce the percolation rate of the soil, thus encouraging erosion by wind and water. Grazing and the collection of firewood reduce or eliminate plants that help to bind the soil. In the past, some have mistakenly believed that drought is another cause of desertification. Global warming caused by increasing greenhouse gas levels in the atmosphere is expected to increase the variability of weather conditions and extreme events. Many dryland areas face increasingly low and erratic rainfalls, coupled with soil erosion by wind, and the drying up of water resources through increased regional temperatures. Now, it is known that well-managed lands can recover successfully after the rains return; however, continued abuse of the land during droughts increases land degradation.

This phenomenon is beginning to cause panic around the world because of the loss of fertile agricultural lands. Estimates suggest that 35 percent of the Earth's land surface is at risk, and the livelihoods of 850 million people are directly affected. Worldwide, desertification is making approximately 12 million hectares useless for cultivation every year. In the Sahel (the semi-arid area south of the Sahara Desert), for example, the desert moved 100 km southwards between 1950 and 1975. The demand, therefore, for desert agriculture and new growing techniques is rising. Growers nowadays are looking for new methods that will allow them to exploit arid regions and grow crops on otherwise "dead" lands. In order to survive the harsh conditions in deserts, judicial selection of crops must be made and careful management of all water resources instigated.

3.7 Some world desert environments

As already stated, more than one-third of the total surface area of the Earth is classified as various categories of drylands; some of them are extreme and barely populated, whereas others are currently degraded and in the process of desertification. Below, some of the most well-known world deserts are described. Although some sort of agricultural husbandry has been practiced either inside these desert regions or in their peripheral or transitional areas, which are prone to desertification, it has not necessarily involved olive cultivation; however, it is very important to know the basics of these areas for the effective implementation of advanced modern biotechnologies in order to improve agricultural production, including that of olives. The practice of agriculture in these deserts is mostly dependent on the availability of water, which is the main limiting factor. Experience has shown that groundwater is the most suitable water source in the desert environment. The efficient use of water will be discussed in detail in Chapter 5.

As the prime objective of this book is to demonstrate the possibility of expanding olive oil cultivation into various world deserts, the Negev Desert in Israel is the first to be described. Here, eco-friendly sustainable desert olive oil cultivation, based on advanced biotechnologies, has successfully been established over recent decades.

3.7.1 The Negev Desert

The Negev Desert in Israel covers over 12,500 square kilometers – more than half of Israel's surface area. It is a triangular area of land that forms part of the great Saharo-Arabian desert belt that extends from the Sahara and the Atlantic seaboard on the west side of Africa, across the Arabian Desert, to the desert of Sind in India in the East (BIDR, 2000) (Figure 3.3). The Negev Desert has arid, semi-arid and hyper-arid climates, characterized by strong solar radiation, extreme temperatures, and high levels of water evaporation from the ground surface. Rainfall in the Negev occurs only in the winter, and varies annually. The semi-arid zone lies north of Beer-Sheva, and includes the northern Negev, the upper reaches of the Judean Desert, the northern Jordan Valley, the Hula Valley, and the Sea of Galilee (Kinnerot) region. The mean annual precipitation in this zone is 300–500 mm, while the mean annual evapotranspiration ranges from 1500 to 1700 mm.

The region is characterized by a typical Asian biota, mixed with both Mediterranean and Sahara–Arabian desert elements. The arid zone, known as the central Negev, is located below the semi-arid zone, and extends from south of Beer-Sheva to Mitzpe Ramon, and east to the lower reaches of the Jordan Valley. The mean annual precipitation in this region does not exceed 100–300 mm, while the mean annual evapotranspiration ranges from 1700 to 1800 mm. The region is characterized by a typical Saharo-Arabian desert biota, mixed with Asian and Mediterranean elements, especially in its northern and higher-elevation areas. The hyper-arid zone is located south and east of the arid zone, and includes the central and southern Negev, from Mitzpe Ramon to Eilat, the Dead Sea Basin, and the Arava Valley. The mean annual precipitation in this zone does not exceed 30–90 mm, and the mean annual evapotranspiration ranges between 1800 and 2800 mm. The hyper-arid region is a typical Saharo-Arabian desert, with an acacia savannah in the Arava Rift Valley (Protnov and Safriel, 2006).

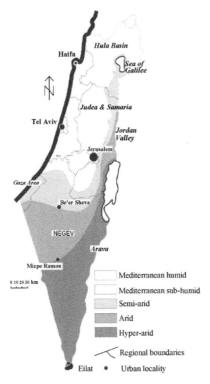

Figure 3.3: Israel and the Negev Desert.

Temperatures in the Negev are extreme. There are sharp differences in temperature between day and night, and between summer and winter. Solar radiation and evaporation rates are high throughout the year, and relative humidity and cloudiness remain low. Summers are hot and dry, with average temperatures ranging between 25 °C and 35 °C in July and August, and peak temperatures of 42–43 °C. Solar radiation is very strong, especially during June and July when the solar rays strike the desert surface vertically. In addition, relative humidity is low during most of the day, between 25 and 30 percent, but rises considerably during the night, when the ambient temperature drops sharply. The daily temperature fluctuation in the summer is about 18 °C, and the average temperature is within the range of thermal comfort. Winters are cold, sometimes rainy, and uncomfortable. The average temperature in January is 10 °C and the average minimum daily temperature is 3 °C, while temperatures may drop below freezing at night. Floods are not unusual during the few winter rain events.

In summary, of the exposed surfaces in the Negev and adjacent areas, 60–65 percent are rocky deserts, approximately 15 percent are sedimentary plains or plateaus covered by desert pavements, about 5–10 percent are loessial plains and depressions where loess-derived soils are common, some 10 percent are sand fields (including shifting sand dunes), and finally about 1 percent is composed of highly saline sabkhas. As in other deserts, the Negev suffers from lack of water. To conserve Israel's water resources, exploited almost to the limit, several measures have been taken by the Water Commission of Israel. One of them is the use of saline water. The Negev possesses considerable reserves of saline underground water with different concentrations of salts. It was found that, for certain crops, saline water can be used for irrigation in place of fresh water, and this has been implemented for several decades to grow tomatoes, watermelons, grapes, jojoba, olives and other crops.

Israel is also considered to be one of the natural homelands of olives, and this crop has been cultivated there since time immemorial. The traditional area of olive cultivation in Israel is around Jerusalem and/or the north of the country. However, with an increasing population and the scarcity of cultivable land, the planting of olives in Israel has gradually moved southwards towards the Negev Desert. Today, most of the new olive plantations in Israel are concentrated in central and southern Negev around Beer-Sheva, and even further south in the Ramat Negev and Neot Smadar highlands (Figure 3.4). Cultivation has become successful with the implementation of modern techniques, developed by Israeli scientists, for olive cultivation in drylands, and with selective genotypes specifically chosen for the desert environment. This successful olive cultivation in the different parts of the Negev Desert might provide a good model for other world deserts.

3.7.2 The arid region of Southern Spain

Andalusia is a large territory of 87,561 square kilometers in southern Spain. This enormous terrain comprises around 17 percent of the country. Andalusia is composed of eight provinces – from east to west, Almería, Granada, Jaén, Córdoba, Málaga, Seville, Cádiz and Huelva. The plain is bordered in the north by the Sierra Morena and the Bética Mountains, in the south-west by the Atlantic Ocean, and in the east and south-east by the Mediterranean Sea. Most of Andalusia is considered to be mountainous, with some 50 percent of the region lying more than 600 m above sea level (Figure 3.5).

Andalusia is widely known as one of the most arid regions in Europe, with a significant shortage of water. The hydrographical network consists of five river basins: the Guadalquivir, Guadiana and Guadalete-Barbate basins each flow into the Atlantic Ocean, while the South and Segura basins flow into the Mediterranean Sea.

The economic development of Andalusia has led to an increase in the demand for water over time. In turn, this has led to greater pressure on the scarce water resources. The agricultural sectors absorb 90 percent of the available water resources because of the very large amount of irrigated land. Spain ranks third in the world and first in Europe in terms of irrigated land, while Andalusia has 23.3 percent of irrigated land in Spain – a very high percentage. Together with agricultural activities, tourism has raised the demand for water to such an extent that the coastal areas are mostly irrigated using underground water.

Figure 3.4: Successful olive orchards: (a) the Neot Smadar area of the southern Negev; (b) the Bet Gurvrin area of the northern Israeli Desert.

Figure 3.5: The Andalusia region of Spain.

In spite of its geo-climatic situation, this region produces 75 percent of the total olive oil production of Spain, and Spain produces more than 30 percent of the world's olive oils. Although Andalusia has already become a world reference point for intensive agricultural technologies ranging from plastic structures to genetic engineering of seeds and plants, integral water-cycle technologies for agricultural use, hydroponic cultivation, and subsequent processing of the product (Casas *et al.*, 2006), there is no doubt production could be increased if there was proper implementation of the recently developed techniques for desert olive oil cultivation.

3.7.3 Australian deserts

After Antarctica, Australia is the second driest continent in the world. The Australian deserts listed below are distributed throughout the western plateau and interior lowlands (Figure 3.6). They make up about 18 percent of the total mainland area of the continent, but a full 70 percent of Australia receives less than 500 mm of rain annually, making it arid or semi-arid. Only 3 percent of the Australian population lives in these dry areas; the rest is concentrated in the north, east, southeast and southwest, where precipitation is adequate to support vegetation that significantly protects the land surface from weathering.

The air over these desert regions is usually very dry. Annual average relative humidity in the mid-afternoon is below 30 percent over most of the area, and falls to as low as 20 percent in parts of inland Western Australia. Days with humidity falling below 10 percent are not unusual, especially in late winter and in spring.

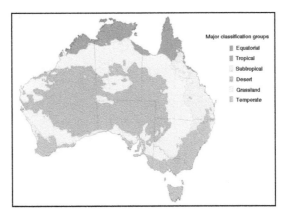

Figure 3.6: Climatic classification of Australia.
Source: Australian Bureau of Meteorology (http://www.bom.gov.au/).

The scarcity of water in Australia has been a major limitation to agricultural development over the past 200 years. Periods of drought have placed great stress on human and animal welfare, economic production and resource sustainability. Rainfall is distributed unevenly, both geographically and seasonally, across the country. Most of this rainfall does not run off into the river systems, with less than one-sixth of the rain that falls in Australia (less than 3 percent in the drier areas and up to 24 percent in the wetter areas) ending up in its rivers. Most of the rainwater evaporates, is used by trees and plants, or ends up in lakes, wetlands or the ocean. Because of this, Australian rivers are characterized by relatively low and variable flows. Most lowland rivers are not only occasionally dry, but also flood from time to time. Rivers in southern Australia are extensively dammed and regulated to provide year-round security of supply for urban and domestic use, and irrigation water for use during the summer. In accordance with the need, Australia stores up to seven years' worth of water in the major dams. Permanently flowing rivers are found mainly in the eastern and southwestern regions, and in Tasmania. The reduced availability of surface water has resulted in far greater groundwater use in recent years over much of Australia, and this has resulted in declining groundwater levels in many areas.

While surface water is unconditionally renewable as long as it rains, groundwater is only conditionally renewable. Some groundwater supplies used for irrigation and urban needs are thousands of years old, and response times to changes in recharge are also very long. This means that rapid use (on a scale of decades) makes the resource practically non-renewable. Groundwater levels are declining because of overuse of many major aquifers, and minute quantities of rainfall.

In spite of these shortages, agriculture is the dominant form of land use in Australia today, occupying around 59 percent of the country's land. Agriculture is also the largest consumer of water in Australia, using 70 percent of the water extracted. Because of climate variability, irrigated agriculture is extensive and growing in area. Though the majority of irrigation water

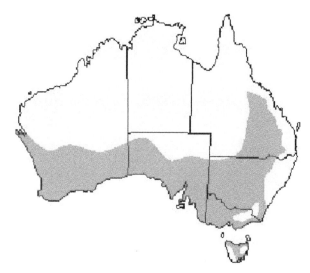

Figure 3.7: Map of Australia. The dark areas represent those areas with a climate similar to that of the traditional olive-growing regions in Europe.
Source: Australian Olives Association (www.australianolives.com.au).

is used for pasture, and lesser amounts are used for some of the higher-value crops and for horticulture, with the growing awareness of olive oils and the availability of additional land the area dedicated to olives has increased rapidly in Australia. An estimate shows that more than 8 million olive trees have been planted in Australia in the past 10 years. There are still parts of the least populated areas of Australia that are very similar to the traditional olive-growing areas of Europe, and arid areas where olive cultivation can be increased effectively in the future (Figure 3.7).

Furthermore, Australia is considered to be the "cleanest" olive-growing area from the point of view of attack by the olive fly (*Bactrocera olea*). It is reported that *B. olea* damages between 5 and 30 percent of the total olive production annually, and Australia is still remote from this insect (Augustinos *et al.*, 2002).

3.7.4 Argentinean deserts

Argentina is the world's eighth largest country, occupying an enormous surface of 2.8 million square kilometres, 3300 km long and about 1200 km wide at its widest point. Argentina is positioned below the equator on the southernmost part of South America; most of it is remarkably flat and 63 percent of the total area is less than 500 m above sea level, with altitudes ranging from 2150 m (~7000 feet) in the Andes in the west to sea level in the east. It extends between latitudes 22 and 55° S, and about two-thirds of this vast territory is associated with arid and semi-arid rangeland ecosystems, from the hot deserts and semi-deserts in the north to the frozen

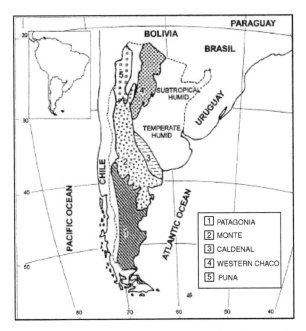

Figure 3.8: Arid and semi-arid rangelands of Argentina.

arid zones in Patagonia. Argentina is divided into four major regions. There are five main arid and semi-arid regions in Argentina, each with its own characteristics (Figure 3.8).

In Argentina, the olive-growing areas have increased in recent decades to 13 million plants. Most of the olive-growing areas in Argentina are in the arid regions. The Mendoza, San Juan, La Rioja, Catamarca, Cordoba and Buenos Aires provinces are the main areas where olive trees are currently planted in Argentina; however, there is tremendous potential to increase the area of olive plantations in other dryland regions (Barreara *et al.*, 2003).

3.7.5 The California Desert

The California Desert stretches over 25,000 square miles (~64,700 sq km), and is divided into two basic zones: the Mojave, or high desert, and the Colorado, or low desert. Each of these areas contains unique fauna and flora, as well as other natural resources.

The Mojave Desert

The Mojave Desert is a rain-shadow desert that occupies a significant portion of southern California and parts of Utah, Nevada and Arizona. It functions as a transitional desert between the Great Basin Desert to the north and the Sonoran to the south, mainly between latitudes 34° and 38° north (Figure 3.9). It is found at elevations of 925–1850 m (3000–6000 feet), and is considered to be a "high desert." The Mojave has a typical mountain-and-basin

Figure 3.9: North American deserts.
Source: http://www.friendsofsaguaro.org/mapdeserts.gif.

topography, with sparse vegetation. Sand and gravel basins drain to central salt flats, from which borax, potash and salt are extracted. Silver, tungsten, gold and iron deposits are also mined intensively.

The Colorado Desert

The Colorado Desert is a low, flat depression surrounded by mountain ranges, situated in the south-eastern corner of California. The Colorado Desert region encompasses approximately 7 million acres, reaching from the Mexican border in the south to the higher-elevation Mojave Desert in the north, and from the Colorado River in the east to the peninsular mountain range in the west. The Colorado Desert is the western extension of the Sonoran Desert that covers southern Arizona and northwestern Mexico. It is a desert of much lower elevation than the Mojave Desert to the north, and much of the land lies between 70 m (230 feet) below sea level and 675 m (2200 feet) above sea level. Although the highest peaks of the Peninsular Range reach elevations of nearly 3075 m (~10,000 feet), most of the region's mountains do not exceed 925 m (3000 feet). These ranges block moist coastal air and rains, producing an arid climate. Common

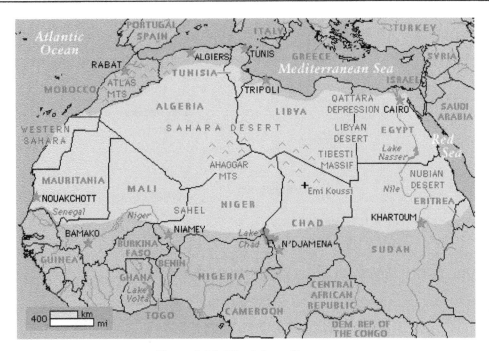

Figure 3.10: The Sahara Desert.
Source: http://library.thinkquest.org.

habitat includes sandy desert, scrub, palm oasis, and desert wash. The soil types include Aridisols and Entisols with hyperthermic soil temperature regimes and aridic soil moisture regimes.

3.7.6 The Sahara Desert

North Africa is considered to be one of the driest regions on Earth. Nearly two-thirds of the region is desert. The Sahara Desert, covering most of North Africa, is the largest non-polar desert in the world, covering approximately 9 million square km. The desert includes most of the Western Sahara, Mauritania, Algeria, Niger, Libya and Egypt; the southern regions of Morocco and Tunisia; and the northern areas of Senegal, Mali, Chad and Sudan. The eastern Sahara is usually divided into three regions: the Libyan Desert, which extends west from the Nile valley through West Egypt and East Libya; the Arabian Desert, or Eastern Desert, which lies between the Nile valley and the Red Sea in Egypt; and the Nubian Desert, which is in north-eastern Sudan (Figure 3.10).

Although olives have been grown in the northern part of the Sahara Desert, where there is the influence of Mediterranean climates, there have been no reports of olive cultivation in the southern part of this desert, called the Sahel; however, with selected cultivars the possibility cannot be ignored. Agricultural development has been the only method so far with the potential

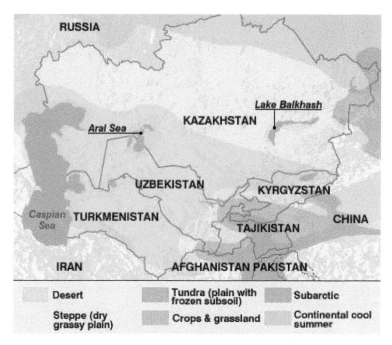

Figure 3.11: The Turkestan Desert.

to improve African economies in this region, but it requires extensive knowledge and advanced biotechnologies to succeed.

3.7.7 The Turkestan Desert

Turkestan, which was once inhabited mainly by Turkish races and is also called Central Asia, is a large, historic region that lies between Siberia to the north, Tibet, India, Afghanistan and Iran to the south, the Gobi Desert to the east and the Caspian Sea to the west. It covers approximately 2,600,000 square kilometers (Figure 3.11). At present, this area includes Turkmenistan, Uzbekistan, Tajikistan, Kyrgyzstan and the southern part of Kazakhstan, and East Turkestan (now the Uighur Autonomous Region of the Sinkiang area of China). The Turkestan Desert, which occupies around 192,000 sq km (~75,000 square miles), is situated in southern Russia to the east of the Caspian Sea. The Turkestan Desert consists of the great plain of the Caspian and Aral Seas and the hilly districts on its eastern border, formed by the western branches of the Hindu Kush Mountains.

In this area, the climate is extremely arid and continental. The desert consists of lowland and piedmont plains which rise to heights of around 925–1850 m (3000–6000 feet) above sea level. Precipitation is rare in the desert lowland region, which is considered to be one of the driest areas in the world. Average annual precipitation is 7–20 cm (~3–8 inches). The piedmont, on the other hand, qualifies as semi-arid, receiving annual precipitation of around 30–40 cm (12–16 inches). More than half the precipitation falls in the spring, and around 35 percent in the winter. July is the hottest month, with mean monthly temperatures ranging from about 27 °C along the Caspian Sea to about 32 °C in the interior and south. Mean January temperatures in the desert lowland range from −2° to +4 °C.

Although there is limited availability of information regarding olive cultivation in this region, several unique and promising olive genotypes have been found growing in the historic homeland of the olive tree. Bearing in mind that most of the cultivars of olives currently growing either in the traditional olive-growing area of the Mediterranean basin or in other new places originated from Mediterranean species, the unique genotypes available in the Turkestan Desert could play a vital role in expanding olive production to other similar parts of the world if judicial selection follows. Furthermore, there are still huge areas that are suitable for olive cultivation in this region that remain fallow, and new olive plantations can expand into these areas.

3.7.8 The Gobi Desert

The Gobi Desert is one of the world's greatest deserts, and the largest desert in Asia. It covers 1,300,000 square km of northern China and southeast Mongolia, and is surrounded by mountains (Figure 3.12). It is bounded by the Da Hinggan Ling (Greater Khingan Range) to the east, the Altun Shan and Nan Shan Mountains to the south, the Tian Shan Mountains to the west, and the Altay and Hangayn Nuruu (Khangai) Mountains and Yablonovyy Range to the north. A large area of the Gobi Desert is bare rock rather than sand. The Gobi is formed by a series of small basins within a larger basin rimmed by upland. The elevations of these basins range from

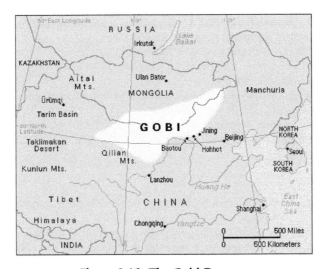

Figure 3.12: The Gobi Desert.

900 m (3000 feet) above sea level in the east to 1500 m (5000 feet) in the west. The basins are divided by low, flat-topped ranges and isolated hills that are the result of faulting action. The floors of the basins are unusually flat and level, and are formed of a desert pavement of small gravel resting on granite or metamorphic rock. There are, however, large areas of sedimentary rock, and some lava beds. Much of the sand and fine material has been blown away, but tall sand dunes rise along the desert's southern edge.

The Gobi Desert has sufficient grass to support scattered herds of sheep, goats and camels. Almost 99 percent of the Gobi Desert and desert steppe – arid and semi-arid zones – is used as natural pasture. In recent years, overgrazing and over-plowing around the desert's borders have resulted in erosion and loss of plant cover, which has caused the Gobi to spread, especially in northern China. This growing desertification has worsened dust and sand storms, which can blanket parts of East Asia and have serious environmental and economic effects. Chinese authorities began working to reverse the loss of vegetation in the early twenty-first century.

The desert is crossed from north to south by traditional trade routes and, alongside them, a railroad from Ulaanbaatar, Mongolia (extending north to the main Trans-Siberian Railroad) to Jining on China's main east–west line in the north.

References

Augustinos, A. A., Stratikopoulos, E. E., Zacharopoulou, A., & Mathiopoulos, K. D. (2002). Polymorphic microsatellite markers in the olive fly *Bacrocera olea*. *Mol Ecology Notes*, 2, 279–280.

BIDR (Blaustein Institute for Desert Research) (2000). The First National Report on the *Implementation of the United Nation Convention to Combat Desertification*. Midreshet Ben-Gurion: Jacob Blaustein Institute for Desert Research (available at http://www.unccd.int).

Bongi, G., & Palliotti, A. (1994). Olive. In: B. Schaffer, P. C. Andersen (Eds.) (pp. 165–187). *Handbook of Environmental Physiology of Fruit Crops. Temperate Crops*, 1. CRC Press, Inc, New York, NY.

Casas, J. J., Gessner, M. O., Langton, P. H., Calle, D., Descals, E., & Salinas, M. (2006). Diversity of patterns and processes in rivers of eastern Andalusia. *Limnetica*, 25, 155–170.

Maas, E. V. (1986). Salt tolerance of plants. *Appl Agric Res*, 1, 12–25.

Protnov, B. A., & Safriel, U. N. (2004). Combating desertification in the Negev: dryland agriculture vs dryland urbanization. *J Arid Environ*, 56, 659–680.

Rengasamy, P. (2002). Transient salinity and subsoil constraints to dryland farming in Australian sodic soils: an overview. *Aust J Exp Agric*, 42, 351–361.

UNEP (United Nations Environment Program) (2007). Environment for Development, Global Desert Outlook. (available at http://www.unep.org/geo/gdoutlook/016.asp, accessed 30 September 2007).

Further reading

Colley, C. C. (1983). The desert shall blossom: North African influence on the American Southwest. *W Historical Q*, 14, 277–290.

Dietzenbacher, E., & Velázquez, E. (2006). *Virtual water and water trade in Andalusia. A study by means of an input–output model*. Universities of Groningen and Pablo de Olavide, New York, NY.

Gat, J. R. (1979). Isotope hydrology of very saline surface waters, *Isotopes in Lake Studies* (pp. 151–162). IAEA, Vienna.

Horita, J. (2005). Saline waters. In: P. K. Agrawal, G. R. Gat, K. F. O. Froehlich (Eds.), *Isotopes in the Water Cycle: Past, Present and Future of a Developing Science* (pp. 271–287). International Energy Agency, Paris.

Rengasamy, P. (2006). World salinization with an emphasis on Australia. *J Exp Bot*, 57, 1017–1023.

Vanderlinden, K., Giraldez, J. V., & Meirrenne, M. V., et al. (2004). Spatial assessment of the average annual water balance in Andalusia. In: X. Sanchez-Vila (Ed.), *GeoENV IV. Geostatistics for Environmental Applications* (pp. 151–162). Kluwer Academic Publishers, Netherlands.

CHAPTER 4

The history of olive oil cultivation in the desert

4.1 Cultivation of olive oil in dry highland areas in ancient times

The exact place where olive cultivation began has not yet been discovered and is the subject of much debate (Loukas and Krimbas, 1983). However, it is commonly suggested that the drylands extending from the southern Caucasus to the Iranian plateau, and the Mediterranean coasts of Syria to Israel, are the original home of the olive tree (Zohary and Spiegel-Roy, 1975; Kiristsakis, 1998). It is also accepted that olive cultivation developed in these two semi-arid regions more than 5000 years ago, spreading from there to the island of Cyprus and on towards the drylands of Anatolia, or from the island of Crete towards Egypt. There is no doubt that the olive tree has distant roots in the history of mankind, and has played an important role in ancient civilizations (Galili *et al.*, 1997). The Bible contains up to 200 references to olive trees and olive oil. Fossils dating back to the Tertiary period (1 million years ago) prove the existence of an ancestor of the olive tree in southern Italy (Boskou, 1996). Two wild olive species, *Olea cuspidatae* and *Olea glandulifera*, have been found in the southern Himalayan range of Western Nepal (Burtolucci and Dhakal, 1999) and some of them are more than 15 m tall. This region lies in comparatively dry land in the rain-shadow of the Himalayas. The discovery shows that there is no doubt that the ancient homeland of olives stretched along the dryland areas from the Mediterranean basin to the Middle East to Central Asia to near the Himalayas.

Almost all the information available today regarding olive trees and olive oil is based on findings collected from the Mediterranean region. Apart from a limited number of reports mainly by Iranian authors (Hosseini-Mazinani *et al.*, 2004; Omrani-Sabbaghi *et al.*, 2007), no systematic data are available regarding the development of olive cultivation in the Iranian plateau and Central Asia.

Galili and colleagues (1997) reported finding thousands of crushed olive stones with olive pulp at the Kfar Samir prehistoric settlement off the Carmel coast south of Haifa. Observations at this site, and at other Late Neolithic to Early Chalcolithic offshore settlements in this region, record olive oil cultivation along the Carmel coastal plain as early as 6500 years ago. These findings have helped to define the technology of olive oil production and refine the chronological definition of cultural units along the southern Levant coast during the seventh millennium BCE – a time of major transition between the end of the Neolithic age and the beginning of the Chalcolithic age. Additional archeological findings suggest that olive cultivation spread among Syria, Israel and Crete between 5000 and 1400 BCE. Commercial networking and the application of new knowledge then brought it to dry areas of Southern Turkey, Cyprus and Egypt. Until 1500 BCE, Greece – and particularly Mycenae – was the area where olives were most heavily cultivated. However, with the expansion of the Greek colonies, olive cultivation reached southern Italy and northern Africa in the eighth century BCE, and then spread into southern France. Olive trees were planted throughout the entire Mediterranean basin during Roman rule. According to the historian Pliny, Italy had "excellent olive oil at reasonable prices" by the first century CE – "the best in the Mediterranean," he maintained (Di Giovacchino, 2000).

According to Jewish mythology, in the drylands of Israel, Kings David and Solomon placed great importance on the cultivation of olive trees. King David even had guards watching over the olive groves and warehouses, ensuring the safety of the trees and their precious oil. Even today, in many places in northern Israel very old trees and olive groves can be seen. Some of them are considered to be more than a thousand years old (Figure 4.1).

Olive trees dominated the rocky Greek countryside and became pillars of Hellenic society. They were so sacred that those who cut one down were condemned to death or exile. In ancient Greece and Rome, olive oil was the hottest commodity. Large, technologically advanced ships were built for the sole purpose of transporting it from Greece to trading posts around the Mediterranean (Frankel *et al.*, 1994).

In the sixteenth century BCE the Phoenicians started disseminating olives throughout the Greek islands, later introducing them to the Greek mainland between the fourteenth and twelfth centuries BCE, where their cultivation increased and gained great importance. In the fourth century BCE, Solon issued decrees regulating olive planting (de Graaff and Eppink, 1999).

Greek mythology records that when Athena, the Goddess of wisdom and peace, struck her magic spear into the Earth, it turned into an olive tree; thus, the location where the olive tree first appeared and grew was named Athens in honor of the Goddess. Local legend dictates that the original olive tree still stands at the ancient sacred site. Citizens claim that all Greek olive trees originate from rooted cuttings that were grown from that original olive tree. The great ancient Greek writer and philosopher, Homer, claimed in his writings that the ancient olive tree growing in Athens was already 10,000 years old. Homer stated that the Greek courts sentenced

Figure 4.1: Ancient olive tree growing in Northern Israel.

people to death if they destroyed an olive tree. In 775 BCE at Olympia, Greece, at the site of the ancient Olympic stadium, athletes competed and trained, and winners were triumphantly acclaimed and crowned with a wreath made of olive twigs. Ancient gold coins that were minted in Athens depicted the face of the Goddess Athena wearing an olive leaf wreath on her helmet and holding a clay vessel of olive oil.

It is documented that the Greeks began olive cultivation in 700 BCE (Malcolm, 2006). Among the Greeks the oil was valued as an important part of the diet, as well as for its external use. The Romans employed it widely in food and cookery, and the wealthy used it as an indispensable adjunct to grooming. In the luxurious days of the later Roman Empire it was said that long and pleasant life depended on two fluids – wine and olive oil. Pliny describes 15 varieties of olive cultivated in his day, the Licinian being the most esteemed, and the oil obtained from Venafrum in Campania being the finest known to Roman connoisseurs. The produce of Istria and Baetica was then regarded as second only to that of the Italian peninsula.

4.2 Expansion of olive oil cultivation in the Mediterranean region

Cultivation of the olive was (and remains) a key characteristic of Mediterranean mixed farming, and played a large part in the economic development of ancient Greece because of the suitability of olive oil as an export crop. For instance, Attica, a region of Athens, imported grain and exported olive oil from early historic times. The Athenian pottery industry was stimulated largely by the demand for containers in which to export olive oil.

From the sixth century BCE onwards the olive spread throughout the Mediterranean countries, reaching the dry areas of Tripoli, Tunis, and the island of Sicily. From there, it moved to southern Italy. Presto, however, maintained that the olive tree in Italy dates back to three centuries before the fall of Troy (1200 BCE). Another Roman annalist (Penestrello) defends the traditional view that the first olive tree was brought to Italy during the reign of Lucius Tarquinius Priscus the Elder (616–578 BCE), possibly from Tripoli or Gabes (Tunisia). Cultivation moved from south to north, from Calabria to Liguria. By the time the Romans arrived in northern Africa, the Berbers knew how to graft wild olives and had developed its cultivation widely throughout the territories they occupied.

The Romans continued the expansion of the olive tree to the countries bordering the Mediterranean, using it as a peaceful incentive in their conquest to settle the people. It was introduced in Marseilles around 600 BCE, and spread from there to the whole of Gaul. The olive tree made its appearance in Sardinia in Roman times, while in Corsica it is said to have been brought by the Genoese after the fall of the Roman Empire.

Olive growing was introduced into Spain during the maritime domination of the Phoenicians (1050 BCE), but did not develop to a noteworthy extent until the arrival of Scipio (212 BCE) and Roman rule (45 BCE). After the third Punic War, olives occupied a large stretch of the Baetica valley and spread towards the central and Mediterranean coastal areas of the Iberian Penisula, including Portugal. The Arabs brought their varieties with them to the south of Spain, and influenced the spread of cultivation to such a degree that the Spanish words for olive (*aceituna*), oil (*aceite*) and wild olive tree (*acebuche*), and the Portuguese words for olive (*azeitona*) and for olive oil (*azeite*), have Arabic roots.

In 1100 CE, olive groves again begin to propagate in Italy, with Tuscany becoming the primary center for their cultivation. Strict laws were made during that time regulating the cultivation of the olives and the commerce of oil, and many of those same laws still stand today.

Venice and Genoa also began to trade oil, and each year an increasing quantity of oil was produced by the countries of the Mediterranean river basin. At the beginning of 1300 CE, Apuglia became an enormous olive grove and olive trees were planted in other regions of Italy, making it the center of olive oil production throughout the world.

The wars of 1400 CE marked a critical time for oil production, but it was only a short while before olive oil, especially Italian olive oil, recovered and again became the lead player on the sumptuous Renaissance tables of Europe.

4.3 Olive oil cultivation beyond the Mediterranean region

With the discovery of North America in 1492, olive farming spread beyond its Mediterranean confines. The first olive trees were carried from Seville to the West Indies, and later to the North American continent. By 1560, olive groves were being cultivated in dry areas of Mexico, and subsequently in Peru, California, Chile and Argentina, where one of the plants brought over during the Conquest – the old Arauco olive tree – lives to this day.

Soon after the discovery of the Americas, Spanish settlers began to cultivate olive trees. The olives flourished as well as in their native lands of the Mediterranean, and olive oil of fair quality is known to be produced in South America. Olives were carried to Peru at a later date, but have not prospered as successfully there. They were taken to Mexico by the Jesuit missionaries of the seventeenth century, and to northern California, where the industry stagnated under the careless management of the later English-speaking culture. Olive cultivation has also been attempted in the south-eastern states, especially South Carolina, Florida and Mississippi.

Olive oil production leveled off during the 1600s due to high taxes, and did not begin to grow until the early 1700s, with the development of the free market and the exemption of taxes for olive groves. In the 1800s, olive groves continued to flourish thanks to the monetary incentives guaranteed by the Papal State.

Recent findings of morphological changes in the endocarps of olives under domestic cultivation in both Mediterranean geographical areas and timescales have provided new criteria for the identification of ancient developed olive cultivars. These changes have allowed the determination of the origins of cultivated forms created and/or introduced in the north-western Mediterranean regions, and understanding how human migrations affected the rest of the western Mediterranean regions. Based on these findings, a model of the diffusion of olive cultivation has been proposed. It shows evidence of an indigenous origin of the domestication process, which is currently recognized in the north-western area, during the Bronze Age.

4.4 Olive oil cultivation in Central Asia and in the eastern hemisphere

In more modern times the olive tree has continued to spread outside the Mediterranean region, and today is farmed in places as far removed from its origins as southern Africa, Australia, Japan and China. As the French writer George Duhamel said, "The Mediterranean ends where the olive tree no longer grows," which can be capped by saying that "Where the sun permits, the olive tree takes root and gains ground."

In the eastern hemisphere, the olive has been established in many inland districts that would in ancient times have been considered ill-adapted for it. It was known comparatively early historically in Armenia and Persia. The tree has been introduced into Chinese agriculture and has become an important addition to the crops of Australia's farmers; there are probably few coastal districts in Australia where the tree would not flourish. In Queensland and in South Australia, near Adelaide, the climate has been found to be especially suitable for olive cultivation, and it has also been successfully introduced into some drylands of South Africa and New Zealand.

In the 1970s, American scientists published several studies citing the correlation between the Mediterranean diet and a lower incidence of health problems. Based on fresh vegetables, seasonal fruits, grains and olive oil, the Mediterranean diet was confirmed to be not only delicious but also good for health.

Although evidence has shown that the central Asian region, extending from Iran to near the Himalayan region, is the natural homeland of the wild olive, and numerous other species belonging to the family oleaceae are widely found in this region, modern olive oil cultivation generally only began there when it had reached its full potential in the Mediterranean region (Burtolucci and Dhakal, 1999). Today the area of land dedicated to olive growing is expanding, as in Turkmenistan, where there are new groves in the south-east zone in particular. The significant amount of olive oil produced by this region has already started to appear in the international olive oil market.

4.5 Olive oil uses and trade in the ancient Middle East

Olive oil is considered to be the oldest alimentary oil after sesame oil. In ancient times, olive oil was used as a food, as a medicinal product, and in religious ceremonies. When the history of the olive and its cultivation began along the coastal regions of the Mediterranean basin more than 5000 years ago, it was immediately used as food (the fruit) and for medicinal purposes (the oil, which was extracted using rudimentary methods).

The Greek philosopher Homer called it "liquid gold." In ancient Greece, athletes ritually rubbed it all over their bodies. Its mystical glow has illuminated history. Drops of it have seeped into the bones of dead saints and martyrs through holes in their tombs. Olive oil has been more than mere food to the people of the Eastern Mediterranean; it has been medicinal, magical, an endless source of fascination and wonder, and the fountain of great wealth and power. The olive tree, a symbol of abundance, glory and peace, gave its leafy branches to crown the victorious in friendly games and bloody war, and the oil of its fruit has anointed the noblest of heads throughout history. Olive crowns and branches, emblems of benediction and purification, were ritually offered to deities and powerful figures – some were even found in Tutankhamen's tomb.

Phoenicians, who were the great oil traders of ancient times, then brought the precious olive oil to Greece, where it was revered for its splendor. In addition to it being food, the Greeks used

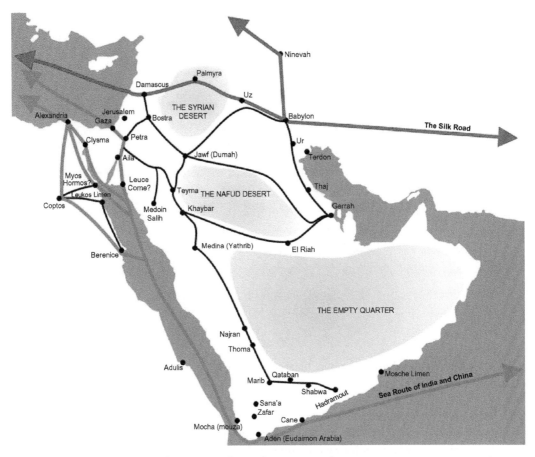

Figure 4.2: The Nabataean trade route.
Source: http://www.nabataea.net/trader.html.

olive oil as fuel for lamps, in sacred initiations, and as a massage oil to prepare the skin before fighting in the arena.

It was a common and widespread belief that olive oil conferred strength and youth. In ancient Egypt, Greece and Rome, it was infused with flowers and grasses to produce both medicines and cosmetics; a list has been excavated in Mycenae enumerating the aromatics (fennel, sesame, celery, watercress, mint, sage, rose and juniper, among others) added to olive oil in the preparation of ointments.

The sacred lamp that was used in ancient Greek culture for lighting dark rooms at night was fueled by olive oil. Aged olive oil was also used in consecrated anointing rituals at weddings and baptisms. Herodotus wrote in 500 BCE that the cultivation and export of olives and olive oil were so sacred that only virgins and eunuchs were allowed to tend orchards of olive trees.

Figure 4.3: Shivta, a Nabataean town located in Israel's Negev Desert in the center of the Nabataean trade route.

The importance of olive oil continued to increase throughout the Roman Empire to the point where the Empire's southern regions were organized around oil provinces. However, the collapse of the Roman Empire signaled the end of the cultivation of the olive tree. The ancient plant survived only around convents and in the fortified regions high in the hills of Tuscany.

The Nabataeans were an ancient (200 BCE–630 CE) trading race who built a brilliant, sophisticated civilization based on agriculture and commerce, which at its peak stretched from modern-day Yemen to Damascus, and from western Iraq into the Sinai Desert. They traded goods from one place to another, selling them at local markets, thus forming a solid link between eastern goods and western markets, by using their knowledge of sea and caravan routes. Amazingly, they managed to take their caravans through the desert, unaffected by the local tribes, by developing water collection systems along trading routes (Figure 4.2) that provided them with water in the desert at places unknown to others. They built prosperous cities at the crossroads of the ancient world, at points of transit between India, Yemen, Egypt, Greece and Rome. The Nabataeans began to focus their attention on agriculture. To survive and prosper in this arid region, the Nabataeans had to develop and maintain a relatively advanced water-management capability. They were skilled at collecting, transporting and storing water, and irrigated their land with an extensive system of dams, canals and reservoirs. With this complicated system of water conservation, they were able to supply fresh water to their cities and for the agriculture that sustained them. They also built terraces, and had enough water to grow crops and raise livestock in the surrounding countryside. Large ranches and horse-breeding programs were developed. Ranching and the cultivation of olives and grapes appear to have become their focus (Figure 4.3).

The Negev Desert is an example of a unique system that can be used to demonstrate the skill of the Nabataeans in developing new hydrological systems, and irrigation and agricultural methods, for sustainable desert cultivation. This involved relatively simple technological manipulation of the natural landscape in a manner that dealt simultaneously with the problems of both water and nutrients. Steep lands were left bare to encourage runoff during the brief, intense rainstorms characteristic of the region. Small "catchment runoff farms" were constructed in catchment areas located on slopes, and in cultivated areas in the drainage bottomlands below these catchments. The ratio of catchment to farm plots varied according to the amount of runoff. The farm plots were constructed with rock dikes across the water courses, thus accumulating and conserving soil inside the plots. The catchment slopes were modified to maximize runoff. Stone conduits were built to carry water to various parts of the bottomland farm plots as required. The cropping systems varied according to the size of the watershed and its drainage channels.

References

Boskou, D. (1996). History and characteristics of the olive tree. In: D. Boskou (Ed.), *Olive Oil: Chemistry and Technology* (pp. 1–11). AOCS Press, Champaign, IL.

Burtolucci P, Dhakal BR (1999). *Prospects for Olive Growing in Nepal*. Field Report-1. FAO TCP/NEP/6713 Kathmandu, Nepal.

de Graaff, J., & Eppink, L. A. A. J. (1999). Olive oil production and soil conservation in southern Spain, in relation to EU subsidy policies. *Land Use Policy*, 16, 259–267.

Di Giovacchino, L. (2000). Technological aspects. In: J. Hardwood, R. Aparicio (Eds.), *Handbook of Olive Oil. Analysis and Properties* (pp. 17–59). Aspen Publications, Aspen, CO.

Frankel, R., Avitsur, S., & Ayalon, E. (1994). *History and Technology of Olive Oil in the Holy Land*. Olearius Editions, Arlington, VA, p 208.

Galili, E., Stanely, J. D., Sharvit, J., & Wienstein-Evron, M. (1997). Evidence for earliest olive oil production in submerged settlements off the Carmel Coast. *Israel. J Archaeol Sci*, 24, 1141–1150.

Hosseini-Mazinani, S. M., Mohammadreza Samaee, S., Sadeghi, H., & Caballero, J. M. (2004). Evaluation of olive germplasm in Iran on the basis of morphological traits: assessment of 'ZARD' and ROWGHANT' cultivars. *Acta Hort*, 634, 145–151.

Kiristsakis, P. (1998). *Olive Oil. From the Tree to the Table*, 2nd edn. Food & Nutrition Press, Inc., Trumbull, CT.

Malcolm P, (2006). History of olive trees. *EzineArticles*, http://ezinearticles.com/?History-Of-Olive-Trees&id=368070 (accessed 17 September 2007).

Omrani-Sabbaghi, A., Shahriari, M., Falahati-Anbaran, M., Mohammadi, S. A., Nankali, A., Mardi, M., & Ghareyazie, B. (2007). Microsatellite markers based assessment of genetic diversity in Iranian olive (*Olea europaea* L.) collections. *Sci Hort*, 112, 439–447.

Zohary, D., & Spiegel-Roy, P. (1975). Beginning of fruit growing in the old world. *Science*, 175, 319–327.

Part 2
Biotechnologies and Olive Oil Cultivation

CHAPTER 5

Environmentally friendly desert agro-biotechnologies

Over the past three decades, several environmentally friendly advanced desert biotechnologies (Ag-Environ biotech) have been developed for intensive agricultural uses. Prime among them are advanced techniques regarding soil and irrigation management. In this chapter, these two basic and essential topics are reviewed, with special emphasis on olive oil cultivation. Application of these environmentally friendly agro-biotechnologies will be discussed, using data from studies carried out in the Israeli Negev Desert as a model. These technologies have been successfully practiced in the Negev for a number of years, and could be implemented worldwide in similar desert circumstances. In order to understand more precisely the environmentally friendly desert agro-biotechnologies that have been applied in the Negev Desert, this chapter is divided into three main sections: (1) The soils of the Negev Desert; (2) Irrigation in the Negev Desert; and (3) Advanced biotechnologies for olive cultivation under saline conditions. The former two sections will aid in understanding the nature of the soil and water sources of the Negev Desert, which will help in selecting olive-cultivable land in other desert regions; the latter section will provide guidance regarding cultivating olives in the desert.

5.1 The soils of the Negev Desert

5.1.1 Geomorphology and mapping of Negev soils

As described in Chapter 3, the Negev Desert lies in the central southern part of Israel. Most of the land of the Negev is hyper-arid, arid or semi-arid, with extreme temperatures and low rainfall. Owing to the limited rainfall and therefore reduced weathering, the soil in the desert is considered to be newer than in other areas. Most of the Negev terrain consists of marine sedimentary rock, primarily limestone and chalk, ranging widely in age. The landscape consists of rocky hills and mountains, gravel-paved plateaus, coarse sediments and sands. The typical Negev soil is loess, a buff-colored, fine-grained deposit of desert dust originating from the Sinai Desert. Though unstable and highly erodible by wind and water, loess has some water-retaining capabilities that lead the soil to become highly productive if water is available.

5.1.2 Analysis of Negev soils

Two soil groups are commonly found in the Negev area. The Aeolian soil group is found mostly in the plains and the depression in the Negev, and is calcareous and slightly saline. The Reg soil group is closely associated with the desert parts of Israel, and is characterized by desert pavement. Since the soil is saline and stony, it is not used for agriculture as such.

5.2 Irrigation in the Negev Desert

Israel's population and economy have grown significantly over the past 60 years. This significant increase in population combined with sustained economic growth has placed enormous pressure on the country's scarce water resources. Since its establishment in 1948 as an independent state, research into water resources, planning, development, control, management and monitoring has become a national priority regarding the legislative basis, institutional set-up, budgetary investment, R&D into water-related technology and agronomy, and the development of monitoring data and tools. Rigorous enforcement of policies, institutional and legal measures has been very effective (World Bank, 2006). The total annual volume of fresh water available in Israel is 1800 million cm^3 (Table 5.1). In 2005, approximately 45 percent of this was used for municipal consumption and 50 percent for agriculture. Groundwater development started early in Israel's history, with archeological findings showing that hydro-geological research was carried out as early as 1000 BCE. In the twentieth century aquifers have been mapped and knowledge about them has developed; over 65 percent of the total water supply of the country is now pumped from the aquifers.

5.2.1 National water delivery plan for Israel

Owing to the geographical conditions and the rainfall regime, approximately 80 percent of the water resources are located in the northern part of the country. Conversely, 65 percent of the

Table 5.1: The long-term potential for renewable water from main sources in Israel

Source	Replenishable amount (million cm^3/yr)
Coastal aquifer	320
Mountain aquifers	370
Lake Kinneret	700
Additional regional resources	410
Total	1800

Source: World Bank, 2006.

country's arid land is located in the south. Hence, large quantities of water must be conveyed from the north to the south. The National Water Carrier (NWC) interlinks the three main water sources (Lake Kinneret, the Yarkon-Taninim aquifer and the coastal Aquifer) with branching regional water grids. The NWC originates from the Sapir pumping station, located on the shore of Lake Kinneret, and flows via gravitation across the Yizre'el Valley to a terminal station near Yarkon Springs. From here, the water is transmitted southward for over 95 km to the arid Negev region (Figure 5.1).

From the very start, it was apparent that the limiting factor regarding agriculture in the Negev is scarcity of water. Awareness that successful modern agriculture hinges upon irrigation, which requires a reliable supply of water, led to the launching of a series of exploratory studies. Attempts were made to drill wells and draw underground water near the settlements, and to build dams and reservoirs to collect seasonal floodwaters, but these failed because of the large annual fluctuations in water quantity, the intensity of floods, and technical difficulties. It was concluded that the only way to secure a dependable and sufficiently large supply of water for desert agriculture was to transport fresh water from northern sources via pipes. In 1964 The National Water Carrier (NWC) was initiated, joining local water factories.

The National Water Carrier distributes 400 million cm^3 of water from Lake Kinneret (also known as the Sea of Galilee) in the north to points of demand in central Israel and the south. Most of the regional water systems are incorporated into the National Water Carrier to form a well-balanced network in which water can be shifted from one line to another according to conditions and demands (Gvirzman, 2002). This integrated network of pumping stations consists of piping systems, aqueducts, open canals, tunnels, reservoirs, and large-scale pumping stations. The NWC delivers water as far as Mizpe Ramon in the northern Negev.

5.2.2 Hydrological mapping of Negev aquifers

The Negev region of Israel consists of two main aquifers, the north Negev Eocene aquifer and the Arava aquifer. Together they provide a complementary supply of groundwater for agriculture

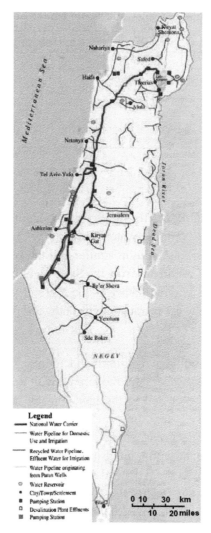

Figure 5.1: National water delivery plan, Israel.
Source: http://ww.jewishvirtuallibrary.org.

and industrial use where water is sparse, or where there is no extension of The National Water Carrier. The groundwater levels and direction of flow as described by Issar and Nativ (1988) are displayed in Figure 5.2.

Eocene aquifer

The Eocene aquifer, which lies in the north of the Negev Desert, consists mainly of chalk and contains little water, with the annual recharge estimated to be between 10 and 15 million cm^3/y (Gvirzman, 2002). High evaporation rates, a minute amount of rain and the low filtration rate of

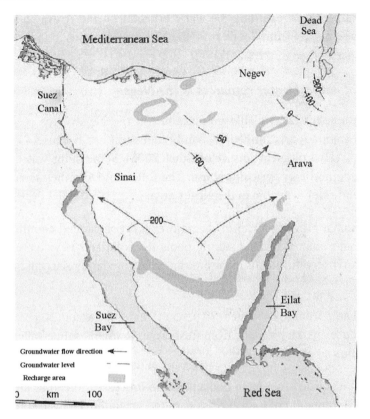

**Figure 5.2: Groundwater levels and direction of flow, Negev, Israel.
After Issar and Nativ (1988).**

the soil determine the low recharge of the aquifer. This reservoir is already becoming polluted – especially in the Ramat Hovav area, where more than 20 chemical factories and Israel's national waste storage are located.

Arava aquifer

Since the Arava Valley forms a rift valley, thick layers of alluvium have accumulated in it. As the precipitation on the valley proper is negligible, the main recharge of the alluvial aquifers comes from the floods that flow from the Negev and Edom Mountains. Part of the floodwater percolates through the alluvial fans into the body of groundwater. Limestone aquifers are located at the north-western edge of the Arava Valley. The water is stored in limestone rocks, and forms a semi-fossil aquifer that extends westwards under the Central Negev. Today, water from this aquifer is pumped for the potash industry of the Dead Sea. The Nubian sandstone (fossil) aquifer is a regional aquifer containing several hundred billion cm^3 of brackish fossil water, extending from the Sinai through the Negev and down to the Arava Valley. The fossil water in this aquifer was recharged tens of thousands of years ago, when the region was more humid (Tsur *et al.*,

1989). The current estimate regarding the water potential of the Arava aquifer is 40 million cm^3/y (Haim, 2002), mostly from floods from the Negev Mountains in the west and the Edom Mountains in the east.

5.2.3 Analysis of natural water resources in the Negev

Despite the large amount of available groundwater in the Negev area at a depth of about 700–1100 m, its exploitation is limited to mainly industrial use because 43 percent is saline (over 600 mg Cl^-/l) (Hydrological Service of Israel, 2005). Some of the water can be reclaimed and used later for domestic or agricultural use. The following describes some of the chemical properties of groundwater available in the Negev area.

The location of Ramat Hovav on top of the aquifer has contributed significantly to the low quality of the groundwater. Damaged storing pools for pollutants have caused constant leaks over the past years. Observation drills in the area have indicated irregularly high levels of copper, chrome and lead.

Both aquifers recharge from fossil water (70 percent) and floodwater (30 percent) (Adar *et al.*, 1992). The chloride levels vary widely, from fresh water to highly saline water (Table 5.2), and increase with depth.

Oren *et al.* (2004) identified two main contamination processes linked to human activity: (1) salinization due to circulation of dissolved salts in the irrigation water itself (mainly chloride, sulfate, sodium and calcium); and (2) direct input of nitrate and potassium, mainly from

Table 5.2: Chloride levels (mg/l) in the Arava Valley

Region	Aquifer	
	Kurnub group	Judea group
Neot Hakikar		
Hazeva-Idan	715	
En Yahav	600–700	
Zofar	3500	480–900
Paran Menuha Hiyon	580–750	400–1700
The Arava divide		
Yahel Yaalon	400–1350	150–850
Qetura Yotvata	350–3500	
Samar Timna Be'er Ora	3000–3500	200–800
Eilat		

Salination of groundwater is controlled by both present-day and ancient base levels, namely the Dead Sea in the north and the Gulf of Elat in the south (Kafri and Goldman, 2007).

fertilizers. The nitrate concentrations in local shallow groundwater range from 100 to 300 mg/l, and in the upper sub-aquifer they are over 50 mg/l. In addition, a water survey conducted by the Hydrological Service in 2005 revealed irregular levels of iron, manganese and sulfur, which must be treated prior to use.

5.2.4 Rainwater analysis

Recharge of the three main aquifers in the Negev area is partly from rainwater. The composition of these rainwater recharged aquifers depends on desert dust carried in the wind, and evaporation of sea water. In arid zones, its composition is mainly determined by desert dust carried by rain drops containing calcium carbonate ($CaCO_3$), dolomite ($CaMg(CO_3)_2$) and quartz (silicon dioxide). Sparse rainwater in the Negev area is relatively alkaline (pH 5.3–7.6) (Mamane *et al.*, 1987) due to a lower density of industrial facilities in the surrounding area (some of the acidity is neutralized by the mineral grain calcium carbonate).

5.2.5 Water-harvesting technologies in the Negev

Floodwater from excess rain can be collected for use; this is called water harvesting. Floodwater harvesting involves several methods to improve the use of ephemeral rainwater at a specific site, before it leaves a geographical region. Thus, the effect of temporal shortages of water supply during dry spells can be mitigated.

In the Negev, the amount of floodwater is quite high – for example, the average floodwater potential of the Besor stream is estimated to be 14.3 million cm^3/y. Floods occur in the area on average five times a year in the northern Negev, and once a year in the southern Negev.

5.2.6 Floodwater catchments in the Negev

The first catchment in Israel was established to capture the Shiqma stream. In the winter of 2002/03, 12.8 million cm^3 was harvested. The Besor stream is the largest stream draining into the Mediterranean Sea, with a drainage area of 3400 km^2 and a peak flow of 1000 m^3/s (Israeli Water Association, 2003). Combining reclaimed water with the Besor-harvested floodwater provides a constant supply of water for agricultural use (Figure 5.3). Another catchment is located in the Arava Valley, and consists of six separate structures with a total water-holding capacity of 8 million cm^3, mainly for agricultural use. The salinity and quality of the water from the different reservoirs differ, so dilution and treatment of the floodwater must be applied before diverting it for irrigation.

5.2.7 Economic aspects of underground water utilization

The development of groundwater catchments entails a large initial investment, including a feasibility study, the construction of drilling wells, installment of pumping facilities, etc. Since

Figure 5.3: Ephemeral Beer-Sheva stream in the winter of 2003.
Source: http://www.sviva.gov.il/Environment/Static/Binaries/index_pirsumim/p0367_1.pdf.

most of the underground water or fossil water is below 1000 m, an effective and long-term plan should always be the first step for any underground water utilization project. It is important to plan not only for the initial stage, but also for the long run. Planning for the future involves unavoidable uncertainties. The decision to develop a given fossil groundwater stock is therefore twofold: first, a decision must be made on whether to develop the aquifer, and second, the scale of the project and the allocation of the groundwater over time must be planned.

As in any other production process, the operation of pumping water from the ground entails two basic types of costs. The first is a fixed cost, which includes costs not affected by the extraction rate – such as facilities, infrastructure, and equipment required by the groundwater extraction technology. The second is a variable cost (VC), which is the operational cost of extraction. This depends mainly on the extraction rate and the volume of remaining stock. As the stock decreases the water table declines; therefore the water must travel a longer distance to reach the surface, and hence the variable cost increases (Tsur *et al.*, 1989).

Two basic extraction technologies are available. In the first, long galleries driven from deep shafts are constructed, and water is pumped through these galleries (Issar, 1985). The second is conventional drilling technology.

Since the cost of installation and operation of the groundwater pump is high anywhere, including the desert, it is almost impossible for individual growers to implement such a project unless they are very rich, and cultivate a large area on a long-term contract basis. Therefore, the concerned government or regional authority or any other public company should be involved. One of the reasons such deep wells are operating successfully in Israel is the government's involvement in the project. In order to reduce the cost of the production of such groundwater, multiple uses such as greenhouse heating, fish farming and irrigation can be served. The details of such uses will be discussed in later chapters.

5.2.8 Reuse of wastewater: an Israeli experience

Sustainability of agriculture is directly linked to the efficiency of water use. As fresh water becomes scarce, and competition with other sectors (i.e., urban, industrial and environmental) increases throughout the world, and particularly in dryland areas, farmers are finding themselves relying more and more on the utilization of marginal water resources (recycled and saline water). Reuse of wastewater for desert agriculture can enhance the available water supply in times of water shortage. Interest in this water resource developed in the early 1980s, after the construction of a centralized sewer system and centers for various water treatments (Biswas, 1990). For most arid and semi-arid countries, where conventional sources are almost fully committed, reclaimed water may have a greater impact on future water availability than any other technological solution aimed at increasing water supply, such as water harvesting, weather modification and desalination. Treated water can be used for irrigation, industrial purposes, and groundwater recharge (Biswas and Arar, 1988). Furthermore, wastewater irrigation is an environmentally sound wastewater disposal practice.

Israel has placed wastewater reuse high on its list of national priorities. This is because of a combination of a severe water shortage, the threat of pollution to the diminishing water resources, and a concentrated urban population with high levels of water consumption and wastewater production. Indeed, relative to its size and means, Israel has devoted more effort to wastewater reuse than any other country. Data have shown that about 470 million cm^3 of wastewater is produced every year in Israel, of which about 430 million cm^3 (almost 90 percent) is treated. About 70 percent (340 million cm^3) of effluent (treated wastewater) is reused every year for agriculture, and only 26 percent (110 million cm^3) of non-reused effluent is returned to the environment (Mekorot, 2007). This is reflected in the fact that Israel has the highest percentage in the world of wastewater effluent being reused for agricultural irrigation, and of wastewater reuse per capita, and is second (after California) in overall wastewater reuse. Extensive experience has therefore been gathered in this field, and a multitude of technologies and approaches are continuously being practiced and tested in Israel (Shelef, 2006).

Israel's national policy calls for the gradual replacement of the freshwater allocation to agriculture by reclaimed effluent. In 1999, treated wastewater constituted about 22 percent of consumption by the agricultural sector. It is estimated that effluent will constitute 45 percent of the water supplied to agriculture by 2010 (www.sviva.gov.il). Rules governing the treatment of wastewater designated for irrigation of different crops were established by the Ministry of Health in 1981 under the Public Health Ordinance. They require municipalities or industries whose sewage exceeds an amount equivalent to 10,000 residents to treat sewage to a minimum baseline level of 20 mg/l BOD and 30 mg/l suspended solids.

According to the plan, the agricultural water supply, which constituted approximately 72 percent of the overall water resources in 1985, will diminish to approximately 50 percent of overall resources in 2010, and 36 percent of this agricultural water supply will be reclaimed wastewater (Table 5.3). It should be noted that during drought years, the water supply to the agricultural sector is severely diminished. During such times, reclaimed wastewater constitutes a significant part of the agricultural water supply (approximately 40 percent in 1999/2000). Farmers connected to reclaimed wastewater conduits received almost the full amount of their annual water supply, even during drought years, whereas those who relied on fresh water sustained painful cuts.

About one-fourth of Israel's total wastewater undergoes treatment in the Dan Region Wastewater Treatment Reclamation Plant, which produces high-quality effluents. The Dan Region Treatment and Reclamation Plant (Shafdan) is located south of Tel-Aviv, and is the largest wastewater treatment plant in Israel, serving a population of 2.1 million with an annual flow of 120 million cm^3/y. It is also the largest reclamation plant in the wastewater reuse scheme in Israel, and one of the largest in the world. It serves the city of Tel-Aviv, and 12 other municipalities of the Dan Region. The layout of the scheme, including the treatment, effluent spreading sites, recovery wells after soil-aquifer treatment, and the "third-line" conveyance pipeline to the Southern and Negev districts to which 115 million cm^3 of reclaimed water was delivered for unrestricted irrigation in the year 2000 as shown in Figure 5.4.

The system consists of facilities for collection, treatment, groundwater recharge and reuse of municipal wastewater from the Dan metropolitan area (population 2.1 million, average municipal wastewater flow 270,000 cm^3/day). It is based on a modern biological-mechanical activated sludge plant with nitrogen removal (Figure 5.5). Following treatment, the effluent is recharged into the regional aquifer by means of spreading basins. A separate zone is created within the regional aquifer that is centered beneath the recharge basins and is dedicated to treatment and seasonal storage of the effluent (SAT – soil aquifer treatment). SAT consists of controlled passage of effluent through the unsaturated zone of the aquifer, mainly for purification purposes, as

Table 5.3: Wastewater reuse by reclamation as part of the overall water supply and the agricultural supply in Israel

Hydrological year (Oct–Sep)	Total water supply (million cm³)	Agricultural supply (million cm³)	Reused wastewater		
			Quantity (million cm³/y)	% of total supply	% of agriculture
1985/86	2050	1490	80	3.9	5.4
1988/89	2050	1280	160	7.8	12.5
1990/91	1430	750	190	13.3	25.3
1997/98	2040	1230	255	12.5	20.7
1999/00	1585	720	285	18.0	39.6
2010/11	2400	1190	430	17.9	36.1

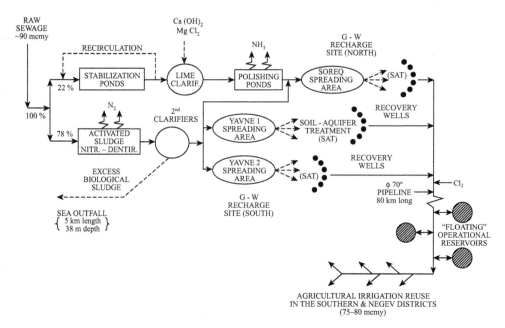

Figure 5.4: The Dan Region (Greater Tel-Aviv area of Israel) Wastewater Treatment and Reclamation Scheme.
Source: http://www.biu.ac.l/Besa/waterarticle3.html).

Figure 5.5: A view of the largest wastewater treatment plant in Israel.

well as for seasonal and multi-annual storage. The recharge operation is carried out by means of spreading basins, which are surrounded by adequately spaced recovery wells that permit separation of the recharge zone from the rest of the aquifer.

The major purification processes occurring in the soil aquifer system are slow-sand filtration, chemical precipitation, adsorption, ion exchange, biological degradation, nitrification, denitrification and disinfection. Water quality control in the recharge zone is virtually complete, and the very high quality of reclaimed water obtained after SAT is suitable for a variety of non-potable uses – and especially for unrestricted agricultural irrigation. Over a period of five years, about 400 million cm^3 of reclaimed water was supplied to the south of the country for unrestricted irrigation (i.e., in the Negev Desert).

Following the success of the Dan Region Treatment and Reclamation Plant, another treatment plant was built in the Haifa region. The Greater Haifa region (including the city of Haifa – the third largest city in Israel – and its neighboring municipalities) has a population of 400,000 and considerable industry. Currently, raw wastewater flows at approximately 35 million cm^3/y.

5.3 Advanced biotechnologies for olive cultivation under saline conditions

5.3.1 Mixing the irrigation water

Although the olive is a glycophyte with intermediate tolerance to salinity, it can be cultivated in saline condition when other fruit trees cannot grow (El-Gazzar *et al.*, 1979; Klein *et al.*, 1994). However, the underground water pumped from aquifers (as is the case in the Negev) is too highly saline for olive cultivation, at around 8–10 dS/m, so it must be mixed with fresh water to reduce the salinity to less than 5 dS/m in order to achieve olive cultivation (Wiesman *et al.*, 2002).

5.3.2 Irrigation practices

Irrigation management practices are a key step for the efficient use of saline water for olive cultivation in desert. This can be achieved by maintaining salt accumulation in the root zone at levels lower than the threshold values. Above the threshold values, salt will not only reduce the yield but the plant might also collapse. In high salt conditions, salt toxicity symptoms appear in the leaves. Typical toxicity symptoms are dead edges on the leaves, leaf drop, and necrosis of the stem tip (Figure 5.6) (Chartzoulakis, 2005).

Improved management practices for olive cultivation under saline conditions include the selection of proper irrigation methods and schedules, and efficient leaching of salts from the olive root zone.

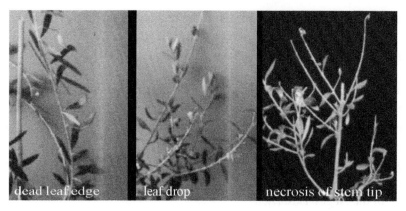

dead leaf edge   leaf drop   necrosis of stem tip

Figure 5.6: Typical salt toxicity symptoms in the olive tree.
Source: Chartzoulakis (2005).

Irrigation methods and scheduling

The irrigation method applied for olive cultivation under saline conditions may have a significant influence on the accumulation and distribution of salt in the soil profile (Paranychianakis and Chartzoulakis, 2005). Flood irrigation is impossible in desert conditions, due to the lack of water resources, and is completely unsuitable for olives. Sprinkler irrigation with saline water may cause injury due to high rates of foliar salt absorption, and the risk of injury will be much higher if such irrigation is practiced during the daytime when the evapotranspiration rate is high. Drip or trickle irrigation is the best and only option recommended for olive growing in the desert environment under saline conditions. A drip irrigation system keeps the soil moisture at a continuously high level at the root zone, and maintains a low level of salt concentration. Common problems that are associated with drip irrigation include the accumulation of salt at the wetting surface and clogging of drippers; however, in most cases the use of sub-surface drip irrigation (SDI) and leaching can overcome these problems.

Irrigation scheduling, which includes both the estimation of the irrigation requirements and application of the appropriate irrigation intervals, is very important in saline water irrigation. It is much more complicated to establish the appropriate irrigation schedule under saline conditions than with freshwater irrigation. The plant growth response under saline irrigation is a function of the salt's concentration in the soil solution (particularly Na^+ and Cl^-) and the matrix potential of the soil (Paranychianakis and Charzoulakis, 2005). To maintain adequate soil-water availability it is essential to restrict salt accumulation; this can be achieved by increasing the irrigation frequency. After irrigation, the moisture content of the soil is high and the salt concentration or osmotic pressure of the soil solution approaches its minimum. On other hand, in olive trees water stress can produce major changes in fruit set and fruit drop, fruit growth pattern and final

Table 5.4: The impact of water stress on the various growth stages of the olive tree

	Olive growth stage	Impact of water stress
1	Shoot growth	Reduced growth
2	Flower and bud development	Reduced flower formation
3	Blooming period	Incomplete flowering
4	Fruit setting stage	Poor fruit set and alternate bearing
5	First stage fruit growth	Reduced fruit size due to decreased cell division
6	Second stage fruit growth	Reduced fruit size; reason unknown
7	Last stage of fruit growth	Reduced fruit size due to decreased cell expansion
8	Oil accumulation stage	Reduced fruit oil content

Source: GSA (2006).

fruit size, fruit ripening and oil content on a dry weight basis. The critical stages of olive growth in relation to water stress are illustrated in Table 5.4.

Soil washing/leaching

Salinity is one of the most severe abiotic factors limiting agricultural production worldwide. The high rate of population growth and global warming and the consequent limitation of freshwater for irrigation, especially in dryland areas such as the Mediterranean, has led to salinity becoming more of a problem in these areas than ever before. It is well known that irrigation development has caused numerous cases of salinization of land and water resources. Of the 270 million hectares of irrigated land in the world, about 110 million hectares (roughly 40 percent) are located in this arid zone. The remaining 60 percent of irrigated land is irrigated under more humid climatic conditions, providing enough leaching to prevent the harmful accumulation of salts (Lambert and Karim, 2002). Irrigation development is likely to create new geo-hydrological flow regimens under which part of the large fossil/primary salt storage in the basin is being mobilized. In the Negev, where most of the crop water requirements are supplied through irrigation, and water often contains large amounts of dissolved salts, salinity control is a major objective of irrigation management. In the desert environment, when underground or brackish water is used for olive production, the problem of salt deposition in the root zone increases tremendously. In order to sustain olive productivity, the EC of the soil in the active root zone must be maintained at a level lower than 6 dS/m. The effects of irrigation and fertilization in arid zones not only cause salinization of the soil and groundwater below the fields (Stigter *et al.*, 1998), but also increase the nitrate concentrations in groundwater (Allaire-Leung *et al.*, 2001). Irrigation in arid zones causes an increase in the salt concentrations in the root zone due to high evapotranspiration (Oren *et al.*, 2004).

The accumulation of soluble salts that leads to salinization transforms fertile and productive land to barren land. Salinity adversely affects non-halophytes by inducing injury, inhibiting growth, and altering the plant morphology and anatomy. Injury is induced not only by the osmotic effects of salts but also by the accumulation of Cl^- and Na^+. When irrigating with saline water, the sodium ions displace other more useful ions, such as calcium and magnesium. Over time, the sodium in the soil becomes concentrated (Wiesman *et al.*, 2004).

Leaching is the process of applying more water to the crop than can be held by the root zone, so that the excess water drains below it, carrying soluble salts with it. When saline water is supplied to crops, leaching become indispensable in order to reduce salt levels in the root zone. Therefore, a proper leaching methodology throughout the entire year has been developed specifically for olive cultivation in arid areas. The amount of water (in terms of a fraction of the applied water) that must be applied in excess in order to control salts is referred to as the "leaching requirement" (LR), and can be calculated, for drip irrigation, from the following formula (Ayers and Westcot, 1985):

$$LR = ECw/(5ECe - ECw)$$

where ECw is the electrical conductivity (dS/m) of the irrigation water and ECe is the electrical conductivity (dS/m) of the saturation extract. ECe is the average soil salinity tolerated by the crop. Depending on the crop and salinity of the water and soil, a 15–20 percent leaching fraction is commonly recommended (Paranchianakis and Chartzoulakis, 2005). In areas such as the Ramat Negev region of the central south Israeli desert, it has been found that an additional $1000\,cm^3$/ha of moderate saline water (4.2 dS/m) in March (at the end of the winter season) and in November (at the beginning of the next winter) enables reduction of the salt level in the developing root zone to a level lower than 6 dS/m (Weissbein, 2006). This leaching regimen, together with accurate weekly drip irrigation, has overcome the problem of high levels of soluble salts in the active root zone, and can keep the soil conditions set for a productivity rate and yield very similar to that achieved with fresh-water irrigated trees of the same variety.

5.3.3 Soil quality improvement in the Negev

Soil quality has been defined as the capacity of a soil function, within ecosystem boundaries, to sustain biological productivity, maintain environmental quality, and promote plant and animal health (Doran and Parkin, 1995). Agricultural production can reduce soil quality. Tillage, for instance, initiates processes that may damage the natural soil ecosystem. It alters many of the soil physical properties, including bulk density, pore space and pore-size distribution, water-holding capacity, soil-water content, and aggregation (Spedding *et al.*, 2004). Degradation of soil quality through water and wind erosion is manifested as organic matter and nutrient depletion, soil-penetration resistance, soil acidity, and decreased microbial activity (Bezdicek *et al.*, 2003). The following are some of the key practices that help to improve soil quality by natural methods.

Soil solarization

Though soil-borne pests can be controlled in fruit trees by pre-planting application of pesticides, including the fumigants methyl bromide, chloropicrin and metam sodium (Katan, 1984), the use of these chemicals is often undesirable owing to their toxicity to animals and human beings, their residual toxicity in plants and soils, their complexity, and the high cost of treatment. Furthermore, restrictions on the use of soil-applied pesticides appear to be imminent as future environmental legislation is being implemented. To overcome this problem and remove soil-borne pests, solarization is considered the best technique (Di Primo *et al.*, 2003).

Soil solarization offers a simple organic solution to the problem. Solarization can control *Verticillium*, *Rhizoctonia solani*, *Fusarium oxysporum* f. sp. *vasinfectum* or *melonis*, *Orobanche*, *Sclerotium rolfsii*, *Pratylenchus* and others, as well as weeds, and can also increase yield.

By placing transparent plastic sheets over moist soil during periods of high ambient temperature, the sun's radiant energy can be absorbed and trapped by the soil, thereby heating the topsoil layer. Solarization during the hot summer months can increase the soil temperature to high enough levels (solar energy in the Negev area is estimated to be 195–201 kcal/cm^2 per year) to completely eradicate pathogens, nematodes, weed seeds and seedlings. It leaves no toxic residues, improves soil structure, and increases the availability of nitrogen (N) and other essential plant nutrients.

Soil solarization is a basic technique for Negev farmers, especially for off-season vegetable and flower production for the export market. It offers a simple, cost-effective, non-pesticidal treatment. This technique is equally useful in olive production, and can also increase the productivity of olive trees in an arid environment.

Mulching

Mulching is a common practice that helps to improve soil aeration. Mulch can be defined as a material used on the surface of the soil primarily to prevent the loss of water by evaporation. Soil aeration is one of the most important determinants of soil productivity. As the oxygen supply in the soil is limited, the rate of growth of most crops slows down, and stops entirely when the oxygen concentration gets below 2 percent. The oxygen supply is constantly exhausted by roots and microbes. Without sufficient oxygen in the soil, the normal functions of most crop plants and of aerobic microbes come to a standstill. Anaerobic bacteria use oxygen in organic and inorganic compounds, reducing them to sulfides, nitrites, ferrous compounds and other reduced compounds that are toxic to the plants. An excess of oxygen in the soil is also undesirable, because it leads to the organic matter being oxidized too rapidly. Soil surface conditions can be altered by mulching in order to aerate the soil better. Plants are sensitive to the aeration status of the soil. It is generally agreed that gaseous exchange between the soil atmosphere and the aerial atmosphere is primarily accomplished by the process of diffusion. The oxygen diffusion rate (ODR) is used as an index of the soil aeration status in the cropped field. Mulching also alters the soil temperature. The ODR is temperature dependent, and an increase in temperature decreases the solubility of oxygen and increases the diffusion through both gases and liquids (Khan and Mohsin, 1976).

Weeds in the tree rows can also be controlled using mulches. Organic mulches (cereal straw, green waste, composted wood chips) or synthetic mulches (polyethylene, polypropylene or polyester) can be used around young trees. A 30-gauge polythene sheet (transparent or black) is quite commonly used for mulching in arid and semi-arid areas. Shredded tree prunings also make a good mulch. However, mulches may also provide a good habitat for gophers, voles, field mice and snakes, or be a source of new weed seeds that came with the mulch.

Composting

Composts are stabilized organic residues used in arable soils to improve fertility through increasing soil organic matter content, thereby enhancing microbial activity and improving the physical properties and nutrient-supplying capacity of the soil. The rate of decomposition of compost incorporated into soil is usually less than 35 percent in the first year (Herbert et al., 1991; Cheneby et al., 1994). A major concern regarding the continuous use of compost is the load of organic nitrogen (N) imposed upon the soil; this will eventually mineralize, and nitrates in excess of crop requirements will be leached into and contaminate groundwater. To minimize this hazard, the amount of compost applied should be in accordance with the crop's ability to take up available nitrogen, and complemented or replaced with inorganic fertilization. Using composted municipal solid wastes can increase the available nitrogen- and phosphorus-supplying capacity of the soil without exceeding the crop's ability to take up nitrogen (Hadas et al., 2004).

Although the use of compost has many beneficial effects, to date it has mainly been used in organic olive cultivation in the southern Negev.

References

Adar, E., Issar, A. S., Rosenthal, E., & Batelaan, O. (1992). Quantitative assessment of flow pattern in the southern Arava Valley (Israel) by environmental tracers in a mixing cell model. *J Hydrology*, 136, 333–354.

Ayers, R. S., & Westcot, D. W. (1985). Water quality for agriculture. FAO Irrigation and Drainage Paper 29 (Rev. 1). Rome: Food and Agriculture Organization (FAO) of the United Nations.

Allaire-Leung, S. E., Wu, L., Mitchell, J. P., & Sanden, B. L. (2001). Nitrate leaching and soil nitrate content as affected by irrigation uniformity in a carrot field. *Agric Water Manag*, 48, 37–50.

Bezdicek, D. F., Beaver, T., & Granatstein, D. (2003). Subsoil ridge tillage and lime effects on soil microbial activity, soil pH, erosion, and wheat and pea yield in the Pacific Northwest. USA. *Soil Tillage Res*, 74, 55–63.

Biswas, A. K. (1990). Conservation and management of water resources. In: A. S. Goudie (Ed.), *Techniques for Desert Reclamation* (pp. 251–267). John Wiley & Sons Ltd, Chichester.

Biswas, A. K., & Arar, A. (1988). *Treatment and Reuse of Wastewater*. Butterworth, London.

Chartzoulaki, K. S. (2005). Salinity and olives: growth, salt tolerance, photosynthesis and yield. *Agric Water Manag*, 78, 108–121.

Cheneby, D., Nicolardot, B., Godden, B., & Penninckx, M. (1994). Mineralization of composted 15N labeled farmyard manure during soil incubation. *Biol Agric Hort*, 10, 255–264.

Di Primo, P., Gamiliel, A., Austerwell, M., Steimer, B., Peretz-Alon, I., & Katan, J. (2003). Accelerated degradation of metam-sodium and dazomet in soil: characterization and consequences for pathogen control. *Crop Protection*, 22, 635–646.

Doran, J. W., & Parkin, T. B. (1995). Defining and assessing soil quality. In: J. W. Doran (Ed.), *Defining Soil Quality for a Sustainable Environment*. (pp. 3–21). Soil Science Society of America, Madison, WI.

El-Gazzar, A. M., El-Azab, E. M., & Sheata, M. (1979). Effect of irrigation with fraction of sea water and drainage water on growth and mineral composition of young grapes, guavas, oranges and olives. *Alexandria J Agric Res*, 27, 207–219.

Gvirzman H (2002). *Israel Water Resources* [in Hebrew]. Jerusalem: Yad Yizhak Ben Zvi.

Hadas, A., Agassi, M., Zhevelev, H., Kautsky, L., Levy, G. J., Fizik, E., & Gotessman, M. (2004). Mulching with composted municipal solid wastes in the Central Negev, Israel: II. Effect on available nitrogen and phosphorus and on organic matter in soil. *Soil Tillage Res*, 78, 115–128.

Haim G (2002). *Israel Water Resources* [in Hebrew]. Jerusalem: Yad Yizhak Ben Zvi.

Herbert, M., Karam, A., & Parent, L. E. (1991). Mineralization of nitrogen and carbon in soils amended with composted manure. *Biol Agric Hort*, 7, 349–361.

Hydrological Service of Israel, (2005). *Development of Use and Current Status of Water Resources in Israel* [in Hebrew]. Jerusalem: HSI, 404 pp.

Israeli Water Association, (2003). *Proceedings of the Conference on Arid Zones Hydrology*. Ness Ziona: IWA.

Issar, A. (1985). Fossil water under the Sinai-Negev peninsula. *Scientific Am*, 253, 104–110.

Issar, A., & Nativ, R. (1988). Water beneath deserts: keys to the past, a resource to the present. *Episodes*, 11, 256–261.

Kafri, U., Goldman, M., & Levi, E. (2007). The relationship between saline groundwater within the Arava Rift Valley in Israel and the present and ancient base levels as detected by deep geoelectromagnetic soundings. *Environ Geol Intl J Geosci*, 54, 1435–1445.

Katan, J. (1984). Soil solarization II International Symposium on Soil Disinfestations. *Acta Hort*, 152, 227–236.

Khan, A. R., & Mohsin, M. (1976). Oxygen diffusion meter for use in soils. *Res Industries*, 21, 89–92.

Klein, I., Ben-Tal, Y., Lavee, S., De Malach, Y., & David, I. (1994). Saline irrigation of cv. Manzanillo and Uovo di Piccione trees. *Acta Hort*, 356, 176–180.

Lambert, S., & Karim, S. (2002). Irrigation and salinity: a perspective review of the salinity hazards of irrigation development in the arid zone. *Irrigation Drainage Syst*, 16, 161–174.

Mamane, Y., Dayan, U., & Miller, J. M. (1987). Contribution of alkaline and acidic sources to precipitation in Israel. *Sci Total Environ*, 61, 15–22.

Mekorot (Israel National Water Company) (2007). http://www.mekorot.co.il/mekorot/download/Mekorot_EN.pdf.

Oren, O., Yechieli, Y., Bohlke, J. K., & Dody, A. (2004). Contamination of groundwater under cultivated fields in an arid environment, Central Arava Valley, Israel. *J Hydrology*, 290(3/4), 312–328.

Paranychianakis, N. V., & Chartzoulakis, K. S. (2005). Irrigation of Mediterannean crops with saline water: from physiology to management practices. *Agric Ecosyst Environ*, 106, 171–187.

Shelef G (2006). Wastewater treatment, reclamation and reuse in Israel. Available at http://www.biu.ac.il/Besa/waterarticle3.html.

Spedding, T. A., Hamel, C., Mehuys, G. R., & Madramootoo, C. A. (2004). Soil microbial dynamics in maize-growing soil under different tillage and residue management systems. *Soil Biol Biochem*, 36, 499–512.

Stigter, T. Y., Van Ooijen, S. P. J., Post, V. E. A., Appelo, C. A. J., & Carvalho Dill, A. M. M. (1998). A hydrogeological and hydrochemical explanation of the groundwater composition under irrigated land in a Mediterranean environment, Algarve, Portugal. *J Hydrology*, 208, 262–279.

Tsur, Y., Park, H., & Issar, A. (1989). Fossil groundwater resources as a basis for arid zone development? An economic inquiry. *Intl J Water Res*, 5, 191–201.

Weissbein S, (2006). Effect of moderate saline water irrigation on vegetative growth and oil quality on different varieties of olive grown in Ramat Negev region of Israel. MS. Thesis, Albert Katz International School for Desert Studies, Ben-Gurion University of the Negev, Sedeboker, Israel.

Wiesman, Z., De Malach, Y., & David, I. (2002). Olive and saline water – story of success. *Intl Water Irrigation*, 22, 18–21.

Wiesman, Z., Itzhak, D., & Ben Dom, N. (2004). Optimization of saline water level for sustainable Barnea olive and oil production in desert conditions. *Scientia Hort*, 100, 257–266.

World Bank (2006). World Bank Analytical and Advisory Assistance (AAA) Program, China: Addressing Water Scarcity. Background Paper No. 3, June.

Further reading

Ben-Zvi, A., & Meirovich, L. (1997). Direct probabilistic description of arrival times of runoff events within the year. *Stoch Hydrol Hydraul*, 11, 511–521.

Dan, J., & Bruins, H. J. (1981). Soils of the southern coastal plain. In: J. Dan, R. Gerson, H. Koyumdjisky, D. H. Yaalon (Eds.), *Aridic Soils of Israel – Properties, Genesis and Management* (pp. 143–164). Division of Scientific Publications, The Volcani Center, Bet Dagan.

Dan, J., Yaalon, D. H., & Koyumdjisky, H. (1972). Catenary soil relationships in Israel, 2. The Bet Guvrin. catena on chalk and Nari limestone crust in the Shephela. *Israel J Earth Sci*, 21, 29–49.

Dan Y, Marish, S, (1980). *Arava Valley: Survey of Soil for Agricultural Use* [in Hebrew]. Jerusalem: Ministry of Agriculture.

GSA (Government of South Australia) (2006). *Olive Irrigation in Drought Conditions*, Fact Sheet No: 17/06. Adelaide: GSA.

Hillel, D. (1982). *Negev: Land, Water, and Life in a Desert Environment*. Praeger Publishers, Westport, CT.

Khan, A. R., Chandra, D., Quraishiand, S., & Sinha, S. R. K. (2000). Soil aeration under different soil surface conditions. *J Agropec Crop Sci*, 185, 105–112.

Muñoz, A., Nunes, J. M., López-Piñeiro, A., Albarrán, A., & Coelho, J. (2007). Changes in selected soil properties caused by 30 years of continuous irrigation under Mediterranean conditions. *Geoderma*, 139, 321–328.

Naor H, Granit Y (2000). Central Arava: updating the hydrological and operational situation of the aquifers and wells, and forecasts until 2004 [in Hebrew]. Tahal: Water Section – Hydrology Department.

Narvot J, Ravikovitch, S, (1972). Trace elements in soil profiles of Israel. Internal Report. Rehovot: The Hebrew University of Jerusalem, Faculty of Agriculture.

Ravikovitch, S (1953). The Aeolian Soils of the Northern Negev Desert. Jerusalem: Research Council of Israel, pp. 404–433.

Singer, A. (2007). *Soils of the Negev, The Soils of Israel*. Springer, New York, NY, pp. 47–83.

Smedema, L. K., & Shiati, K. (2002). Irrigation and salinity: a perspective review of the salinity hazards of irrigation development in the arid zone. *Irrigation Drainage Syst*, 16, 161–174.

Kanarek, A., & Michail, M. (1996). Groundwater recharge with municipal effluent: Dan Region Reclamation Project. *Israel. Water Sci Technol*, 34, 227–233.

CHAPTER 6

Biotechnologies for intensive production of olives in desert conditions

The main purpose of this chapter is to discuss the latest biotechnological aspects of olive growing in desert olive plantations, which have developed over the years in the Negev Desert in Israel, and elsewhere. For clarity, the chapter is divided into three main sections:

1. General description of the olive plant

2. Establishment of olive orchards in desert conditions

3. Cultural practices for the growth and development of olive trees in desert conditions.

6.1 General description of the olive plant

The common olive (*Olea europaea* L.), which belongs to the family oleaceae, is an evergreen tree that grows up to ~12 m in height with a spread of about 8 m. However, many larger olive trees are found around the world, with huge, spreading trunks. The trees are also tenacious, easily sprouting again even when chopped to the ground. Sometimes it is difficult to recognize which is the primary trunk. The tree can be kept at a height of about 5 m with regular pruning. The olive tree has a very attractive appearance, with its graceful and billowing shape. In an all-green garden, its grayish foliage serves as an interesting accent. The gnarled branching pattern

Figure 6.1: Typical old olive tree in the Middle Eastern dryland region.

is also quite distinctive. Olives are long-lived, with a life expectancy of greater than 500 years. Some trees grown in Israel and Middle Eastern regions are believed to be more than 1000 years old (Figure 6.1).

Olive leaves are feather-shaped, and grow opposite one another. Their skin is rich in tannin, giving the mature leaf its gray-green appearance. The leaves are replaced every two or three years, leaf-fall usually occurring at the same time as new growth appears in the spring and/or in the autumn. Leaves have stomata only on their lower surfaces.

Mature olive trees produce huge numbers of flowers, but the fruit set is normally less than 5 percent (Martin *et al.*, 1994a). Olive flowers are small, fragrant and cream-colored, and are largely hidden by the evergreen leaves; they grow on a long stem arising from the leaf axils. The olive produces two kinds of flowers: a perfect flower containing both male and female parts, and a staminate flower with stamens only. Flowers are borne along the shoot in auxiliary panicles (Figure 6.2a). Generally, there are 12 to 20 flowers in one panicle. Perfect flowers consist of both pistillate and staminate parts, normally a small calyx, four petals, two stamens and filaments supporting large pollen-bearing anthers, and a plain green pistil, short and thick in shape, with a large stigma.

The flowers are largely wind-pollinated, with most olive varieties being self-pollinating, although fruit set is usually improved by cross-pollination with other varieties. Most of the olive varieties are self-compatible; however, it has been suggested that there are also a few self-incompatible varieties that do not set fruit without other varieties being nearby, and that there are varieties that are incompatible with certain others. It is reported that extreme high temperature might cause incompatibility in olives – for example, the Manzanillo cultivar is usually self-pollinating, but under hot conditions its pollen develops slowly, resulting in little or

Figure 6.2: Flowering olive branches (a) and fruiting branches at various stages of ripening (b–d).

no fertilization (Ayerza and Coates, 2004). However, spraying with additional micronutrients and/or hormones dramatically minimizes this type of loss (Talaie and Taheri, 2001).

The olive fruit is termed a *drupe* botanically, similar to almonds, apricots, cherries, nectarines, peaches and plums, which are green in color at the beginning and generally become blackish-purple when fully ripe. A few varieties are green even when ripe, and some turn a shade of copper brown (Figure 6.2b–d). Olive fruits consist of a carpel, and the wall of the ovary has both fleshy and dry portions. The skin (exocarp) is free of hairs and contains stomata. The flesh (mesocarp) is the tissue that is eaten, and the pit (endocarp) encloses the seed. Olive cultivars vary considerably in size, shape, oil content and flavor. The shapes range from almost round to oval, or elongated with pointed ends. Raw olive fruits contain an alkaloid that makes them bitter and unpalatable. A few varieties are sweet enough to be eaten after sun-drying. In general, the trees reach bearing age at about four years.

Olive cultivars are basically classified into "oil olives" and "table olives," and oil cultivars predominate. Olive cultivars are also classified according to the origin of the cultivar – for example, Spanish, Italian, Greek, Syrian, Moroccan, Israeli, etc. The most popular cultivars are: Picual, Arbequina, Cornicabra, Hojiblanca and Empeltre in Spain; Frantoio, Moraiolo, Leccino, Coratina and Pendolino in Italy; Koroneiki in Greece; Chemlali in Tunisia; Ayvalik in Turkey; Nabali, Suori and Barnea in Israel and The West Bank; Picholin in France; Mission in California; and various varieties in Australia. The table olive cultivars include Manzanilla and Gordal from Spain; "Kalamata" from Greece; "Ascolano" from Italy; and "Barouni" from Tunisia (Jacoboni and Fontanazza, 1981; Weissbein, 2006). More detailed descriptions of olive varieties can be found in Chapter 7, which is dedicated to olive genetic material.

6.2 Establishment of olive orchards in desert conditions

6.2.1 Production of planting materials

Owing to the need to conserve the selected olive varieties, almost none of the common olive varieties have been commercially propagated by seeds for at least the past century. Olive vegetative plant production has been intensively studied and developed from the beginning of the second half of the twentieth century (Hartmann *et al.*, 1990; Wiesman *et al.*, 1995a, 1995b; Fabbri *et al.*, 2004). Based on intensive studies, biotechnologies concentrating on the rooting of cuttings and tissue culture have been established for optimal growing and cultivation environments. More recently, additional Ag-Environ biotechnologies have been developed that also allow efficient and cost-effective olive plant propagation in desert conditions. There is the possibility of using grafted material, but this is uncommon due mainly to the lack of availability of superior olive rootstocks. However, the latter methodology can be achieved using grafting or chip budding with material from desired cultivars, and in some varieties bark-grafting or top-working can be performed to renew the orchard. Propagation by using suckers is also possible. This type of propagation is done only in mature trees when there is a need to replace the grafted cultivar. If suckers develop from the lower part of the trunk, they are usually grafted with a new selected scion. Cuttings are the most commonly practiced method of propagation in the olive industry, although lately tissue culture techniques are also being applied to produce seedlings from selected mother stock plants (Fabbri *et al.*, 2004). The tissue culture technique is usually more expensive, but helps to produce more healthy and virus-free olive plants.

Propagation of olives from cuttings

For successful propagation of olive cuttings in the dry desert climate, it is very important to use specific biotechnologies that consider the following aspects.

Selection of plant materials for propagation

It is important to select semi-hard cuttings containing adequate energy reserves in the form of starch in their pith tissue, as demonstrated in KI assay and SEM micrographs (Wiesman and Lavee, 1995). Owing to the extended vegetative growing seasons of the olive canopy in most desert climates, it is essential to choose the best time for removing cuttings from the mother plant. Usually, mid-winter and summer, when a relatively low rate of vegetative growth occurs, are the best periods for preparing cuttings. It has been reported that various growth retardants, such as paclobutrazol, significantly increase the rate of rooting of olive cuttings (Wiesman and

Lavee, 1995). During the rooting process, the rate of photosynthesis of the cuttings is very low; therefore, only two or three leaves should be left on an average 15 cm long cutting. The starch reserve of the cutting is supposed to supply the energy needed during the rooting process, which on average takes three to four weeks, depending mainly on the olive varieties used (Wiesman and Lavee, 1995; Wiesman *et al.*, 1995a).

Root formation stimulation

Usually, olive varieties are divided into three main groups in terms of their rooting ability. In the vast majority of varieties, about 50 percent of the cuttings form roots in a period of about four weeks (Wiesman and Lavee, 1994a, 1995). Another significant group of varieties are known as "easy to root," and more than 80 percent of the cuttings form roots within four weeks. The third group of varieties consists of those difficult to root, with less than 20 percent of them forming roots. Hormonal balance regulates the process of root formation and elongation, with auxin dominating the initiation of roots (Wiesman *et al.*, 1989; Hartmann *et al.*, 1990). Owing to the fact that, in olives, the natural supply of auxin from the apical buds of the cutting through the base is limited in most varieties, there is a need to supply exogenous auxin (Wiesman *et al.*, 1994a). The most common auxin for root stimulation of olive cuttings is indole-3-butyric acid (IBA). However, some additional plant-growth regulators, including growth retardants and others, have also been found to contribute additively to the rate of root formation accelerated by IBA application (Wiesman and Lavee, 1995).

As a follow-up to a basic study of the interaction of auxins with growth retardants, cytokinins and mineral nutrients in relation to the development of plant root systems (Wiesman *et al.*, 1994a, 1995b), a novel product to stimulate rooting of cuttings has been developed by the current author's research group. Granular encapsulated controlled-release hormone-fertilizer formulations have been designed and engineered, and are particularly suitable for various horticultural uses: they are relatively cheap and easy to use, environmental contamination is relatively small and, most significantly, they facilitate good uptake and efficacy of the applied biomaterials (hormones, fertilizers and/or fungicides). Selected hormone-fertilizer formulations have been intensively and successfully tested in large-scale plant production. This product has already been tested successfully and approved by commercial nurseries in Israel, America, Australia and New Zealand. International patents for the concept, technology and applications have been registered in many countries throughout the world (Wiesman *et al.*, 2002a, 2003; Wiesman and Markus, 2004). As it takes three or four weeks for olive cuttings to root, the controlled-release hormone-fertilizer technology is of special interest for the rooting of olive cuttings with an emphasis on dry and low-humidity desert conditions.

Environmental conditions

Atmospheric humidity, soil temperature, light and hygiene are the key factors that control the rooting of the cuttings. For maintaining the proper "incubator" environmental conditions needed for efficient root formation in the base of olive cuttings, it is very important to carefully control the levels of atmospheric and rooting-media humidity and moisture. In particular, in dry desert conditions rapid changes in the rate of humidity occur. Recent developments in controlled and isolated greenhouse technologies are making great contributions to solving this problem. Today several types of greenhouses with digitally controlled environment are available; these greenhouses also control the rate of temperature change and light around the clock (Hartmann *et al.*, 1990; Linker *et al.*, 1999).

Novel cutting incubation facilities including sensitive mist and fumigation technologies enable control of the microclimate and moisture directly around the olive cuttings. Such facilities enable maintenance of the proper moisture level in the rooting media, where the cuttings are kept throughout the entire rooting process. These facilities also maintain a high rate of hygiene in the incubating environment, so that important phytosanitary aspects do not limit rooting of the cuttings (Fabbri *et al.*, 2004).

Application of water to the cuttings

For rooting purposes, a fogging system of irrigation that maintains a proper atmospheric humidity level in the propagation facility is accepted as being superior to most other conventional misting technologies (Fabbri *et al.*, 2004). Fog can be applied by a microjet fogger in short pulses of 5 seconds each, followed by intervals of 20–30 seconds, to provide the required humidity level (which must be adapted to specific local conditions). Water application for growth of the cutting (mainly after the roots are already formed) will require an additional sprinkler system.

Nutrition

The nutritional requirements of the rooted olive cuttings are relatively low. However, it has been found that adding and maintaining a relatively low level of nutrients to the rooting medium and maintaining the level significantly increases the quality of the rooted olive plants (Wiesman *et al.*, 1995b). Continuous application of nutrients has also been found to improve the relatively slow rooting process of olive cuttings in the nursery (Wiesman *et al.*, 1995b, 2002a). The novel controlled-release biotechnology, recently adapted by Wiesman and colleagues (2004) from the pharmaceutical drug field to the agricultural industry, was found to significantly improve the rooting of olive cuttings. This technology is based on the encapsulation of a root-stimulation hormone mix including IBA and paclobutrazol (growth retardant) together with fungicides and a proper balance of fertilizers. It consists of a multilayered coating of extruded polymeric materials containing the bioactive material. The active compounds needed for the stimulation of root formation and for root elongation are then slowly released over various periods of time. Using this easily applied technology, it is possible to maintain the right hormone balance for the whole period of rooting. It is also possible to stimulate further root development and significantly to reduce the chances of soil fungi infestation (Wiesman *et al.*, 2002a).

Micropropagation of olives

Micropropagation of olives is a new method of propagation that has not yet replaced the traditional propagation systems (rooted cuttings and grafting), mainly because of the high costs. However, recent development of low-cost mass-scale techniques may enable this technique, which is highly suitable for olive culture – especially in dry areas. As described by Fabbri *et al.* (2004), micropropagation includes five stages: establishment, multiplication, rooting, transplanting, and acclimatization. Establishment starts with the collection of the parts used for propagating material – perhaps microcuttings with one or more buds. Sterilization of this material follows, to prevent infestation to the medium and ensure that plants are completely free from pests and diseases. It is crucial that the plant material is from a known genetic source of good quality. Collection of this material should be performed in the spring, when vegetative activity is high. A growth chamber set at 20–24 °C is used, with a 2000- to 4000-lux light intensity, and a lighting period of 16 hours or so.

There are some particularly critical phases in micropropagation of olives that must be taken into consideration. The most significant of these is the first stage of *in vitro* adaptation, where it is important to balance the auxin/cytokinin levels to promote emission of axillary buds. Once this first step has been overcome, the rooting phase is more straightforward. According to their response to *in vitro* conditions, olive cultivars can be divided into three main groups: cultivars that are very recalcitrant regarding *in vitro* immission, such as Leccino and Picholine; cultivars that show a medium level of difficulty, such as Pendolino and Frantoio; and finally the readily adaptable cultivars, such as Arbequina, Barnea and Hojiblanca (García-Férriz *et al.*, 2002). For cultivars with easy and medium levels of adaptation to *in vitro* conditions, micropropagation can be applied for mass-scale production by a commercial outfit, and may represent a good alternative methodology to the traditional nursery techniques based on rooting of cuttings. This technique provides plants that have only a short unproductive period and do not exhibit juvenile traits (Briccoli Bati *et al.*, 2002). With proper care, most micropropagated olive trees can start to flower in their third year.

Hardening and acclimation of young olive plants

Seedlings produced from either rooted cuttings or micropropagation must be hardened off before planting in the main field; this will not only increase their survival rate in the field, but also increase their productivity at a younger stage. Due to the large-scale demand for coverage of large areas of desert land with olive plantations throughout the world, and in order to improve the production cost/efficiency, a new methodology has recently been developed in the field of olive nurseries. In this industrial methodology, rooted olive plants are hardened and grown, after the initial phase of the indoor rooting process, in large netting houses using relatively small pot sets and arranged in trays. The young olive plants remain here for a period of several months, without repotting, before being planted in the field. This highly intensive practice requires a great deal of care, with an emphasis on irrigation and fertilization, and on phytosanitary aspects.

Figure 6.3: Different stages of olive plants in the hardening nursery.

During this hardening and growth-stimulation phase, it is essential to ensure that there is proper root system development, which is necessary for water uptake, and leaf cuticle membrane development, in order to reduce and control the rate of water loss and stimulate the development of the hydraulic system. These biological systems are very important for the stimulation of root elongation and development via the balance of the shoot/root ratio, and underpin the whole rooted plant development (Wiesman *et al.*, 1995a, 2002a). Furthermore, these systems are particularly significant for the cultivation of olive plants in desert conditions. Fully controlled and automated irrigation systems specifically designed for nursery purposes, based on sprinklers or dippers, have been well-developed for more than a decade. The nutrient and phytosanitary aspects are being addressed by the new controlled-release encapsulated hormone-fertilizer technology, which enables the sustainable delivery of the balance of elements needed for whole rooted plant growth and for the next phase of hardening. The granules of the encapsulated products are mixed into the growing medium before potting and planting of the rooted olive plants, and the active ingredients are released and activated by the irrigation water (Wiesman *et al.*, 2002a, 2004). In this way, any single small potted olive plant gets all the necessary elements for its development. Furthermore, since common practice is to plant the hardened olive plants in the field directly from their small pots, the well-balanced and enriched medium and essential elements increase the survival rate of these plants, in particular in the adverse desert conditions – indeed, this methodology has been found to be highly effective in many desert areas and elsewhere all over the world.

The different stages of olive plants in the hardening nursery are shown in Figure 6.3.

6.2.2 Design and preparation of olive orchards
Selection of the cultivation site

The initial step toward desert olive oil cultivation is correct selection of the location for establishment of a plantation. The following aspects should be taken into account when selecting a site for an olive grove.

Soil survey

A soil survey must be performed to determine the suitability of the soil for olive production. The survey should include the soil type, structure, drainage capacity, acidity and fertility; the presence of any residual chemicals which may affect future organic registration. Common desert soil types are sandy or loam. In many desert environments where there is a high percentage of silt or clay in the soil, the water drainage capacity is limited; this is important, because water-logging is a limiting factor for olive trees. The availability and level of various nutrients in the soil should be checked. Soil acidity should also be determined; it is known that olive trees develop best in soils with a pH of 7.0–8.0. Slopes and rocky soil structure may interfere with cultivation practices with an emphasis on bio-automation.

Soil analysis

Before planting olive trees in an arid environment, soil analysis must be performed in order to correct any nutrient deficiencies prior to planting. The lack of documented standards for olive trees leads to the use of mineral analysis in the foliage to obtain the full picture. Living cells depend on minerals for their proper structure and function and, like vitamins, minerals function as coenzymes, which are required for growth and healing.

To perform the analysis, an inexpensive package can be purchased that measures soil pH (acidity or alkalinity) and salinity (salt level), lime levels, texture, class, and plant-available phosphorus and potassium.

Assessment includes:

- soil analysis – texture, pH, $CaCO_3$, organic matter, macro-nutrients (at least P, K and Mg) and micro-nutrients (e.g., Boron);

- thorough elimination of sources of disease inoculums, especially *Verticillium dahliae*; the absence of this fungus should be verified, and the area avoided if previous crops have hosted *Verticillium*;

- elimination of perennial problem weeds.

In particular, large-scale meliorations (e.g., excavations and land-fills) should be assessed critically with respect to their environmental impact and the destruction of a diversified topography and existing ecological compensation areas.

The soil texture is determined by the distribution of particle sizes of the soil, with sand being the largest particle and clay the smallest. Testing of the soil texture can provide an estimate of the percentages of each particle size. The texture can indicate the stability, strength and drainage of a soil, and these are important characteristics when pre-planting fruit trees.

Sandy soils (sand, loamy sand, sandy loam) have lower water- and nutrient-holding capacities, whereas soils with a high level of clay (clay, silty clay, clay loam, silty clay loam) tend to be poorly drained and are subject to compaction.

Climate evaluation

Desert climates are usually characterized by cool winters and hot summers. Olives favor such conditions, but data should be collected regarding temperature changes throughout the year to ensure that there are enough cold periods for olive flower induction. Strong winds are known to interfere with olive flowering; therefore, historical data regarding wind events should also be collected. Olives are ideally adapted to the Mediterranean climate, but even there they can sometimes suffer severely if the winter is harsh. They do, however, require a cold spell of several months during which temperatures fall to below $9\,°C$, but never lower than $-9\,°C$, in order to set fruit. They will produce in other areas as long as they have the correct chilling requirement (winter temperatures fluctuating between $1.5°$ and $18\,°C$) and summers that are long and warm enough to ripen the fruit. The trees can suffer severe damage at temperatures less than $-5\,°C$.

Rainfall

Olive trees are subtropical species from semi-arid climates. They can cope with several months of little or no rain, but do not tolerate prolonged spells of cold, wet weather. Hot, dry winds or rain at pollination in late spring can reduce fruit set. Significant rain at harvest-time can reduce the extractability of oil from the fruit due to its high water content, especially in fruit normally grown for table olives.

Annual rainfall must also be taken into account. In low-rainfall areas (200–300 mm) olive yield is satisfactory in soils with good water-retaining capacity if irrigation is applied. In high-rainfall areas (400–600 mm) olive yield is good on condition that adequate drainage is provided, since water-logging reduces the yield. Experience has shown that, with proper irrigation, olives can be cultivated in areas with virtually no rain. There are some areas in the Israeli desert where the average annual rainfall is less than 50 mm but, with proper irrigation management, olives have been grown successfully. However, the irrigation system is severely affected by rain when plants are irrigated by saline water. Extra leaching or over-irrigation should be managed during the rainy period to protect against salt entering the root zone.

Wind

Wind is essential for pollination, but can also cause poor fruit set, floral abortion, fruitlet drop, fruit shrivel, windmarks, sandblasting, and wind-rocking of young trees. It is therefore imperative to select varieties less vulnerable to wind, such as Mission, when such conditions exist. Frost is also a major consideration when selecting the location for olive trees.

Available water reservoir

For intensive large-scale desert olive oil cultivation, it is very important to ensure the availability of a large enough suitable water resource for olive irrigation. Saline water reservoirs are common in most world desert environments, and it is important to analyze the quality of such water, with emphasis on the composition of the salts and the electrical conductivity. For cost-effective production analysis, it is also important to evaluate the cost of the irrigation infrastructure and the power needed for irrigation.

History of previous crops

Olives are known to be sensitive to various soil fungi and nematodes that may have been hosted by previous crops cultivated in the same site, so it is important to know the crop history of the area in order to manage crop husbandry better.

Availability of labor

Some olive practices still need manpower, such as pruning and harvesting. Although machine harvesting has been developed and is commonly practiced, there is still the requirement for a large amount of human labor during the harvesting period. If olives are being grown for table purposes, or if the farming is organic, then more labor is needed.

Wildlife and fire risk

Some desert environments are rich in wildlife that could damage olive trees. Data regarding such wildlife, and significant local fire risks, may be of value. Before considering the location of the plot, one criterion for the selection of a possible area is the presence of undamaged olive trees in the vicinity for at least five years.

Planting density and design

As olives are long-lived permanent tree crops with high adaptability, a small mistake at the time of planting could have a serious effect in the long run – so significant precautions should be taken when designing the planting of new olive groves. The design of the site depends on the land topography, the cultivars and their type, and the purpose of the plantings (i.e., for table olives or for oil). Denser planting is adopted to give a higher yield per hectare from younger trees, but the design of a desert olive grove must provide maximum sunlight exposure to the planned number of trees during all stages of growth and at maturity. The olive grove must also allow for easy and efficient operation of cultivation equipment. The traditional system of spacing was quite wide, with 10–12 m between plants and rows, and even more in semi-desert areas – up to 24 m between plants and rows. These days closer spacing, such as 6×5 m or 6×6 m, to accommodate about 300 plants per hectare, is common. The most common olive tree densities are 250 trees per hectare for olive varieties of moderate vigor (8×5 m) and 123 trees per hectare for more vigorous olive varieties (9×9 m). Plant density is related to plant productivity. Olive trees remain unproductive for the first three or four years after planting, and the increase in production through the years depends upon the plant density. A greater plant density produces an earlier yield, but the economical life of the orchard is shorter. These higher-density plantings have also shown less alternate-bearing behavior (Beede and Goldhamer, 1994). New olive orchards are often high density.

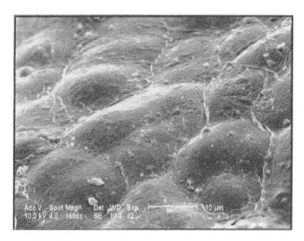

Figure 6.4: Typical scanning electron microscope (SEM) view of Souri leaf upper surface, demonstrating the massive cuticle membrane formed for reduction of water loss.

Super-high-density olive planting has also been developed recently. In this system, 1500–2250 olive trees are planted per hectare; they are then trellised as in vineyards, pruned to remain short in stature, and harvested with a conventional mechanical grape harvester. The system has already been successfully implemented in Spain, and has been recently introduced in California, but it is highly intensive and needs special technical know-how and technology (UC Cooperation Extension, 2007). Current reports are that this system has so far provided a higher yield and may have great implications for the future, but it is still in the study phase and is therefore not a priority in this book.

Irrigation management

Irrigation in olive groves
Traditionally, the majority of olive orchards around the Mediterranean are not irrigated. Due to the reduced rate of water loss owing to adaptation during the evolution of the olive species, olive cultivation was traditionally developed and established in areas receiving about 500 mm rain annually. A relatively well-developed and massive cuticle membrane can be found in most of the traditional olive cultivars common in the Mediterranean region, as demonstrated in Figure 6.4. The small olive leaf, well coated with cuticle membrane, leads to a significant reduction in the rate of transpiration and water loss, and makes it possible to cultivate most of the common olive cultivars in desert environments.

However, in arid desert regions, olive trees without supplemental irrigation, especially during the hot summer season, are poorly developed and have a very low olive yield. Studies published in recent decades (Weissbein *et al.*, 2006, 2008) have clearly shown that proper intensive irrigation could stimulate most of the olive cultivars to be very vigorous and highly productive in desert conditions (Figure 6.5).

Figure 6.5: Typical extensive, non-irrigated olive orchard (a) and intensively cultivated olive orchard (b), both in the same Negev Desert environment.

The high economic importance of the olive industry to the traditional olive-producing countries, led by Spain, Italy and Greece, where the vast majority of olive trees are not irrigated, means that until recently most of the technological efforts in this field were directed towards the development of technologies for supplemental irrigation application to traditionally cultivated trees. Before going further into intensive olive desert irrigation technologies, some aspects of supplemental irrigation are therefore discussed here.

Several attempts to develop a supplemental irrigation strategy have been carried out in recent years, and it has finally been confirmed that application of irrigation, even in traditional olive groves, significantly increases both the number and size of the fruits, leading to an increased final olive yield. It has also been suggested that the ratio between root and leaf dry weight is usually greater in non-irrigated than in irrigated plots. Dichio and colleagues (2004) showed that roots explored a soil volume ranging from $0.5\,m^3$ in the first year to $16.8\,m^3$ in the seventh year in irrigated plots, and from $0.5\,m^3$ to $13.4\,m^3$ in non-irrigated plots. Moreover, in non-irrigated plots the canopy growth is drastically reduced while root growth is restricted to a lesser degree, probably as a defense strategy against the water deficit most common in the desert olive origin environment, making for a better root/leaf ratio and consequently greater water availability to the leaves.

Although olive trees are normally cultivated under dry farming conditions, with the application of small amounts of water it is possible to increase their production. However, if maximum yields are desired, greater amounts of water will be needed (Goldhamer *et al.*, 1994). Unfortunately, large amounts of water are not always available, especially in the most common growing areas of the Mediterranean basin. In the driest environment, water stress accounts for major changes in the fruit growth pattern and fruit size, reduced fruit set and reduced oil content on a dry weight basis, and enhanced fruit drop and advanced fruit ripening, but does not affect the oil accumulation pattern and oil characteristics.

However, in areas where irrigation water is carefully controlled by the correct application of regulated deficit irrigation (RDI), it is possible to attain maximum yield while limiting the amount of water applied (Alegre *et al.*, 1999). In this system, the available water is utilized efficiently to regulate oil production and oil quality. Particular developmental periods of olive trees (such as the blooming period) that are highly sensitive to the low moisture level in dry regions, which may cause excessive fruit drop and alternate bearing, are noted and irrigated judiciously. Insufficient irrigation, even in the summer, reduces shoot growth and carbohydrate production. In the RDI system, the olive tree is irrigated on the basis of its seasonal sensitivity to water stress as an indicator of total and seasonal water requirement (Beede and Goldhamer, 1994). RDI has also been found to increase the polyphenol concentration and heat stability in olive oils in comparison with levels in regular olive trees (Motilva *et al.*, 2000).

Critical periods for water stress – phenological stages of the olive

A comprehensive description of the various phenological stages related to water regimen has been suggested by Beede and Goldhamer (1994):

- Stage 1 is characterized by flower bud development, blooming, fruit set and shoot growth. At this developmental stage, massive events of cell division and some cell enlargement take place. Low soil-water availability reduces flower formation and results in incomplete flowering, reduces fruit set, increases alternate bearing, and decreases shoot growth.

- Stage 2 is characterized by initial fruit growth due to cell division, and shoot growth. Water stress in this stage results in a small fruit size, increased fruit drop and decreased shoot growth.

- Stage 3 is characterized by fruit growth due to cell enlargement, and shoot growth. Water stress in this stage will result in a final small fruit size, fruit drop, and decreased shoot growth.

Studies conducted in California, Spain, Italy, Israel and other locations have concluded that water application can be reduced and moderate stress applied during the period of pit-hardening that takes place in Stage 2. Therefore, RDI during the pit-hardening period (Stage 2 of fruit development) is suggested to be a useful strategy for applying regulated stress if required. RDI should be used as an irrigation strategy when there is a shortage of water. Applying more water in the pit-hardening period will not reduce oil quality.

Monitoring of water status

Tree orchard canopies are more complex than those of most agricultural crops. Therefore, it is much more complicated to describe any transpiration or photosynthesis involved. However, as detailed knowledge of water use and carbon assimilation is critical for optimizing irrigation, several research studies have been conducted on this issue (Diaz-Espejo *et al.*, 2005). The leaf water potential is used in many research projects to describe the olive tree's water status. In most cases, the pre-dawn leaf water potential is used as a comparable parameter between irrigation treatments. A comparison between the water status of olive trees under dry farming and drip farming conditions was made by Fernández *et al.* (1993), and the value of leaf water potential was shown to be very sensitive to atmospheric demand. Lower values were found in September than in August.

Excess water

Olive trees are very sensitive to over-irrigation, and will not perform well in waterlogged soil. Waterlogged soil, often a result of poor drainage, causes poor soil aeration and root deterioration, and can lead to the death of olive trees (Beede and Goldhamer, 1994). Trees in saturated soils are more susceptible to varying weather conditions and soilborne pathogens such as *Phytophthora* and *Verticillium*.

Irrigation of young olive trees

A simple useful rule of thumb for the application of irrigation to young olive trees in desert conditions is based on the percentage of the grove floor area shaded by the tree canopy at midday. If less than 25 percent of the grove floor is shaded by the olive tree canopy, it is common to apply about 50 percent of the water volume generally applied to mature trees in the same environment. If more than 50 percent of the grove floor is shaded, 100 percent of the water volume is applied.

Since saline water is the prime source for this region, proper irrigation management should be established in order to achieve successful olive cultivation. The technology developed over the years for olive cultivation in the Negev Desert can be utilized equally well in other desert environments throughout the world.

Negev Desert saline irrigation technology

The increased demand for healthy sources of fat for human consumption, along with the fact that olives are considered to be well adapted to semi-arid conditions, has led to an intensive wave of olive planting in many places worldwide in the past decade (Food and Agricultural Organization (FAO), http://www.fao.org). Traditionally olives have been cultivated in the Mediterranean region, and many superior cultivars have been selected in most countries in this area. However, because of the increasing shortage of tap water, mainly in semi-arid areas, the irrigation of most new olive plantations is based on available low-quality sources of water, all of which are characterized by relatively high salinity (Wiesman *et al.*, 2004). The relationship between saline water and olive cultivation has been intensively studied for many years, and significant progress has been made in understanding this topic (Bernstein, 1964; Mass and Hoffman, 1977; Munns, 1993; Gucci and Tattini, 1997; Wiesman *et al.*, 2004; Aragues *et al.*, 2005). It is generally well established that saline conditions limit the vegetative and reproductive development of olives, mainly due to interference with the osmotic balance in the root system zone and detrimental effects caused by specific toxic accumulation of Cl^- and Na^+ in the leaves (Benlloch et al., 1991; Hassan et al., 2000). Salinity is known to be a common limiting factor in semi-arid areas, even when tap water is used, due to the high rate of water evaporation (Shalhevet, 1994).

The productivity of the olive tree in traditional cultivation systems is relatively low. It is well established that the production potential of olive trees could be several-fold higher if they were grown under intensive, irrigated conditions (Goldhamer et al., 1994). In many olive-growing regions, and particularly in desert conditions, water availability is the most significant limiting growth factor; therefore, in order to use water effectively, it needs to be delivered and applied properly.

Human beings knew about the potential for growing olives in drier and marginal areas hundreds of years ago; however, because of the lack of a proper irrigation system for such regions, the possibility of extending olive cultivation into desert marginal environments was limited. Although other irrigation systems, such as hand watering, flood irrigation and sprinkler irrigation, were developed long ago, it was only with the development of drip or trickle irrigation that the possibility of expanding olive growing into desert lands, using saline water, became a reality. The drip irrigation technology developed during the past two decades, and mainly sub-surface drip irrigation, has helped to expand olive growing into this region, using the saline water that is generally available in desert areas.

Hand watering and flood irrigation are not suitable for large-scale desert olive cultivation because hand watering is labor-intensive and flood irrigation provides uneven water application and is very wasteful. Micro-sprinkler irrigation/overhead irrigation is more advanced than the former two irrigation systems, and is used in olive nurseries; however, the high set-up costs and evaporation losses mean that this system is inefficient in the olive orchard.

Drip irrigation on the ground surface and especially sub-surface drip irrigation (Figure 6.6), which can be buried underground and reduces evaporation losses, are the easiest irrigation methods to manage, and the most efficient for saline water. In surface drip-irrigated orchards, different applications can be used. In most cases, one dripline is placed on the ground along each planted row of trees. The use of two driplines per row is also common.

In some orchards, the dripline is hung on the trees in order to enable criss-cross cultivation (Figure 6.7). Similar applications to those used in drip irrigation can be found in olive orchards irrigated by mini-sprinklers – meaning that in some cases the pipes and mini-sprinklers hang on trees and in other cases the sub-main pipes lie on or under the ground, with one or two mini-sprinklers per tree. The sprinkler system is not so efficient, but some growers have found this to be more suitable in organic olive plantations due to the problem of weed control. However, in recent years the use of sub-surface drip irrigation systems has increased. A sub-surface irrigation

(a) (b)

Figure 6.6: Typical ground upper surface (a) and sub-surface (b) drip irrigation systems.

Figure 6.7: Typical olive orchard hung irrigation system.

system consisting of thick-walled lines with compensated drippers has recently been developed and specifically engineered for desert olive cultivation; this has some advantages over most of the other available irrigation systems in that it:

- prevents water loss due to low rates of evaporation and drainage;

- decreases weed spread infestation;

- reduces problems of mechanical harvesting;

- reduces system maintenance;

- improves the uptake of nutrients;

- increases the root-zone air status;

- eases plot cultivation.

Soil management

Soil analysis is the most important part of soil management. This practice is particularly important in desert conditions. Soil analysis should be performed before planting olives, as it is easier to make corrections before the trees are planted, and then adjust the soil periodically. As well as ensuring a correct balance of trace elements, a combination of nitrogen, phosphorous and potassium (NPK) fertilizers should be applied, half in the autumn and half in the spring. The following are the main parameters that influence the production of olive plants, and which should be monitored regularly.

Soil pH

Olive trees can grow in a range of pH levels between neutral and moderately alkaline. The soil pH can easily be modified by adding Ag-lime in the case of acidic soils. Samples should be extracted from the topsoil and sub-soil for complete analysis.

Total salt concentration

The more concentrated the salts (electrolytes) in the soil solution, the more likely it is that clay will flocculate. This is the "electrolyte effect." All soluble salts have this effect, including common salt (sodium chloride) and gypsum (calcium sulfate). When the salt concentration of the soil solution is greater than that between the clay platelets, water will move from between the platelets into the soil solution by osmosis. The clay will flocculate and be very stable. In contrast, if the soil solution has a low salt concentration, water will tend to move from the soil solution into the zone between the clay platelets; this will cause swelling and dispersion unless the cations there are strongly charged. Salts such as gypsum, lime (calcium carbonate which, when dissolved, forms calcium salts with any anions present), ammonium or nitrate salts in fertilizer or manure, various forms of phosphate, potassium salts, and trace elements such as zinc sulfate are all salts that are added to the soil in agriculture. Each adds to the total salt (electrolyte) concentration of the soil solution.

Salinity

Saline soil is a major factor in the process of plot selection. It originates from soluble salts in the soil. A common source of salts in irrigated soils is the irrigation water itself. Salinization transforms fertile and productive land into barren land. Salinity limits the vegetative and reproductive growth of plants by inducing physiological dysfunctions and causing widespread direct and indirect harmful effects, even at low salt concentrations. The salt tolerance of plants is difficult to quantify because it varies appreciably with environmental factors (e.g., soil fertility, soil physical conditions, salinity components (ions), distribution of salt in the soil profile, irrigation methods, and climate) and plant factors (e.g., stage of growth, variety, and rootstock). It is well established that saline conditions limit olive vegetative and reproductive development, interfering in the osmotic balance between the soil and the root system, causing toxic accumulation of Cl^- and Na^+ in the leaves.

Soil salinity refers to the concentration of salts in the soil solution; these salts are the positive and negative ions that are dissolved in the soil water. Any ions present in the soil water will add to soil salinity. Soil needs to have a low level of salinity, as a soil without salt would contain no plant-available (soluble) nutrients, and therefore when referring to salinity, we usually mean *excessive* salinity. The degree of soil salinity is most conveniently assessed by measuring the electrical conductivity (EC) of the soil solution. The more salt there is dissolved in the soil water, the more readily it will conduct an electrical current. The current flow is measured with an electrical conductivity meter, and the result is usually expressed as deciSiemens per meter (dS/m). The effective EC is calculated according to the soil type, and is referred to as the ECe.

Salt occurs naturally in many soils in desert areas. Sea spray enters the atmosphere and can be carried long distances inland on the wind. Minerals in rocks contain elements that can form salt, and when those minerals decompose by weathering, the salts are released. Sedimentary rocks such as sandstone may contain high levels of salt if they were laid down when the sea or salt lakes originally covered that area of land. Normally, the salt is deep in the soil. The water table can rise and bring dissolved salts up with it. Moreover, when the water table is within 1–2 m of the soil surface, the water and salts rise to the surface by capillary action, causing the root zone to become saline.

Water is the principal agent of salt transport, and evaporation and transpiration by plants affects salt concentration. Salinity works against plant water uptake – the greater the concentration of salts in the soil solution, the harder the plant has to work to take up water.

Excessive amounts of sodium chloride (common salt) also dilute and displace plant nutrients, reducing soil fertility. At high concentrations, chloride is toxic to plants. Plants vary in their tolerance to salinity, but even low levels of salt leave fewer options for crop choice.

Sodicity

Soil sodicity is the accumulation of sodium salt relative to other types of salt cations, especially calcium. An increase in soil pH and decreases in calcium and magnesium usually accompany this process. Sodic soils are characterized by a poor soil structure and low infiltration rate, are poorly aerated and are difficult to cultivate. They contain clay that swells and disperses when wet; dispersion shows up as a fine suspension of clay in the soil water. Soils with sandy topsoil and dense clay subsoils may have severe problems at depth without any surface signs. The clay disperses because of an excessive proportion of sodium in the exchangeable cations attached to the surface of the clay. Soils with 6 percent or more of sodium as a percentage of the total exchangeable cations are sodic. The measure of soil sodicity is known as the Exchangeable Sodium Percentage, or ESP.

Sodicity in soils has a strong influence on the soil structure of the layer in which it is present. A high proportion of sodium within the soil can result in dispersion. Dispersion occurs when the clay particles swell strongly and separate from each other on wetting. On drying, the soil becomes dense, cloddy and without structure. This dense layer is often impermeable to water and plant roots. In addition, scalding can occur when the topsoil is eroded and sodic subsoil is exposed to the surface, increasing erodibility. Thus, sodic soils adversely affect the plants' growth.

Improvement of sodic soils requires reduction of the amount of sodium present in the soil. This is done in two stages. First, chemicals such as gypsum (calcium sulfate), which are rich in calcium, are mixed with the soil. The calcium replaces the sodium, which is leached from the root zone by irrigation water.

Organic matter

Soil organic matter consists of living roots and organisms (including earthworms, bacteria and fungi), and decomposing plant, animal and microbial residues. Most soil organisms are understood very poorly, and only a fraction have been identified. It is estimated that 80–90 percent of soil biological activity is performed by bacteria and fungi. Humus is the dark, relatively stable end-product of organic matter breakdown. It may be thousands of years old. Some carbon (as much as 50 percent) may be present in the relatively inert form of charcoal. Unless the soil is being regularly supplied with fresh organic matter, the amount of useful (labile) material in it will decline. Humus particles are negatively charged, and act in much the same way as clay particles in the exchange of positive ions such as ammonium, calcium, sodium, magnesium, zinc, copper and iron. They make a very significant contribution to the cation exchange capacity (CEe) of a soil. Some humus particles or parts of particles are positively charged, and can adsorb and store on their surface the negative ions found in the soil solution. Metal chelation is the formation of complexes of metal ions with organic matter. The metal ions in these complexes are in a form that is available to plants. If these elements were not protected in these molecules, they would be readily precipitated into a form that would be unavailable to plants. Metals that are chelated in the soil by this means include copper, iron, manganese and zinc.

The breakdown of organic matter by soil microbes (mineralization) provides plant nutrients and is one of the major sources of nitrogen for plants. The ammonium form of nitrogen is the first produced when organic matter is broken down. Plants can use this form of nitrogen as well as nitrate.

6.3 Cultural practices for the growth and development of olive trees in desert conditions

6.3.1 Training and pruning

Proper pruning is important for the olive for better growth and development of the trees. Pruning is necessary to adjust the trees to the climatic conditions of the area and increase the plantation's productivity. Pruning not only regulates production; it also shapes the tree for easier harvesting. The trees can withstand radical pruning, so it is relatively easy to keep them at the desired height. The problem of alternate bearing can also be reduced with careful pruning every year. It should be kept in mind that the olive never bears fruit in the same place twice, and usually bears on the previous year's growth. To leave a single trunk, any suckers and branches growing below the point where branching is desired should be pruned. For the gnarled effect of several trunks, basal suckers and lower branches should be staked out at the desired angle.

Young trees need to be pruned to encourage them into the correct shape (usually vase-shaped or conical) to optimize the efficient removal of olives by mechanical shaking. Once this is achieved, the tree should be pruned every year to maintain its shape and health by allowing air and light to enter and circulate through the canopy. If the trees have been planted at high density specifically for mechanized straddle harvesting, protruding limbs that can obstruct the harvester need to be controlled. Olive trees are biennial bearers, and pruning at the correct time during "on" years will encourage more shoots and subsequent fruit growth in the following "off" years. There are three main types of pruning, described below.

Regulated pruning

Regulated pruning aims to develop the tree's frame, and is of great importance in the first years of its life. This pruning into shape facilitates cultivation, spraying and especially harvesting. At this early stage, very severe pruning should be avoided because it delays trees from entering the fruiting period. The most common shaping system is the "free cup" (Figure 6.8a). To form this shape, one-year-old trees are cut back to 60–80 cm above the ground when they are planted. In the first year, the main focus is to create side branches around the central axis to a height of 30–60 cm from the ground. In the following years, pruning is very mild, aimed at the removal of broken shoots or of shoots that intersect one another. After the tree has developed well, three to five main branches are chosen around the central axis, with a 20–30 cm distance between them. When the tree enters the fruiting period, and if no severe pruning is performed, it gradually takes a free spherical shape.

In intensive cultivation, where trees are densely planted, short pruning shapes are desired – namely, the "short cup" and the "bush" (Figure 6.8b, 6.8c). In the "short cup," branching is allowed very close to the ground, at a height of 30–40 cm, while in the "bush" no pruning is carried out for the first five or six years. After this, only weak shoots and top branches

Figure 6.8: Three different systems for training young olive trees: (a) free cup; (b) short cup; (c) bush.
Source: TDC-Olive (2004).

exceeding 3 m in height are removed. The bush shape has certain advantages for intensive cultivation systems, including an earlier fruiting period, higher yields per hectare compared with other pruning shapes, and lower labor costs because the fruit can be harvested at ground level, without using ladders – although for mechanical harvesting, this is not a good shape.

Pruning for fruiting

The aim of pruning for fruiting is to induce productive branches to form fruits, leaving the structural branches unaffected. Additionally, it maintains uniform production in terms of yield and quality – a feature that is particularly important for table olive varieties. Olive trees produce fruits on the previous year's branches, but very vigorous branches are not productive (they have only vegetative buds) and weak branches produce few fruits. Therefore, the aim of pruning is to induce branches to form fruits, ensure good lighting conditions and maintain an active and vigorous fruiting zone. The above goals are difficult to achieve in densely planted trees, because of the reduced lighting to the crown. In this case, the fruiting zone is restricted to the top branches and certain areas of the south part of the canopy, where there is more light. The productivity of these trees is greatly reduced when their tops are pruned to give a shorter shape, because a significant part of their canopy is removed.

In the productive stage, it is suggested that mild pruning be performed every year to remove dead and dense branches from the fruiting zone. This is necessary because the fruiting zone has the tendency to produce short and dense shoots with time. The aim of this mild pruning is to improve the length of the shoots and ensure good lighting throughout the fruiting zone. It must be noted that such pruning must be severe for trees growing in arid and infertile soils, to reduce the surface area of the canopy and thus save nutrients and water for the new fruiting growth.

Conversely, trees growing in fertile soil with good fertilization and irrigation need less severe pruning, because there are adequate nutrients and water for both the present vegetation and the development of new fruiting growth. Severe pruning of these trees results in the development of sucker shoots.

In the case of table olive varieties, pruning is also used to improve the size of the fruit. For this reason, it is suggested that excessive fruits be thinned immediately after fruit setting, especially in high-yield years. Proper pruning can also reduce the alternate bearing of the olive tree. In this case, severe pruning is suggested in the winter following the year of high yield, cutting off low-vigor shoots.

Renovating pruning

Renovating pruning aims to stimulate sprouting in order to rejuvenate senescent trees. The main characteristic of the olive tree is its longevity, because it has the ability to produce new shoots from almost any part of its wood; this makes it possible to renovate senescent or frost-damaged trees. Old or low-yield trees can be rejuvenated by cutting off their trunk at a low height, or at the point of ramification. For partial renewal or reduction of the canopy surface in densely planted trees, pruning is performed at the branches or their first ramifications at a desirable height. New vivid shoots will develop from the cutting points, the most appropriate of which are chosen for the new shape of the tree. The new tree will again enter the fruiting period after about two years, depending on the level of care it receives. When frost damage occurs, trees are left unattended for one year to estimate the real extent of the damage. New branches will form from the newly developed shoots and all the damaged parts should be cut off.

When should pruning be carried out, and how severe should it be?
Olive trees can be pruned immediately after harvest. For table olive varieties, pruning begins in November–December for green olives, or February–March for black olives. In general, pruning can be performed from autumn to the first months of spring, but it should be delayed in areas with a high risk of frosts.

When deciding upon the severity of pruning, the following parameters must be taken into account:

- the level of rainfall in autumn and winter;

- the yield of the previous year;

- the vegetative condition (vigor) of the tree when pruning;

- the end product (table olives or olive oil);

- planting density and the pruning system applied.

6.3.2 Weed control

Olive trees can survive in low-fertility soils under semi-arid conditions. Unfortunately, many weed species can adapt to the same conditions and grow faster than the olive trees, providing strong competition for moisture and nutrients.

Weeds, especially perennial species, have almost the same growth pattern as olive trees. However, their adaptability and greater efficiency ensure earlier and larger growth than that of the olive. Competition from weeds can be a major problem for young trees, but is easily managed either by orchard floor management or by using herbicides.

Orchard floor management decisions are significantly influenced by the location, climatic conditions, soil, irrigation practices, topography, and grower preferences. Weeds are commonly controlled either by chemical or mechanical means. The area between tree rows may also be chemically treated, or mechanically mowed or tilled. Alternatively, mulches, subsurface irrigation and flamers can be used. Often, several weed-management techniques are combined. Weed control must be applied four to six weeks before there is visible spring growth in olive trees.

Trees are most sensitive to weed competition during the first few years of growth, and where soil depth is limited. Weedy orchards may require several more years to become economically productive than do weed-free orchards. Regardless of the method used to control weeds, special attention must be paid to ensure that trees do not suffer chemical injury, or mechanical damage to the trunk or roots. As trees become established, competition from weeds reduces as shade from the tree canopy limits weed growth. Some of the more common ways to control weeds in new orchards follow.

To control weeds with herbicides after trees are planted and before fruiting, a pre-emergence herbicide is applied to either a square or a circle around each tree, or as a band along the tree row. Herbicides can also be applied to control weeds after they emerge. Selective herbicides are available for annual grass control and suppression of perennial grasses. Paraquat (1,1'-dimethyl-4,4'-bipyridinium dichloride) can be used to control weeds near young trees protected with shields or wraps. The non-selective herbicide glyphosate (sodium 2-[(hydroxy-oxido-phosphoryl)methylamino]acetic acid) can control broadleaf weeds after emergence, but should be used only around mature trees with brown bark, and must not be allowed to make contact with tree leaves.

Pre-emergence herbicides can be applied alone; in combinations of herbicides in the autumn after the harvest; divided into two applications (autumn and spring); or in the winter with a post-emergence (foliar) herbicide. It may be most beneficial to delay the pre-emergent application in winter until most weeds have germinated. Afterwards, a post-emergence herbicide can be used. This allows longer weed control during the summer, and therefore the tree does not suffer much

Table 6.1: Main herbicides recommended for olive trees

Active ingredients	Comments
Pre-emergence	
Simazine	Apply 3–4 years after tree planting
Diuron	
Oxyfluorfen	Also recommended for young trees
EPTC	
Chlorthal dimethyl	For olive tree nurseries
Post-emergence	
Paraquat	Contact herbicides
Diquat	
Paraquat and diquat	
Glufosinate ammonium	Exert slight systemic activity
Glyphosat	Systemic herbicides
Glyphosat trimensium	
Aminotriazole (amitrole)	
Pre-and post-emergence	
Simazine and paraquat	Effective on germinated weeds
Simazine and aminotriazole	
Diuron and amonitriazole	
Terbuthylazine and glyphosate	

competition from weeds. For safety reasons, herbicide sprays must be directed only at the soil or at the weed foliage, and not at the tree leaves. Some of the main herbicides recommended for olive trees are presented in Table 6.1 (TDC-Olive, 2004).

6.3.3 Root-zone soil buffering

A practical breakthrough in the field of utilization of saline water irrigation of olives in the Israeli Negev Desert area was made about three decades ago with the pioneering studies of De Malaach and Klein, and finalized by Wiesman (Klein *et al.*, 1994; Wiesman *et al.*, 2002b, 2004). These studies addressed the following topics and questions:

- Is it possible to cultivate olive trees with solely saline water irrigation?

- What is the optimal saline water concentration for sustainable olive cultivation?

- What is the effect of saline irrigation water on the vegetative and reproductive response of olive varieties?

- What is the effect of saline irrigation water on olive root-system development?

- What is the effect of saline irrigation water on olive oil quality?

Is it possible to cultivate olive trees with solely saline water irrigation?

Central Negev has a reservoir of saline water with 7.5 dS/m electrical conductivity. Based on the results of a set of long-term basic and practical field experiments performed over the past 20 years, it is clear that it is possible to utilize local available saline water using proper advanced drip irrigation technology specifically developed for the cultivation of olive trees. However, with direct use of highly saline water (7.5 dS/m) pumped from the local reservoir, olive trees grew and survived, but vegetative development and yield were significantly reduced, particularly after about 10 years.

It is very important that soil flushing is performed at least twice a year, forcing excess volumes of saline water through the drippers to push the salts toward the margin of the irrigated soil zone.

Winter rainfall significantly increases the level of salts in the root-system growing area in the soil. As a result of this, many young olive trees were found to collapse and die about a year after planting. To overcome this problem, a field irrigation technique using saline water drip irrigation was developed for use during rains, and this can save the plants by pushing salt towards the margin of the root zone.

Furthermore, late winter and late summer soil-flushing via the drip irrigation system as an annual practice was developed, and this maintains the level of salts in the root zone low enough to assure growth and development of the olive trees (Klein *et al.*, 1994; Wiesman *et al.*, 2002b).

What is the optimal saline water concentration for sustainable olive cultivation?

When water of two different degrees of salinity (7.5 dS/m in the pumped water from the underground source of the Ramat Negev area of Israel, and 4.2 dS/m made by mixing this with fresh water) were used for drip irrigation and compared to control water (1.2 dS/m) in the same olive plot, data collected over 15 years clearly demonstrated the sustainability of olive cultivation in desert conditions using 4.2 dS/m in comparison to control water. No significant vegetative growth reduction occurred, and no reduction of olive yield was observed. The trees irrigated with 7.5-dS/m water demonstrated a clear pattern of reduction of canopy growth and average annual yield (Table 6.2a, b).

What is the effect of saline irrigation on the vegetative and reproductive response of olive varieties?

Comprehensive study of the effect of moderately saline water irrigation (4.2 dS/m), used in drip irrigation technology in the Negev Desert, on the growth and development of a list of 14 olive

Table 6.2a: Effect of saline water irrigation on the trunk circumference of Barnea olive trees grown in the Ramat Negev area of the Negev Desert, Israel

Salinity level (dS/m)	Trunk circumference of the tree (cm)		
	First year	Fourth year	Ninth year
1.2	3.14 ± 0.24	34.17 ± 1.72	62.47 ± 5.12
4.2	2.35 ± 0.13	31.44 ± 1.16	55.61 ± 4.22
7.5	3.32 ± 0.10	26.73 ± 1.17	41.73 ± 6.70

Values are the mean of six plants ± standard deviation.
Adapted from Wiesman et al. (2002b).

Table 6.2b: Effect of saline water irrigation on the yield of Barnea olives grown in the Ramat Negev area of the Negev Desert, Israel

Salinity level (dS/m)	Annual yield (Mt/ha)			Cumulative yield (Mt/ha)	Average annual yield (Mt/ha)
	1999	2000	2001		
1.2	18.04	9.8	17.43	42.27	15.09
4.2	22.05	12.7	20.62	55.37	18.46
7.5	17.43	11.49	17.6	46.52	15.51

Adapted from Wiesman et al. (2002b).

varieties originating from various traditional olive-cultivating countries in the Mediterranean basin clearly demonstrated that there was no significant effect of saline water on the rate of vegetative growth of each of the tested varieties, in comparison to olives irrigated with control irrigation water (1.2 dS/m). Similarly, no significant difference was observed regarding the olive yield, or the percentage oil achieved, between all the varieties of olives tested under control and under moderate saline irrigation conditions. Clear variation between the various olive varieties cultivated in the same plot in the same Negev Desert environment was demonstrated in both irrigating treatments. These sustained and clear results suggest that, with the use of the new saline irrigation technology, almost all common and known olive varieties can be cultivated in the Negev Desert environment (Wiesman *et al.*, 2002b; Weissbein, 2006; see also Table 6.3).

What is the effect of saline irrigation water on olive root-system development?

From the previous sections it is clear that with a proper water management system – and especially maintenance of the leaching fraction – moderately saline water be can used equally as well as fresh water for olive cultivation in desert areas. Since roots are the major sink for photoassimilates and their primary task is water and nutrient uptake, in desert conditions with saline water irrigation the direct impact of the system will be on the roots. It is therefore very important to know and understand the physiological aspects of olive roots under such circumstances (Munns and Termaat, 1986). In order to do this, a root-imaging study using

Table 6.3: Effect of saline irrigation on the vegetative and reproductive response of olive varieties

Variety	Salinity (dS/m)	Average trunk growth (cm/yr) (2001–2005)	Average olive yield (kg/tree per year) 2001–2004	Average oil (%), (2001–2004)
Israeli				
Barnea	1.2	5.1	55	20.4
Barnea	4.2	5.8	52	19.6
Souri	1.2	6.9	32	20.3
Souri	4.2	5.7	32	19.4
Maalot	1.2	6.8	37	24.3
Maalot	4.2	8.7	34	22.6
Italian				
Frantioio	1.2	4.8	39	12.5
Frantioio	4.2	6.2	32	15.9
Leccino	1.2	6.1	46	17.5
Leccino	4.2	6.5	46	16.6
Spanish				
Arbequina	1.2	5.9	43	18.7
Arbequina	4.2	5.3	45	19.3
Picual	1.2	6.3	43	14.7
Picual	4.2	6.4	43	14.7
Picudo	1.2	4.7	32	16.5
Picudo	4.2	6.7	31	15.5
Greek				
Kalamata	1.2	6.3	30	16.6
Kalamata	4.2	7	39	15.4
Koroneiki	1.2	5.2	44	19.6
Koroneiki	4.2	5.4	33	20.9
French				
Picholin	1.2	6.3	54	16.8
Picholin	4.2	6.2	59	14.4
Moroccan				
P. Morocco	1.2	7	37	13.5
P. Morocco	4.2	9.3	35	13.8

Values are the average of 10 trees of each variety in both saline water regimes. The olive plants were planted in 1997 in the Ramat Negev regions of the central Negev Desert, Israel.
Adapted from Weissbein (2006).

Figure 6.9: Three-tube disposition in the minirhizotron olive experiment.
Adapted from Weissbein (2006).

advanced minirhizotron technology is very helpful. This technology provides a non-destructive tool for investigating root-system dynamics, and consists of a miniature video camera installed in clear tubes. Digital images of the roots are provided by means of fiber-optic technology introduced via 90-cm-long pipes located at various distances from the olive tree trunk and dripper. The root images are captured on video and later analyzed using a root-tracing computer program (Box, 1996; Hendrick and Pregitzer, 1996). Early studies have clearly demonstrated the concentration of the roots of olive trees irrigated with saline water in the center of the wetted zone, in comparison to the more spread roots of the control-water irrigated trees. Though calculation of the total number and length of roots of olives under saline water and under fresh water irrigation conditions shows a significant difference between them, in the main root active zone a critical mass of roots can be obtained in the saline-irrigated trees more precisely than in the control-water irrigated trees (Weissbein, 2006). The installation of this minirhizotron system in an olive orchard can be seen in Figure 6.9.

In order to achieve precise results from the system, the clear acrylate tubes should be installed very carefully; in particular, disturbance to the soil surrounding the tube should be minimized. After installing the tubes in the desired location (generally 0.5, 1.0 and 1.5 m from the tree

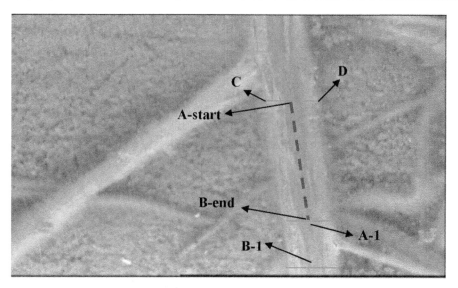

Figure 6.10: Part of the olive root image, obtained by video, of the minirhizotron system under saline irrigation conditions in the desert. The "straight line" method was used for determination of the tip number, root length and root diameter.
Source: Weissbein (2006).

trunk), images can be taken periodically at different depths (generally at 0–30 cm, 30–60 cm and 60–90 cm). The number of roots at different locations can be counted manually in the video images. The root length can be measured using the straight-line method described by Win Rhizotron software (Win Rhizotron, 2003), by drawing in the image exactly as with a paint program. This procedure allows the program to calculate the morphological measurements, length and surface area of the path of the root (Weissbein, 2006). Part of a root image of an olive root under saline water is presented in Figure 6.10.

What is the effect of saline irrigation on olive oil quality?

A few years ago, it was claimed that the oxidative effects of saline water might damage and reduce the quality of olive oil (Raven *et al.*, 1994). However, our study regarding the effect of saline water on olive oil quality of various varieties showed that no significant differences were found with regard to most of the basic IOOC quality parameters (International Olive Oil Council, http://www.internationaloliveoil.org), including free fatty acid composition, peroxide value and fatty acid profile (Wiesman *et al.*, 2002b; see also Table 6.4). The results of this study clearly showed that the quality of the oil produced from the 4.2 dS/m irrigated trees was similar to that produced from the control-water irrigated trees, by all IOOC standards. This suggests that saline irrigation technology, applied practically and exposing the olive trees to a moderate saline stress conditions, does not drastically interfere with the antioxidative mechanisms of the olives that operate to protect the accumulated oil. However, a clear variation between various olive

Table 6.4: Effect of saline irrigation on the quality of oil from Barnea olives grown in the Ramat Negev area of the Negev Desert, Israel

Salinity level (dS/m)	FFA (%)	Peroxide value (meq)	FA profile (%)						UN/SA	Polyphenols (ppm)	Vitamin E (ppm)
			16:00	18:00	18:01	18:02	18:03	20:00			
1.2	0.92	7.6	15.9	2.4	55.8	20.9	0.8	0.7	4	78	41.3
4.2	0.71	9.3	16.3	2.5	59.1	18	0.6	0.1	4.1	86.3	163.7
7.5	0.96	9.8	16.8	2.9	60.4	17.2	0.6	0.6	3.8	147.3	111.2

FFA, free fatty acids; FA, fatty acids; UN, unsaturated fatty acids; SA, saturated fatty acids.
Adapted from Wiesman et al. (2002b).

varieties was obtained. Some interesting results were obtained regarding the effects of saline water on triglyceride composition, specific phenol compounds and phytosterols. These findings clearly suggest that olive oil produced via saline-irrigation is not inferior to oil produced via regular irrigation water; in some cases, it is even superior in some aspects (see Chapter 11 for further details). In addition, saline irrigation water may become a way of differentiating olive oil produced in desert conditions, perhaps in time leading to a new brand name.

6.3.4 Alternate bearing and fruit thinning
Alternate bearing

Alternate bearing is common in olive cultivation (Barranco *et al.*, 1998). Under normal conditions, olive trees produce heavy crops one year (the "on" year) and light ones the next (the "off" year). Experience has shown that large olive yields tend to lead to smaller crops the following year. The potential light crop is predictable from the limited amount of new growth present in the trees. The previous heavy crop limits shoot growth and, because flowers are produced only on the new shoots of the previous season, the tree's potential for olive production is reduced. This alternate bearing is one of the most important limitations of natural extensive olive cultivation.

Generally, it is suggested that after an "on" year, in which most of the olive tree energy is directed towards the growth and development of the olives, and fruit oil accumulation, the energy reserves are significantly reduced. In particular, in extensive non-irrigated and non-fertilized olive cultivation practice, the growth of new shoots is very limited. In intensively cultivated olive trees, the rate of growth and elongation of the new shoots in an "on" year is also limited, mostly due to winter arriving very soon after the olive harvest. However, in intensively cultivated olives in desert conditions, where winter begins late, growth of the newly formed shoots can be boosted by intensive irrigation and fertilization. Indeed, from our experience with a wide range of olive varieties, the rate of growth of new shoots is significantly stimulated at the end of the "on" year and the fluctuations in olive yield are significantly reduced. Furthermore, olive alternate bearing can be moderated by a list of additional intensive cultivation technologies, such as olive thinning in "on" years, pre-blooming potassium foliar fertilization, summer foliar fertilization, and others that are discussed below.

Fruit thinning

Fruit thinning is generally recommended in order to produce a more consistent yearly crop with larger and higher-quality fruit, an earlier maturity date and lower harvesting costs.

Usually olives (and mainly table olives) are valued based on the fruit size. Heavy crops have a preponderance of small fruits that are expensive to pick and of limited value. Furthermore, the alternate bearing of olives is the cause of light crops invariably following heavy crops. Fruit thinning has been shown to reduce the crop load, improve the size of the remaining fruit, and minimize alternate bearing.

Hand thinning is not practical for olives, and generally chemical thinning is practiced. Naptha-lene acetic acid (NAA) is an effective fruit thinner when applied 7–14 days following full bloom (Martin *et al.*, 1994b). The activity of the thinning agent is influenced by the condi-tion of the tree, the air temperature, and the NAA dose. Chemical olive-thinning operations are highly complicated and dangerous. If performed improperly, or under adverse conditions, over-thinning can result. NAA application more than 2.5 weeks after blooming is both illegal and ineffective. Application that is too early may cause over thinning, while late application can yield unsatisfactory results. Application during blooming can even eliminate the whole crop.

6.3.5 Fertilization

Generally, because of their wide distribution in semi-arid environments, olive trees are not considered to be highly nutrient-demanding plant species. However, in recent years it has become well understood that, in order to maintain high productivity and cost-effective management in the olive oil industry, it is very important that the olive trees, especially in desert-like areas and under intensive cultivation practices, are provided with not only a proper irrigation regime but also an accurate fertilization methodology. This can be addressed by periodic monitoring of the status of the soil and the tree foliage.

Fertilization of olives is based on routine annual supply. Olive trees are characterized by a fast response to nitrogen fertilizer, causing intensive green mass growth. Proper nutrition for olive trees encourages new growth, better fruit size, heavier production and more regular bearing. For successful olive growing, nine macro-nutrients (i.e., nutrients needed in relatively high amounts) are required: carbon, hydrogen, oxygen, phosphorus, potassium, nitrogen, sulfur, calcium and magnesium. Of these, nitrogen and potassium are the most important for olive growing. Seven other nutrients that are equally important but required in smaller quantities are iron, manganese, boron, zinc, copper, chlorine and molybdenum. Of these, boron plays the crucial role in olive growing. Nitrogen is usually supplied in the form of ammonium sulfate or similar. Symptoms of low nitrogen levels include small, yellowish leaves, poor shoot growth, sporadic blooming and poor fruit set (Figure 6.11). Trees with adequate nitrogen bloom properly, with a good fruit set.

Based on leaf analysis, it was found that most olive groves in the Mediterranean basin lack potassium. Dead leaf tips or margins, light green coloration and twig dieback are the main symptoms of potassium deficiency (Figure 6.11). An adequate supply of potassium is highly recommended in order to increase the vegetative and reproductive growth and development of olives (Fernández-Escobar, 1999). Generally, no significant shortage of phosphorus was observed; however, boron deficiency is relatively common in olive trees, and this has been known for many years (Hanson, 1991). In some cases, deficiencies of sulfur, calcium, magnesium, iron, copper, manganese and zinc could also be detected.

Low Nitrogen   Potassium deficiency   Boron deficiency twig dieback and dead tip leaves

Figure 6.11: (a) Non-intensively cultivated olive trees suffering from typical nutrient deficiencies; **(b)** intensively cultivated olive trees with well-balanced nutrition; **(c)** low nitrogen levels result in pale olive leaves and a lack of new growth; **(d)** potassium deficiency; **(e)** boron deficiency results in twig dieback and **(f)** leaves with dead tips.
Source: Connell (2006).

In most cases the soil is not short in boron, but the olive tree root uptake of the element is very low. Twig dieback, excessive branching, and leaves with dead tips and a yellow band while still green at the base are the prime symptoms of boron deficiency (Figure 6.11). Premature fruit drop and defective fruits (as in "monkey face" symptoms; see Figure 6.12) are also common when plants suffer from boron deficiency.

Traditionally, fertilization is based on organic or granular fertilizers spread in the olive groves twice a year, usually in early summer and following the harvest, in mid-autumn. In recent years, with the availability of irrigation systems in most new olive plantations and even in many traditional groves, a new technology named "fertigation," consisting of the application of soluble nutrients through the irrigation systems, has become popular among olive growers.

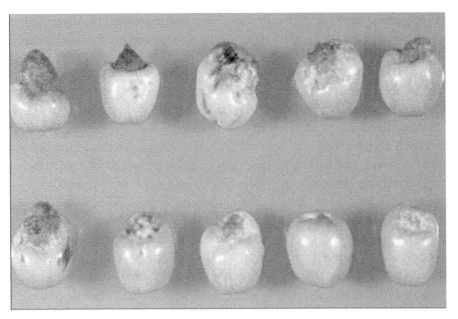

Figure 6.12: Defective olive fruit showing "monkey face" symptoms of vitamin B deficiency. Source: Connell (2006).

Fertigation

A well-designed fertigation system can reduce fertilizer application costs considerably, and supply nutrients in precise and uniform amounts to the wetted irrigation zone around the tree where the active feeder roots are concentrated. Applying timely doses of small amounts of nutrients to the trees throughout the growing season has significant advantages over conventional fertilizer practices. When plants receive conventional widespread applications of granular fertilizer once or twice a year, they receive a large quantity of fertilizer that is in excess of what they require at the time it is applied. This can result in salt damage to roots and excessive fertilizer losses through leaching and volatilization to the atmosphere, and can be toxic to beneficial soil micro-organisms and to fauna such as earthworms. Generally, with the use of conventional fertilizer spreaders much of the fertilizer is placed outside the wetted irrigation zone in areas not exploited by the tree roots. Since fertigation allows precise and uniform application of the nutrients to just the wetted root volume, where the active roots are concentrated, there is no doubt that this method increases the efficiency of application of the fertilizer to a remarkable degree, thus allowing reduction of the quantity of fertilizer required and the production costs whilst also reducing the potential for groundwater pollution caused by fertilizer leaching (Imas, 1999). However, there is no simple formula for determining fertigation rates and frequency for olive groves. Rates usually correspond to those used for granular application, but the frequency of application ranges from weekly to fortnightly to monthly. More frequent application ensures that nutrients are being supplied to the tree in adequate amounts and reduces the problems associated with salt damage and leaching losses. Since fertigation is generally part of the normal irrigation program, there is little extra cost associated with more frequent application.

Leaf analysis should form the basis of any fertigation program. In general terms, the annual fertilizer recommendation for a mature, high-yielding olive grove is about 400 g nitrogen (N), 80 g phosphorus (P) and 400 g potassium (K) per tree. Instead of applying this in one or two doses of granular fertilizer, it can now be applied through the drip irrigation system all year around at various levels, taking into account leaf and soil analysis, and expected growth and development. In general, at least 20 applications of 20 g N, 4 g P and 20 g K spread throughout the spring, early summer and mid-autumn is beneficial. However, in non-irrigated olive groves, or where irrigation systems are not installed, the application of granular or even soluble fertilizers is in many cases not very efficient. Usually, the activity of the applied fertilizer is dependent on rainfall, which generally only occurs in the winter in desert environments, and then in very limited amounts – and in the winter, the olive root system is very inactive due to the low soil temperatures (Hartmann *et al.*, 1966). Extensive olive cultivation, characterized by the irregular supply of water and particularly nutrients, is considered to be one of the reasons for the high fluctuations in olive yield known as alternate bearing (Hartmann *et al.*, 1966; Barranco *et al.*, 1998). In advanced intensive cultivation of olives, based on fertigation technology, the continuous application of nutrients is addressed far better than in traditionally cultivated olive groves. However, even in the modern olive groves alternate bearing is very common, and every two to three years there is frequently a strong "off" year in terms of productivity. Indeed, many genetic, pollination and environmental factors that are not fully understood affect and contribute to the fluctuation of olive yields. However, some studies in recent years have clearly demonstrated the effect of a novel foliar nutrient system applied just before the olive blooms at the beginning of the spring in significantly increasing the olive yield in both "on" and "off" years (Wiesman *et al.*, 2002c, 2002d). It is clear now that the efficient application of balanced nutrients, including nitrogen, potassium and boron, before blooming is highly important in stimulating the flowering and olive fruitlet set. As low soil temperatures maintain the olive tree root system in a non-active phase, and because the foliar application of nutrients (with an emphasis on boron and potassium) is so inefficient, almost all attempts to apply nutrients at the end of the winter or in early spring have been found to have little effect at the field level.

Foliar spray for olive development and yield increase

Experience has demonstrated that olive groves, especially the mature trees, show significant deficiencies in many nutrients. Nitrogen, potassium and boron are the main nutrients in which olive plants are deficient (Fernández-Escobar, 1999), although phosphorus deficiency is also seen. In order to discover the nutrient status of a plant, regular analysis of the leaves should be performed. The levels of nutrients in olive leaves are shown in Table 6.5. For effective analysis, samples should be taken in the winter pre-blooming stage and in July. Samples should be obtained from 100 mature leaves from the middle of non-fruiting shoots.

Table 6.5: Leaf analysis of olives

Nutrients	Deficient	Optimum
Nitrogen	1.4%	1.5–2.0%
Potassium	0.4%	0.8–1.0%
Boron	14 ppm	19–150 ppm

Research has shown that foliar application of suitable nutrients to the olive plants, and especially to desert-grown olives, is required twice a year. The first is during the pre-blooming time, in order to increase the number of olives that are produced and set. During the winter, olive roots are inactive in desert environments because of the comparative coldness of the soil. Moreover, in desert environments olive trees start to bloom earlier than in cold areas because of the comparatively high temperature. The slow growth in winter and the early blooming in desert environments lead to a constant shortage of the nutrients in the plants.

Pre-blooming or early spring foliar spray

It has been known for some time that the uptake of some essential nutritional elements (such as boron) from the soil by the root is very low, and therefore the recommendation has been to apply them to the foliage by spraying (Hartmann *et al.*, 1966; Moyoma and Brown, 1995). However, due to the thickness and complex chemico-physical structure of the cuticular membrane coating and protecting the olive leaves (Jenks and Ashworth, 1999), nutrient penetration into the leaf cells is very low (Weinbaum, 1988).

To overcome the natural difficulty of nutrient penetration through the olive cuticle membrane, a five-year research and development project was carried out in the phyto-lipid biotechnology laboratory at Ben-Gurion University of the Negev from 1997 until 2002.

Initially, the olive cuticle membrane was characterized at the morphological (Figures 6.13, 6.14), chemical and physical levels (Luber, 2001; Wiesman *et al.*, 2002c). The membrane is formed by various lipid compounds consisting of waxes, fatty acids, fatty alcohols, aldehydes, ketones, and some additional sterols and phenols (Jenks and Ashworth, 1999). The penetration of nutrients is possible only via the non-crystalline fraction of the cuticle membrane, designated Zone B (Figure 6.15).

In the second phase of the developing project, a novel controlled-release delivery system that facilitated long-lasting delivery of the encapsulated nutrients through the olive cuticle membrane was developed.

The third phase of the project consisted of laboratory radiolabeled nutrient materials being tested for their capacity to penetrate isolated and intact cuticle membranes. The effect of the novel delivery system was compared to that of many other common and advanced surfactant

Plant skin - Cuticle

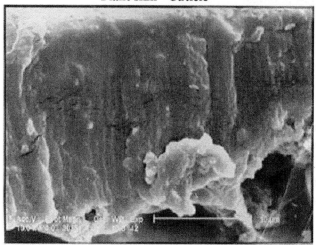

Figure 6.13: Typical thick olive-leaf cuticle membrane as shown by scanning electron microscopy.

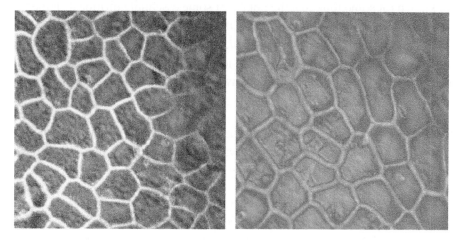

Figure 6.14: Upper surface of olive-leaf cuticle membrane as dyed by fluorescein (left) and rhodamin (right), using laser confocal microscopy.

and delivery systems (Luber, 2001). Following a long and successful testing phase, the fourth phase of large-scale field testing was carried out. This world-wide olive field testing yielded the pre-blooming olive nutrient (PON) treatment (Wiesman *et al.*, 2002c).

The mineral constituents of PON are $N(NH_2)$ 10 percent, P_2O_5 33 percent, K_2O 21 percent, and B, 1.8 percent, coated with the novel controlled-release delivery system providing a long-lasting nutrient penetration effect. Application of this treatment in March, about three weeks before bud opening in olive trees, led to a significant increase in the N, P, K and B levels. Five years'

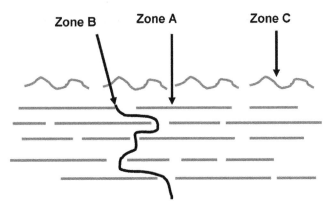

Figure 6.15: Scheme showing cuticle membrane nutrient penetration pathway. Zone A consists of the crystalline lipid fraction; Zone B shows the route of penetration via the gaps in the crystalline fraction; Zone C consists of an amphorous fraction on the upper surface of the cuticle membrane.

monitoring of the fluctuations in olive yield in treated groves demonstrated a significant increase in the average yield, ranging from 20 percent to 50 percent, compared with untreated control treatment. This effect was obtained in both "off" and "on" years. At present, this treatment is well recognized as being the most efficient and recommended foliar olive spring fertilization method.

Summer olive foliar spray

In order to improve the nutritional status of the olive tree during the fruit development period, a summer foliar application of urea was suggested. However, this remained uncommon in practice due to the low effectiveness of the spray and the imbalanced nutrient composition. With recent development of foliar sprays based on an advanced surfactant composition, this practice has become more common. It is generally well accepted that nutrients can be fed much more easily to young olive leaves with a thinner epicuticular wax layer than to mature leaves (Jenks and Ashworth, 1999). However, spraying of young, delicate and less-protected olive leaves significantly increases the foliage burn risks from high concentrations of various mineral elements. The second treatment based on the technology developed in the phyto-lipid biotechnology laboratory at Ben-Gurion University of the Negev addressed the issue of leaf burn risk, which was reduced by the controlled release of nutrients and the long-lasting penetrative effect. The second treatment, named the summer olive nutrient (SON) treatment, comprises coated sodium, phosphorus and potassium (8–16–40).

Application of PON and SON

Spraying PON at 2.0 percent concentration, using 1000–1500 l/ha in the month of March, and SON at the same rate in the month of July significantly increase the yield in olive groves (Table 6.6; Figure 6.16).

Table 6.6: Summary of the effect of PON and SON applications Barnea trees (~50 hectares), 1998–2002

Treatment	Average yield over 5 years (kg/tree)	% Control
Control	24	100
PON	34	142
SON	28	116
PON + SON	38	158

PON, pre-blooming olive nutrient spray; SON, summer olive nutrient spray.

Figure 6.16: PON- and SON-treated olives being weighed and monitored.

Figure 6.17: Olive fruit fly, and damage cause by its larvae.
Source: TDC-Olive (2004).

Figure 6.18: Adult olive moth, and flower damaged by its first-generation larvae.
Source: TDC-Olive (2004).

6.3.6 Pest and disease management

Several insects and diseases limit olive tree growth and olive production. Today, it is accepted that an integrated pest management (IPM) strategy is necessary in order to control and contain the dangers at the lowest possible level, bearing in mind the economic situation and the ecological balance of the surrounding area (Dent, 1995).

Major harmful pests of the olive tree

The olive fly (*Dacus oleae*) is an insect that causes damage by laying its eggs on the olives (Figure 6.17). The number of generations depends on the climate of the area. The damage done by the larvae affects both the quality and the quantity of the olives produced. The olive fly can be found in all Mediterranean olive-growing countries, and extends as far as India in the east and the Canary Islands in the west. However, it has not been found in regions where the olive is an introduced species, such as North and Central America (California, Arizona, Mexico, El Salvador), South America (Argentina, Chile, Peru, Uruguay), Central Asia (China) and Australia. Treatment with products such as dimethate or natural piretrioid, used as insecticides, is recommended when the first fruit shows evidence of viable insect eggs.

The olive kernel-borer or olive moth (*Prays oleae*) is a small moth that attacks the flowers, fruits and leaves of olives (Figure 6.18). It has been known as an olive pest since very early times,

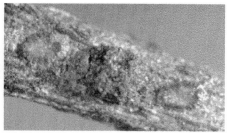

Figure 6.19: Black scale (*S. Oleae*), and exit of black scale, from *Metaphycus helvolus*. Source: TDC-Olive (2004).

as is evident from descriptions in ancient Greek and Roman documents. *P. oleae* is found in all Mediterranean olive-growing countries, and has been identified in areas around the Black Sea, such as the Crimea and Georgia. Its presence has not been reported in Central Asia (Iran, Pakistan, Afghanistan), East Africa (Eritrea), Southern Africa or North or South America. The most effective defense is the use of organophosphorus insecticides. Recently, it has been found that microbiological insecticides based on *Bacillus thuringiensis* provide an effective defense.

Black olive scale (*Saissetia oleae*) causes considerable damage to plants, including leaf drop, shrinkage of shoots and a decrease in blooming. *S. oleae* is widely distributed, extending from Central Asia to Africa. The olive tree is one of a large number of host plants on which *S. oleae* has been found. The preferred habitat is the lower surfaces of olive trees. *S. oleae* damages the olive tree directly by sucking the sap, and indirectly by releasing honeydew onto the leaves. This honeydew is a substrate for the development of different fungi, and olive tree cultivation is thus responsible for the spread of a sooty mold. By coating the leaves, this sooty mold impedes photosynthesis and respiration, and finally induces more or less serious leaf drop (Figure 6.19). In cases of severe infestation, intervention is necessary using selective insecticides.

Of the less important insect pests, some occur in particular areas or conditions at population levels that cause serious damage – for example, *Euphyllura olivine*, *Zeuzera pyrina*, *Aspidiotus nerii* and *Resseliella oleisuga*. Others, although they occur only occasionally, cause serious problems by disrupting the biological balance of the ecosystem – for example, *Parlatoria oleae*, *Leucaspis riccae* and *Philippia follicularis*.

Major harmful diseases of olives

Several diseases can cause serious problems in olives; *Verticillium* wilt, olive leaf spot and olive knot are the most common. Their severity depends on the weather conditions, location, and cultural practices (Fergusan *et al.*, 1994). Foliage diseases and diseases that affect the shoots and branches are caused by pathogens that are spread by wind, splashing water or insects. Injuries such as frost cracks provide entry points for infection. Soil-borne diseases are caused

Figure 6.20: Major diseases of olives: **(a)** *Verticillium* wilt; **(b)** leaf spot; **(c)** olive knot; **(d)** *Phytophtora*.
Source: TDC-Olive (2004).

by soil-inhibiting pathogens that infect roots. Root pathogens are spread by wind, in surface water, and via soil-contaminated equipment.

Verticillium wilt (*Verticillium dahliae* Kleb.)

This disease exists in almost all olive-growing countries. The fungus can survive in the soil for years, either embedded in infected tissues or in the form of sclerotia. It is spread by soil movement during tillage, by irrigation water and by infected tools used for pruning. Symptoms of the disease appear when leaves on one or more branches of the tree suddenly wilt early in the growing season; this process intensifies as the season progresses. Death of mature trees infected with *Verticillium* is also possible (Figure 6.20). Darkening of xylem tissue, a key symptom for distinguishing *Verticillium* wilt in many crops, is frequently not apparent in olives. There are no effective chemical controls for *Verticillium* wilt in established groves, and the most effective management strategies to protect trees from *Verticillium* wilt are those taken before planting. When considering a new site for an olive orchard, it is therefore recommended that land which has been planted for a number of years with crops that are highly susceptible to the disease (such as cotton, eggplant, peppers, potato or tomato) be avoided. Inoculum levels can be reduced before planting by soil fumigation, soil solarization, flooding the fields during summer, growing several seasons of grass cover crops, or a combination of these treatments.

Olive leaf spot (*Spilocea oleaginea*)

Olive leaf spot, a fungal disease, is one of the most common and damaging diseases of olives. This fungus disease infects the leaves with the first fall rains. The fungal spots enlarge and become colored, resembling the "eye spot" of a peacock's tail; they are therefore often referred to as "peacock spots." In spring, infected leaves defoliate, reducing bloom and the subsequent crop. In years of high winter rainfall, untreated trees can lose 50 percent of their canopy due to olive leaf spot infection. Dense trees and humid conditions encourage optimal olive leaf spot infection. Pruning (to eliminate dense clumps of foliage) mitigates infection and enhances the coverage of protective fungicides.

Olive knot (*Pseudomonas syringae* pv savastanoi)

Olive knot, a bacterial disease, requires an opening for infection. Pruning wounds, leaf scars and frost cracks are the most common wounds that become infected with olive knot. Rainfall disseminates the bacteria into these openings. Once infected, tissue grows at an uncontrolled rate, forming a gall ("knot") that girdles and kills the shoots and branches. Severe infections, usually resulting from freezing temperatures that lead to a profusion of wounds, can cause considerable fruit wood death. Growers prune olives in spring to reduce the hazard of opening pruning wounds for infection via rainfall. Pruning out of established knots is recommended in the summer to eliminate inoculum from the grove. Grove floors are managed free of weeds for optimal protection against radiation and advective freezes.

Phytophthora root and crown rot (*Phytophthora* spp.)

Phytophthora species attack olive trees in some areas, but the disease is not considered a major problem. The pathogen enters the tree at the crown near the soil line, at the major roots, or at the feeder roots, depending on the species. Trees affected with *Phytophthora* first show small leaves, sparse foliage, and lack of terminal growth. Infected trees may decline for several years, or die within the same growing season in which the foliage symptoms first appear. *Phytophthora* can survive in the soil for many years and spread and infect the trees during moist cool to moderate temperatures; some infection may occur in the summer, depending on the species. No chemical control measures exist for *Phytophthora*; cultural control measures are based on site selection and preparation, proper irrigation management, and improvement of soil drainage.

Fruit mummification

This disease is caused by the fungus *Gleosporium olivarum*. The fungus can penetrate healthy fruit skin, although existing lesions may facilitate the process. Infected fruits display brownish round spots, which expand in size. Usually, infection begins at the distal end of the olive fruit, where water droplets rest after dew or rainfall. Dissemination of the inoculum is facilitated by rain, since germination occurs only in the presence of water. Fungal conidia may survive for one year in mummified fruit at low temperatures. The disease is common in Mediterranean olive-growing countries, particularly Portugal, Greece and Lebanon (TDC-OLIVE, 2004). The first attacks appear in September, while olive fruits are still green. The combination of rainfall and high relative humidity results in the development of pouches and conidia on the infected fruits, creating secondary infections that result in fruit drop and increased acidity of the extracted olive oil. Occasionally the infection may also spread to vegetal parts, causing leaf drop, shoot death, and the overall weakening of the infected tree. For the control of the disease, preventive fungicide treatment is recommended at the beginning of September, before the rainy period. Application must be repeated later if secondary infections are noticed.

References

Alegre, S., Girona, J., Marsal, J., Arbonés, A., Mata, M., Montagut, D., Teixidó, F., Motilva, M. J., & Romero, M. P. (1999). Regulated deficit irrigation in olive trees. *Acta Hort*, 474, 373–376.

Aragues, R., Puy, J., Royo, A., & Espada, J. L. (2005). Three-year field response of young olive trees (*Olea europea* L. cv Arbequina) to soil salinity: trunk growth and leaf ion accumulation. *Plant Soil*, 271, 265–273.

Ayerza, R., & Coates, W. (2004). Supplimental pollination increasing olive (*Olea europaea*) yields in hot, arid environments. *Exp Agric*, 40, 481–491.

Barranco, D., Fernandez, R., & Rallo, L. (1998). Variedades y partoned del cultivo del olivo [in Spanish]. In: D. Barranco, R. Fernández-Escobar, L. Rallo (Eds.), *Cultivo Del Olivo* (pp. 61–68). Junta de Andalucia, Seville.

Beede, R. H., & Goldhamer, D. A. (1994). Olive irrigation management. In: L. Ferguson, G. Steven Sibbett, G. C. Martin (Eds.), *Olive Production Manual* (pp. 61–68). University of California at Davis, Division of Agriculture and Natural Resources, Berkeley, CA.

Benlloch, M., Arboleda, F., Barranco, D., & Fernández-Escobar, R. (1991). Response of young olive trees to sodium and boron excess in irrigation water. *Hort Sci*, 26, 867–870.

Bernstein, L. (1964). Effects of salinity on mineral composition and growth of plants. In: *Proceedings of the Fourth International Colloquium of Plant Analysis and Fertilizer Problems*, 4, 25–45.

Box Jr., J. E. (1996). Modern methods for root investigations. In: Y. Waisel, A. Eshel, U. Kafkafi (Eds.), *Plant Roots: The Hidden Half* (pp. 193–237) (2nd edn). Marcel Dekker, New York, NY.

Briccoli Bati, C., Godino, G., & Nuzzo, V. (2002). Preliminary agronomic evaluation of two cultivars of olive trees obtained from micropropagation methods. *Acta Hort*, 586, 867–870.

Connell, J. (2006). *Nutrient Considerations for Olives*. UC Cooperatives Extension, Butte County, CA.

Dent, D. R. (1995). *Integrated Pest Management*. Chapman & Hall, London.

Diaz-Espejo, A., Verhoef, A., & Knight, R. (2005). Illustration of micro-scale advection using grid-pattern mini-lysimeters. *Agric Forest Meteorol*, 129, 39–52.

Dicho, B., Xiloyannis, C., Angelopoulos, K., Nuzzo, V., Bufo, S. A., & Celano, G. (2004). Drought induced variations of water relation parameters in *Olea europaea*. *Plant Soil*, 257, 381–389.

Fabbri, A., Bartolini, G., Lambardi, M., & Kailis, S. (2004). *Olive Propagation Manual*. CISRO, Clayton, S. Victoria.

Fergusan, L., Sibbett, G. S., & Martin, G. C. (1994). *Olive Production Manual*. University of California Publications, Berkeley, CA, No. 3353.

Fernández, J. E., Moreno, F., & Martín-Aranda, J. (1993). Water status of olive trees under dry-farming and drip-irrigation. *Acta Hort*, 335, 157–164.

Fernández-Escobar, R., Moreno, R., & García-Creus, M. (1999). Seasonal changes of mineral nutrients in olive leaves during the alternate bearing cycle. *Sci Hort*, 82, 25–45.

García-Férriz, L., Ghorbel, R., Ybarra, M., Marì, A., Belaj, A., & Trujillo, I. (2002). Micropropagation from adult olive trees. *Acta Hort*, 586, 879–882.

Goldhamer, D. J., Dunai, J., & Ferguson, L. (1994). Irrigation requirements of olive trees and responses to sustained deficit irrigation. *Acta Hort*, 356, 172–176.

Gucci, R., & Tattini, M. (1997). Salinity tolerance in olive. *Hort Rev*, 21, 177–214.

Hanson, E. J. (1991). Movement of boron out of tree fruit leaves. *Hort Sci*, 26, 271–273.

Hartmann, H. T., Uriu, K., & Lilleland, O. (1966). Olive nutrition. In: N. F. Childers (Ed.), *Nutrition of Fruit Crops* (pp. 255–261). Rutgers Horticultural Publications, New York, NY.

Hartmann, H. T., Kester, D. E., & Davies, Jr. J. T. (1990). *Plant Propagation: Principles and Practice* (5th edn.). Prentice-Hall, Upper Saddle River, NJ.

Hassan, M. M., Seif, S. A., & Morsi, M. E. (2000). Salt tolerance of olive trees. *Egyptian J Hort*, 27, 105–116.

Hendrick, R. L., & Pregitzer, K. S. (1996). Applications of minirhizotrons to understand root function in forests and other natural ecosystems. *Plant Soil*, 185, 293–304.

Imas, P. (1999). Recent techniques in fertigation of horticultural crops in Israel. Proceedings of the IPI-PRII-KKV Workshop on: Recent Trends in Nutrition Management in Horticultural crops 11–12 February 1999. Dapoli, Maharashtra, India. (available at http://www.ipipotash.org/presentn/rtifohc.html).

IOOC (International Olive Oil Council), http://www.internationaloliveoil.org.

Jacoboni, N., & Fontanazza, G. (1981). *Cultivar*. Rome: REDA L'Olivo, pp. 7–52.

Jenks, M., & Asworth, E. (1999). Plant epicuticular waxes: function, production and genetics. *Hort Rev*, 23, 1–68.

Klein, Y., Ben-Tal, S., Lavee, J., De Malach, Y., & David, I. (1994). Saline irrigation of cv Manzanilo and Uomo di Piccione trees. *Acta Hort*, 356, 176–218.

Luber, M. (2001). Characterization of factors influencing phosphate penetration into *Citrus grandis* leaves. MSc Thesis, Ben-Gurion University of the Negev, Beer-Sheva, Israel.

Martin, G. C., Ferguson, L., & Polito, V. S. (1994a). Flowering, pollination, fruiting, alternate bearing, and abscission. In: L. Ferguson, G. Steven Sibbett, G. C. Martin (Eds.), *Olive Production Manual* (pp. 51–56). University of California Press, Berkeley, CA.

Martin, G. C., Connell, J. H., Freeman, M. W., Krueger, W. H., & Sibbett, G. S. (1994b). Efficacy of foliar application of two napthaleneacetic acid salts for olive fruit thining. *Acta Hort*, 356, 302–305.

Mass, E. V., & Hoffman, G. J. (1977). Crop salt tolerance – current assessment. *J Irrig Drainage*, 103, 115–134.

Motilva, M. J., Tovar, M. J., Romero, M. P., Alegre, S., & Girona, J. (2000). Influence of regulated deficit irrigation strategies applied to olive trees (Arbequina cultivar) on oil yield and oil composition during the fruit ripening period. *J Sci Food Agric*, 80, 2037–2043.

Moyoma, A. M. S., & Brown, P. H. (1995). Effects of the time of B application on almond tissue B concentration and fruit set. *Hort Sci*, 30, 879–884.

Munns, R. (1993). Physiological process limiting plant growth in saline soil: some dogmas and hypotheses. *Plant Cell Envir*, 16, 15–24.

Munns, R., & Termaat, A. (1986). Whole-plant responses to salinity. *Aust J Plant Physiol*, 13, 143–160.

Raven, J. A., Johnston, A. M., Parsons, R., & Kulber, J. (1994). The influence of natural and experimental high O_2 concentrations on O_2-evolving phototrophs. *Biol Rev*, 69, 61–94.

Shalhevet, J. (1994). Using water of marginal water quality for crop production. *Agric Water Manag*, 25, 233–269.

Talaie, A., & Taheri, M. (2001). The effect of foliar spray with N, Zn and B on the fruit set and cropping of Iranian local olive trees. *Acta Hort*, 564, 337–341.

TDC-OLIVE (Technology Dissemination Center for Olive) (2004). *European Commission Priority 5 on Food Quality & Safety*. (available at www.biomatnet.org/pulications).

UC-Cooperative Extension (2007). Sample costs to establish a super high density olive orchard and produce olive oil. Berkeley, CA: University of California Cooperative Extension, 00SV-07.

Weinbaum, S. A. (1988). Foliar nutrition in fruit trees. In: P. M. Neumann (Ed.), *Plant Growth and Leaf-Applied Chemicals* (pp. 81–100). CRC Press, Boca Raton, FL.

Weissbein S, 2006. Characterization of new olive (*Olea europae* L.) varieties' response to irrigation with saline water in the Ramat Negev area. MS. Thesis. Ben-Gurion University of the Negev, Beersheva, Israel.

Wiessbein, S., Wiesman, Z., Ephrath, J., & Zilberbush, M. (2008). Vegetative and reproductive response of olive varieties to moderate saline water irrigation. *Hort Sci*, 43, 320–327.

Wiesman, Z., & Lavee, S. (1994a). Vegetative growth retardation, improved rooting and viability of olive cuttings. *Plant Growth Reg*, 14, 83–90.

Wiesman, Z., & Lavee, S. (1994b). The rooting ability of olive cuttings from cv. Manzanillo F1 progeny plants in relation to their mother cultivar. *Acta Hort*, 356, 28–30.

Wiesman, Z., Riov, J., & Epstein, E. (1989). Characterization and rooting ability of indole-3-butyric acid conjugates formed during rooting of mung bean cuttings. *Plant Physiol*, 91, 1080–1084.

Wiesman, Z., Elgabi, F., & Lavee, S. (1995a). Vegetative propagation of olive. A. Effect of plant material. *Alon Hanotea*, 49, 136–140.

Wiesman, Z., Many, Y., & Lavee, S. (1995b). Vegetative propagation of olive. B. Effect of hormones and propagative agrotechnique. *Alon Hanotea*, 49, 228–233.

Wiesman, Z., Markus, A., Wybraniec, S., Schwartz, L., & Wolf, D. (2002a). Promotion of rooting and development of cuttings by plant growth factors formulated into controlled-release system. *Biol Fertil Soils*, 36, 330–334.

Wiesman, Z., De Malach, Y., & David, I. (2002b). Olives and saline water – story of success. *Intl Water Irrig*, 22, 18–21.

Wiesman, Z., Ronen, A., Ankarion, Y., Novikov, V., Maranz, S., Chapagain, B., & Avramovich, Z. (2002c). Effect of olive Nutri-Vant on yield and quality of olives and oils. *Acta Hort*, 594, 557–562.

Wiesman, Z., Luber, M., Ronen, A., & Markus, A. (2002d). Ferti-Vant – a new nondestructive and long-lasting *in vivo* delivery system for foliar nutrients. *Acta Hort*, 594, 585–590.

Wiesman, Z. Markus, A. (2004). Multi-layer adjuvants for controlled delivery of agro-materials into plant tissues. US Patent No. 10/488,388, registered by BGU.

Wiesman, Z., Itzhak, D., & Ben Dom, N. (2004). Optimization of saline water level for sustainable Barnea olive and oil production in desert conditions. *Sci Hort*, 100, 257–266.

Win Rhizotron (2003). *Manual for Rhizotron Root Measurement*. Regent Instruments. (www.regentinstruments.com).

Further reading

Inglese, P., Gullo, G., & Pace, L. S. (2002). Fruit growth and olive quality in relation to foliar nutrition and time of application. *Acta Hort*, 586, 507–509.

Katzer, G. (2003). Olive (*Olea europea* L.), http://www.uni-graz.at/~katzer/engl/Olea_eur.html.

CHAPTER 7

Desert-suitable genetic material

7.1 Genetic diversity and reproductive characteristics of the olive tree

The olive plant is a member of the *Oleacea* family, the genus *Olea* and species *europaea*. Many different subspecies within the genus *Olea* are found around the world. At least six subspecies of *Olea europea* L have been identified, and each has a specific distribution (Green and Wickens, 1989; Vargas *et al.*, 2001). *Olea europaea* subsp. *europaea* corresponds to the Mediterranean or European olive; *Olea europaea* subsp. *guanchica* is endemic to the Canary Islands; *Olea europaea* subsp. *maroccana* is found in Southern Morocco; *Olea europaea* subsp. *laperrinei* is distributed in the Saharan mountains, Algeria, Sudan and Niger; *Olea europaea* subsp. *cerasiformis* is endemic to the island of Madeira; and *Olea europaea* subsp. *cuspidata* is distributed from South Africa to Iran and China. This latter subspecies is composed of a range of morphological types corresponding to limited geographic areas, with three main species names:

Olea africana Mill from South to East Africa, *Oleaa chrysophylla* Lam from East Africa to Arabica, and *Olea cuspidate* Wall from Iran to China (Besnard *et al.*, 2002). *Olea cuspidata* has also been reported to be distributed in the southern part of the Himalayas in Nepal. All these taxa are supposed to be inter-fertile with Mediterranean olives, and gene exchange between distant populations could have contributed to the evolution of the Mediterranean olive (Besnard *et al.*, 2001).

The cultivated olive, which is believed to have originated from the eastern Mediterranean as early as 3000 BCE, was one of the first trees to be domesticated (Zohary and Spiegel-Roy, 1975) and possibly spread throughout the area following human displacement. Palynological and anthracological (fossil charcoal) evidence supports the occurrence of wild olives in forests of the Mediterranean Basin during the Paleolithic era (Lipschitz *et al.*, 1991). According to anthracological studies, during the Neolithic (10,000–7000 years ago) era, the use of wild olives (oleasters) persisted in several parts of the Basin (Lipschitz *et al.*, 1991; Zohary and Hopf, 1993). Clear signs of olive domestication (olive-oil presses) in the Near East date from the early Bronze Age (the second half of the fifth millennium BCE) (Zohary and Spiegel-Roy, 1975; Lipschitz *et al.*, 1991).

The domesticated olive is a short, squat evergreen tree, and rarely exceeds 8–15 meters in height. It has great longevity, with some specimens known to have lived for more than 1000 years. The crosses between wild local oleasters and introduced selected cultivars led to new cultivars (Besnard *et al.*, 2001). Though the cultivation and harvesting of olives is considered to be quite difficult compared with other oil crops, olive oil consumption is steadily increasing throughout the world because of its high quality and healthy constituents. This has driven cultivation to new areas, such as Australia and Latin America.

The flowers of cultivated olives are small and white, with a four-cleft calyx and corolla, two stamens and a bifid stigma. The flowers are generally borne on the previous year's wood, in racemes springing from the axils of the leaves. Cold winter temperatures are required for the induction of flowers, which are formed on an inflorescence bearing about 15–30 blooms. They may be either perfect or staminate with staminate dominating, and both cross-pollinated and self-pollinated.

The fruit is formed by an epicarp, which usually changes color at physiological maturity, and is joined to a pulp (mesocarp) that contains most of its oil. The endocarp (or stone) is hard, long and pointed at the apex; it encloses a seed in which the embryo and nutritional reserves are found. Alternate bearing of olives is common in most of the cultivated varieties.

Both wild and cultivated olives are diploid ($2n = 2x = 46$) and predominantly allogamous (Green, 2002). Wild olives reproduce sexually by wind pollination, and their seeds are mainly dispersed by birds (Herrera, 1995). Because of the allogamous mode of reproduction, a high level of heterozygosity has been maintained in olives, and domestication has been primarily due to the vegetative multiplication of selected individuals with favorable allele combinations (Zohary and Spiegel-Roy, 1975). Cultivated olives can be both self-compatible and incompatible. Directed breeding is difficult in olives due to their long juvenile phase. Therefore, "open" crosses between individual trees have resulted in shuffling of the genetic background, and this, along with micro-environmental pressure and its long history as a crop, has resulted in numerous cultivars. These cultivars have been selected over the centuries for their qualitative and quantitative traits, and propagated mainly vegetatively.

7.2 Common olive varieties

The cultivated olive tree developed with the exchange of sexual reproduction for vegetative reproduction based on techniques such as grafting of selected scions or propagation of stem cuttings. The criteria used for identification and selection of the best plants are agronomical and technical in nature. The selection of olive varieties has given rise to diverse groups or varieties which today are spread throughout the different olive-oil-producing regions. In most Mediterranean countries where olive cultivation is an ancient art, domestic varieties are almost always used for planting. Today, several thousand olive varieties are reported in all olive cultivation countries worldwide – for example, in Italy alone more than 300 varieties have been enumerated, although only a few are grown to a large extent. In Spain, too, several hundred olive varieties can be found. Similarly, in Greece, Portugal, France and other Mediterranean areas, as well as Middle Eastern countries, several varieties have been recorded. Some of the most common olive cultivars are listed in Table 7.1, along with their respective countries of origin.

The olive varieties available today can be grouped into European (Italian, Spanish, Portuguese, Greek, French), Middle Eastern (Turkish, Syrian, Jordanian, Israeli and Palestinian, Lebanese), North African (Tunisian, Egyptian, Moroccan), Central Asian, and other (Argentinean, Australian, California) varieties. Some of the more widely grown olive varieties are listed below, while many of the olives produced are illustrated in Figure 7.1.

7.2.1 Italian

Leccino

Leccino is a versatile variety, which can adapt to different soils and is extremely resistant to cold and high humidity. Like most of the other Italian varieties, Leccino is also harvested early, and therefore the harvest is spread over a longer period. Leccino is a vigorous tree with distinctively good oil quality, but a low oil content. It is believed that oil content rises as trees perdure, but this has not yet been proven. It can root easily, and is favored because it begins to produce relatively soon after planting and has high and constant productivity. The fruiting potential is high and the response to irrigation is good, causing a marked increase in yield, while the reduction of oil content in the fruits at the optimal stage of maturation is relatively small. The tree is medium

Table 7.1: Common olive varieties of the world

Origin	Cultivar	Origin	Cultivar	Origin	Cultivar
Albania	K.M.Berat	Jordan	Rasei	Spain	Castellana
Albania	Kalinjot	Iran	Mari	Spain	Changlot Real
Albania	Mixani	Iran	Fishomi	Spain	Cornicabra
Albania	U. Bardhei Tirnes	Iran	Zard	Spain	Empeltre
Algeria	Chemlal de Kabylie	Iran	Rowghani	Spain	Farga
Algeria	Sigoise	Iran	Gorgan	Spain	Gordal Sevillana
Egypt	Hamed	Iran	Shengeh	Spain	Hojiblanca
Egypt	Toffahi	Iran	Dezful	Spain	Leccino del Granada
Serbia	Buga	Iran	Shiraz	Spain	Lechin de Sevilla
Serbia	Crnica	Iran	Rashid	Spain	Manzanila de Sevilla
Serbia	Istarska Belica	Iran	Gelooleh	Spain	Manzanilla Cacerefia
Serbia	Lastovska	Iran	Khormazeitoon	Spain	Morisca
Serbia	Oblica	Iran	Dakal	Spain	Morrut
France	Bouteillan	Iran	Khara	Spain	Picual
France	Lucques	Iran	Janoob	Spain	Picudo
France	Picholine	Italy	Ascolana Tenera	Spain	Servillenca
France	Salonenque	Italy	Carolea	Spain	Vera
France	Tanche	Italy	Cellina	Spain	Verdial de Badajoz
France	Verdale	Italy	Coratina	Spain	Verdial de Huevar
Greece	Adramitini	Italy	Drrita	Spain	Verdial de Velez Malaga
Greece	Aggouromanakolia	Italy	Frantoio	Syria	Abu Satel Echlot
Greece	Alorena	Italy	Grappolo	Syria	Daebli
Greece	Amigdaloia	Italy	Itrana	Syria	Dam
Greece	Amigdaloia Nana	Italy	Leccino	Syria	Kaissy
Greece	Chalkidiki	Italy	Leccino del Corno	Syria	Villalonga
Greece	Defnelia	Italy	Moraiolo	Syria	Sorani
Greece	Kalamanta	Italy	Oblonga	Syria	Zaity
Greece	Karolia	Italy	Rosciola	Tunisia	Chemlali
Greece	Koutsourelia	Italy	San Francisco	Tunisia	Chetoui
Greece	Konservolia	Italy	Santa Caterina	Tunisia	Gerboui
Greece	Koroneiki	Italy	Pendolino	Tunisia	Meski
Greece	Lianolia Kerkyras	Italy	Taggiasla	Tunisia	Ouslati
Greece	Mastoidis	Italy	Maurino	Tunisia	Zalmatio
Greece	Mavrelia	Italy	Biancolila	Turkey	Ayvalik
Greece	Megaritiki	Italy	Carboncella	Turkey	Beyaz Yaglik
Greece	Myrtolia	Lebanon	Ayrouni	Turkey	Cakir
Greece	Picrolia	Lebanon	Souri	Turkey	Domat
Greece	Rahati	Morocco	Haouzia	Turkey	Elmacik
Greece	Thiaki	Morocco	Menara	Turkey	Erkence
Greece	Throumpolia	Morocco	Picholine Morocaine	Turkey	Gemlik

Table 7.1: (*Continued*)

Origin	Cultivar	Origin	Cultivar	Origin	Cultivar
Greece	Tragolia	West Bank	Nabali Muhsan	Turkey	Memecilk
Greece	Valanolia	Portugal	Carrasquenha	Turkey	Uslu
Greece	Vasilikada	Portugal	Cobrancosa	Turkey	Izmir Sofralik
Israel	Barnea	Portugal	Cordovial de Serpa	Turkey	Kan Celebi
Israel	Merhavia	Portugal	Galega	Turkey	Kiraz
Israel	Maalot	Portugal	Negrinha	Turkey	Halhali
Jordan	Arabi Altafila	Portugal	Redondil	California	Mission
Jordan	Bathni	Spain	Alfafara	California	Sevillano
Jordan	Kanabisi	Spain	Arbeqina	Chile	Azapa
Jordan	Ketat	Spain	Blanqueta	Argentina	Arauco
Jordan	Kafri Romi	Spain	Canivano Blanca		
Jordan	Nabali Baladi				
Jordan	Nabali Muhasan				
Jordan	Nasouhi Jaba 2				

in size and vigor, is strongly affected by winds, and tends to grow somewhat at an angle. The leaves are medium to large, of an oval-lanciformic and symmetrical shape, with a dull green upper surface and a gray-green lower surface. The drupes mature early and are of a medium size, with an elliptical, rather than symmetrical shape. They require low detachment force, making them very suitable for mechanical harvesting. This variety is known to be tolerant to peacock spot and to olive knot disease.

Frantoio

Frantoio is one of the main Italian varieties, and is favored for its high and constant productivity and its capacity to adapt to different conditions – although it is susceptible to winter cold. It can root easily. The tree is medium in vigor and is generally upright, with relatively long, thin and flexible fruiting branches. The leaves are oval lanciformic, medium in size, with a dark green upper surface and greenish-gray lower surface. The fruit are medium in size and have a somewhat elongated shape, with the widest diameter close to the apex. Color change starts from the distal end and rises to the drupe; at full maturation the fruit has a uniform deep black color. Its oil content is medium to high, and has excellent organoleptic characteristics; it maintains a very stable level of oil, although it does contain high levels of peroxide. The oil is very fruity, with medium bitterness, and strongly pungent. Frantoio is sensitive to olive knot and the olive fly, and tolerates *Verticillium* wilt.

Moraiolo

Moraiolo has a high and constant productivity. The trees can tolerate droughts and marine winds, and put down roots easily, but are susceptible to cold. The fruits mature early, and are of medium

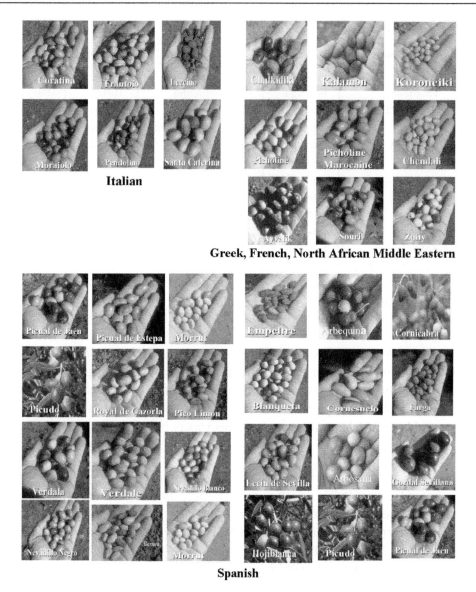

Figure 7.1: Common olive varieties of the world.
Adapted from Vossen (2004).

size. Moraiolo olives are favored for their organoleptic characteristics – the oil is strongly fruity, has a green apple taste, and is slightly bitter and pungent, although this is balanced by a nice sweet sensation in the mouth. The stability of Moraiolo olive oil is considered to be low, and mechanical harvesting of this variety is very difficult. It is sensitive to peacock spot and olive knot.

Coratina

Coratina begins to produce early, and has a relatively constant and high productivity. The tree has vigorous growth with a dense canopy. The leaves are of medium size with an elliptical shape. The drupes are of moderate size, ovoid in shape and slightly asymmetric. Coratina has a stable and high oil content and excellent organoleptic characteristics. Its oil is bitter and pungent. Although Coratina olives mature late, mechanical harvesting is still difficult because they have a high force of detachment. This variety is considered to tolerate cold, but is susceptible to sooty mold and wood root.

Carolea

Carolea is an Italian variety that produces fruits suitable for both oil and pickling. It is grown in Calabria, and in the Enna region of Sicily. It roots vigorously, and is highly tolerant to low temperatures. The drupes are large in size. This variety is favored because of its productivity, high oil content, high-quality oil, and ease of harvesting mechanically. It is susceptible to peacock spot and the olive fly.

Ascolana

Ascolana is the main pickling variety in Italy, owing to its enormous size (8.8 g) and pulp/stone ratio. Its vigorous trees have a very dense crown tending upwards, with slightly drooping fruiting branches. The leaves are regular, elliptical, of medium size, and have an intense green color. The fruit is ellipsoidal in shape and slightly asymmetrical, with a top that is rounded or slightly sub-conical; it is light green in color at harvest. Its flesh, which is milky white in color and represents about 86 percent of the fruit, is very delicate and easily bruised during harvesting. It has a medium oil content. Ascolana is cold tolerant, but demands high-quality soil. Flowering begins late, while fruits mature very early. Ascolana is tolerant to olive knot and peacock spot, but susceptible to the olive fly.

Dritta

Dritta is a tall, vigorous plant that produces large quantities of olives, but a moderate yield of oil. The fruit is of a medium size. The plant is both cold- and parasite-resistant. It is sterile, and is pollinated by Leccino.

Pendolino

Pendolino, which is cultivated in Tuscany and Umbria, is a tree of medium height with good productivity and a moderate oil yield. It is a slow-growing cultivar of limited development with a very obvious weeping habit. The crown is dense and abundant with long, thin, lanceolate leaves of medium size, which are dark greenish-gray in color. The fruits are very small (1.5 g), ellipsoidal, and asymmetrical with rounded tips, and generally mature simultaneously, although the time of maturation is intermediate in relation to other Tuscan cultivars. The oil has a pleasant, delicate flavor. Pendolino has high and constant pollen production, which makes it an ideal pollinator for most Tuscan oil cultivars. It is self-sterile, and prefers Maurino and Leccino as pollinators. It is mildly resistant to cold and, because of its long, flexible branches, is well-suited to manual harvesting.

Taggiasca

Taggiasca is prized for its high production and oil yield. It is widespread in the Ligurian region, especially in the provinces of Savona and Imperia. The tree is of medium vigor, with a pendulant open shape. The crown is very branched and spread out, with numerous small, flexible fruiting branches at well-spaced nodes. The leaves are medium to large, elliptical and elongated in shape, and a shiny dark green on the upper surface. The fruit is medium-sized, elongated and cylindrical in shape, and slightly fatter at the bottom. It matures in the middle to late season (November–December). Taggiasca has a relatively high oil content, which has excellent organoleptic characteristics. The oil is characterized as mildly fruity with low bitterness and pungency, and is considered highly stable. It is self-pollinating, with consistent high production; flowers with aborted ovaries are minimal. It is resistant to all pests and diseases.

Maurino

Maurino is a typically Tuscan oil variety, widely appreciated for its ability to produce significant amounts of fertile pollen and for its compatibility with a wide range of other olive cultivars. It can be cultivated in cold, humid zones that are subject to fog. Its tree is of medium vigor with a pendulous habit, while the fruiting branches are delicate with rising tips. The leaves are lanceolate, of medium size, and grayish-green in color. The fruit is medium, ellipsoidal and slightly asymmetrical in shape and, when fully mature, purple-black in color. Fruit maturation is considered to be relatively early. Fruiting is good, although slightly alternate. Oil content is moderate; the oil itself is delicate and not overly fruity. The cultivar is self-sterile, and the pollinator can be Pendolino, Leccino, Frantoio or Moraiolo. It has a low ovarian abortion rate, generally under 19 percent. It is resistant to peacock spot and olive knot.

Biancolilla

Biancolilla is an oil cultivar, planted prevalently in Sicily. Its production is high, although it tends towards alternate bearing. It tolerates cold and late frosts. The tree is very vigorous, with a dense crown and expansive habit. It has broad branches that produce a large yield of fruit. The leaves are medium-small, lanceolate and dark green. It has symmetrical oval-conical fruit of medium weight (3 g). At harvest the olives are distinctive red-violet in color, and yield a fair amount (15–18 percent) of very aromatic oil of excellent quality. It is partially self-fertilizing, and a good pollinator for other oil varieties. It is resistant to peacock spot.

Carboncella

Carboncella is a typical oil cultivar of central Italy, and is a rustic variety that adapts well to dry limestone soils with good exposure. It tolerates cold and rot. Its constant high production produces fair-quality oil in large quantities. The tree has a good growth rate and a markedly upward habit. It is characterized by long, flexible branches that bend under the weight of the drupes. The leaves are small and dark green in color. The fruit is small in size (2 g), slightly ellipsoidal, symmetrical, and dark violet in color. It matures early (October–November), and all at once. It is resistant to the olive fly and olive knot.

Itrana

Itrana is a rustic variety, good for both table olives and oil, and easily adaptable to many of Italy's olive-producing regions. The tree is highly vigorous, and of notable size and upward growth. The crown is compact and dense, and is covered with fairly large leaves that are elliptical and lanceolate in shape. The fruit is round, asymmetrical and of medium size (3.5 g). At harvest, the drupes are a dark wine color, freckled with a whitish dusting. The oil is of good quality and yield (average 20 percent), pleasing in taste, and highly prized. Maturation is late (November–December). Itrana is self-sterile, very productive, and has a good level of resistance to cold and to most common olive pests.

Santa Caterina

Santa Caterina is an excellent table variety, owing to the large size of its fruits (8 g), and has a regular high production rate. It is very resistant to cold. The tree is of notable size, and very vigorous with a globular, expansive crown. The leaves are elliptical-lanceolate, regular in shape, rather short, and light green in color. The fruit is ellipsoidal and asymmetrical, with a rounded tip and sub-conical base. Ripening occurs early, so harvest generally takes place in September when the olives are still a beautiful intense green color. It is tolerant to peacock spot.

7.2.2 Spanish and Portuguese

Picual

Picual is the main olive variety of Southern Spain, and is favored for beginning production early, and for its high productivity and ease of cultivation. It is a classic dual-purpose variety and, because of its strong tendency to alternate bearing, much of its yield is harvested early for green pickling (early harvesting decreases the negative effect on the next yield). It is generally strong, and adjusts to a wide range of soils and climatic conditions. The Spanish regions in which Picual is most common have cool to cold winters and hot, dry summers. Flowering is in mid-season, and the cultivar is usually self-fertile. Maturation is early, and the fruit should be processed without delay. The tree is vigorous, with a tendency towards an open canopy. Annual growth is medium, and usually lateral. Leaves are uniform, with a shiny dark green upper surface and greenish-gray lower surface; they are somewhat curled inwards on the sides. Drupes are elongated and nearly symmetrical, with a slight bend at the apex. Blackening starts from the apex, and the ripe fruit is uniformly dark black. The fruit is of medium size (4 g) and produces oil that is considered medium in quality, although it stands out for its stability and high polyphenol and oleic acid content. Picual produces nicely aromatic fruity oil that is of medium bitterness and is highly stable when harvested early. The oil content is low – about 14 percent. The yield is relatively high, and the tree responds well to irrigation. Under intensive conditions, an average yield of about 10–12 ton/ha can be achieved. Ripening takes place simultaneously, and it is very suitable for mechanical harvesting. Picual is the variety most susceptible to *Verticillium* wilt, and suffers from Leopard moth. It is particularly resistant to cold and wet soils, slightly tolerant to soils containing lime or calcium, and has relatively high salinity tolerance – although it has not shown clear resistance to salt in our groves. The fruits are susceptible to olive fly, and slightly sensitive to leaf peacock spot.

Arbequina

Arbequina olives are small, brown olives grown in Catalonia. Arbequina is recognized for its aromatic ripe fruitiness, low bitterness and pungency. It is resistant to cold and susceptible to clorosis ferrica (shortage of iron) in highly calcareous terrains. It is favored for beginning production early, its high productivity and excellent oil quality, although it has low stability. The tree is relatively small, with low vigor and somewhat dropping branches of short to medium length. The leaves are small and symmetrical, elliptic-lanciformic in shape, with a dark green upper surface and grayish lower surface. The fruits are very small (1.6 g), generally round, with a slight ovalic and asymmetric tendency. At maturation the fruit is uniformly dark black, but the color change is gradual. Mechanical harvesting is difficult and not very efficient (85 percent fruit removal) because of the fruits' small size. The cultivar is sensitive to peacock spot infection, Black scale and the olive fly.

Hojiblanca

Hojiblanca is another dual-purpose variety, suitable both for oil and pickling. This variety is appreciated for its tolerance to calcareous terrains, cold and droughts, although it is not resistant to salinity. Alternate bearing is of a medium level, and can be reduced by horticultural means. The tree is large and wide with vigorous growth that develops to a wide, open canopy. The branches are of medium length, and their bark has a clear gray color. The leaves are long, lanciformic and rather narrow, with a matt light green upper surface and greenish-gray lower surface. The fruit is medium-large, of uniform elliptic shape, and slightly asymmetrical. When ripe it is a deep violet color; darkening starts at the distal end. The flesh of the fruit is firm and somewhat rough. Even though it has a low oil content, it is very desirable because of the quality of its oil. The oil is fruity, aromatic and mildly pungent, and has low bitterness and stability. It is very resistant to detachment, and thus difficult to harvest mechanically. It is susceptible to *Verticillium* wilt, peacock spot and olive knot.

Empeltre

Empeltre is highly productive, and the fruits produce excellent quality oil whose main trait is its captivating golden-yellow color. The oil tastes mildly fruity, and is pungent with some bitterness. It is considered robust because of its adaptability to various microclimates, as well as freezes and drought. The tree is vigorous, with an upright shape and dense crown that produces an abundance of new wood each year. The leaves are elliptical/lanceolate in shape, medium–small, short and narrow in size, and a matt dark green in color. The tree has low rooting ability, so is usually propagated by graft. The fruit is medium-sized (3 g), black at maturation and slightly freckled, with an elongated oval, slightly asymmetrical shape and a rounded top. The oil content is high. It is ideal for mechanical harvesting, because of its early maturation, low force of detachment and strong branches. Empeltre tolerates *Verticillium* wilt, but suffers damage during cold winters.

Manzanillo

Manzanillo is known for its early productivity and dual purpose use. It has a low oil content, but the oil is of high quality, being fruity, aromatic and herbaceous, mildly bitter, stable, and strongly pungent. It is highly resistant to cold and to weather changes. The tree is vigorous, with a rising habit; the crown has long, pendulant branches, carrying medium-small leaves that are elliptical in shape and bright green in color. The fruit is large in size (4 g), with a slightly asymmetrical shape and a rounded top and bottom. The pulp comprises 85–88 percent of the fruit. Both green and black olives are pickled, because of the large amount of pulp. Since Manzanillo olives mature early and have a low force of detachment, this variety is highly suitable for mechanical harvesting.

Picudo

Picudo is a vigorous variety that is highly adaptive to calcareous terrains. It is favored for its high oil content and excellent organoleptic characteristics. The oil is very aromatic, with a ripe fruitiness, and is mildly pungent and bitter. Its pollen is utilized to pollinate other varieties, owing to its powerful ability to germinate. It matures late and has a high force of detachment, making mechanical harvesting very difficult. Picudo is sensitive to peacock spot and olive knot.

Farga

Farga is highly resistant to winter cold. Its oil content is considered to be relatively high and of good quality, but difficult to extract. It has low rooting ability, although it is sometimes used as rootstock for low-vigor varieties. Bearing is delayed; however, flowering and ripening occur early. Production is high, but alternate. The tree is very vigorous, with a spreading shape and dense canopy. The leaves are elliptical in shape, short in length and medium in width. The drupes are of medium size, and elongated and slightly asymmetric in shape. They mature early, but their high force of detachment makes mechanical harvesting very difficult. Farga is known to be susceptible to peacock spot and resistant to olive knot.

Morrut

Morrut is late in beginning production, and has a low yield. The fruit matures late, and therefore mechanical harvesting is highly efficient. It has moderate oil content, and the oil is unstable. It is able to root easily, and presents early flowering.

Cornicabra

Cornicabra is highly adaptive to harsh conditions such as poor soil, arid climates and cold regions. It has a high oil content, and produces good-quality oil with excellent organoleptic characteristics and stability. Its oil is fruity and aromatic, with medium bitterness and pungency. Its fruity and sweet flavor, reminiscent of apples and almonds, makes a perfect salad dressing, and it is often used in the preparation of sauces. The drupes mature late and have a high force of detachment, so mechanical harvesting is very difficult. Cornicabra is extremely sensitive to peacock spot and olive knot.

Morisca

Morisca is very resistant to drought, and is suitable for production of both oil and pickles, due to the high oil content and size of the fruit. It is easy to propagate vegetatively, but difficult to harvest mechanically because of its relatively high force of detachment. Morisca is susceptible to peacock spot and olive knot.

Verdial

Verdial has low capacity for rooting and lacks productivity; nevertheless, it is found in many areas because it can adapt to humid terrains as well as arid regions. Verdial matures late, and the olives do not blacken. The fruits have a high content of good quality oil. The oil is mildly fruity, bitter and pungent. It is difficult to harvest mechanically because it has a high force of detachment. Verdial is considered to be sensitive to peacock spot and resistant to olive knot.

Gordal

Gordal is used mainly for pickles, since its oil content is very low and the average weight of a fruit is 12.5 g. The fruit is generally oval in shape, with an indent at the stem which can give it an almost heart-shaped appearance. The skin is thin, and speckled with white markings. The flesh is light green, turning purplish-black when ripe, and of a good texture for pickling. It is considered difficult to process because it is easily bruised and has a propensity to split pits. It has a low capacity for rooting, and is thus commercially propagated through grafts. It can tolerate winter cold, is susceptible to olive knot, and is resistant to peacock spot.

Galega

Galega represents around 80 percent of Portugal's olives. This olive variety is used exclusively for the production of olive oil. It is considered to be highly productive but has extremely alternate bearing. Although tolerant to droughts, it is sensitive to cold, salinity and limestone. Its fruit is medium in size (around 2.5 g) and the oil content is low. The oil, from fruits in the middle stage of ripeness, is fruity, with green leaf and grass sensations, and is slightly bitter and pungent with a sweet taste. Its high force of detachment makes it difficult to harvest mechanically. It can resist *Verticillium* wilt, but is susceptible to olive knot and the olive fly.

Arbosana

Arbosana is a high-producing variety grown for oil. It matures late, and has a tendency towards alternate bearing. It starts producing olives after two years, and achieves full fruiting in five years. Its tree is of lower vigor and small in stature, with high yields. It can be planted very densely in orchards. The fruit is small, and high in oil content. The oil has a pungent, fruity taste, and is often blended with oils from other olives to improve their taste and aroma. It is resistant to leaf drop and cold.

7.2.3 Greek

Koroneiki

Koroneiki occupies around 60 percent of the cultivated olive groves in Greece, and is the chief oil variety of the country. It is generally cultivated on the plains, lower hillsides and coastal areas of Crete, where the climate is relatively warm. It can resist droughts, but is susceptible to cold. Productivity is constant and relatively high; it has a very high yield in adverse conditions, whilst with irrigation the yield is lower. The fruits ripen early, avoiding the winter cold. The tree is of medium vigor, with a sparse and spreading canopy. The leaves are short, narrow, and elliptic-lanceolate in shape. The drupes are ovoid in shape and very small (1.1 g); nonetheless, their oil content is high. Moreover, the oil is highly prized because of its excellent organoleptic characteristics, stability and high oleic acid content. It is strongly fruity and herbaceous, mildly bitter and pungent. This cultivar is suitable for producing an early, excellent-quality, premium-price oil. Owing to the small fruit size, mechanical harvesting can be difficult. Koroneiki is peacock spot resistant, but susceptible to olive knot.

Kalamata

Kalamata is a large, black olive, of a smooth meaty taste, named after the city of Kalamata in Greece. It is used mainly for pickling, due to its resistance to treatments and manipulation, its large size (4 g) and high pulp/stone ratio. This variety is intolerant of cold. It has high productivity and alternate bearing, and is characterized by a tree of average height with broad, deep green leaves. The fruit is elongated, with a narrow stone that comes away from the flesh easily. It turns deep black when fully ripe. The oil content is medium, but the oil is of high quality. This cultivar is famous around the world as an excellent table olive. Kalamata is known to be sensitive to peacock spot and *Verticillium* wilt, and tolerant to olive knot and the olive fly.

Konservolia

Konservolia represents more than 75 percent of Greek pickled olive production, owing to the fruit's size (5.5 g), pulp/stone ratio and high quality. It is harvested when green to produce green table olive, or, more rarely, when black for black olives. This vigorous variety is highly productive when cultivated on good soil and properly irrigated; it has relatively good tolerance to the cold, although it undergoes alternate bearing. It ripens relatively early. The tree grows to a great height, with long, average-sized leaves that have a distinctive tip at the end, with a downward bend. The fruit is oval, with a dark, chewy flesh that comes away from the stone easily. It puts down roots easily. This variety is sensitive to *Verticillium* wilt and tolerant to olive knot.

Lianolia Kerkyras

Lianolia Kerkyras is an important oil-producing variety that yields excellent-quality oil. It flourishes even on barren, stony ground, although it has greater demand for moisture (being cultivated primarily in areas with high precipitation). It is a late-ripening variety, and its fruits are harvested late, after the first spring months. This late ripening encourages a high yield. The trees are very large, with vigorous vegetation. Large leaves that fold over upwards are characteristic of this variety, whilst the fruit is small (1.5 g) and elongated, with a slight tip at the top.

Koutsourelia

Koutsourelia has average productivity, and prefers fertile or average soil. It does not grow well at great heights. The fruit is rich, producing a good-quality oil, and ripens relatively early (from the end of October). The tree is of average size, with short shoots. The leaves are small. The fruit is round, and ends in a small, lightly curved tip.

Mastoidis

Mastoidis fruits are spherical in shape (mastoid means breast-shaped), with an average to low yield. Mastoidis demands good soil (deep, lime content) and can be cultivated at great heights. This variety blossoms and ripens late. The leaves are of average size, with a prominent central fiber on the upper surface ending in a pronounced tip. Fruit of 2.5 g is small to medium in size, shaped like a lemon, with a large tip.

Megareitiki

Megareitiki demands little moisture, and can therefore be cultivated in dry areas. Its productivity is considered to be average, and its yield is not high unless it is well tended. The fruit ripens relatively early (November–December), and is typically conical in shape with a narrower base and a peak at the tip, although this can vary. It is large (4.2 g), and is used for both oil and pickling. The leaves are large, and end in a pronounced tip.

Kolovi

Kolovi is considered to be one of the best oil-producing varieties, in terms of both productivity and quality. It has average demands regarding soil and care, and in favorable conditions it can achieve high yields. This variety ripens late (full ripening in February–March), although harvesting starts early (November). The leaves are large, tough and relatively broad. The fruit is characterized by the absence of a tip, and has an oval or spherical shape. It is, however, usually narrower at the base and broader at the top (similar to an acorn). The fruit is of medium size (3.5 g), with a high oil content.

Kothreiki

Kothreiki is a durable variety that is tolerant to dry conditions, cold and strong winds. Its yield is considered to be average, with average demands regarding soil and care. The fruits are large (4.5 g) and spherical in shape, without a nipple, and with a relatively high oil content (around 25 percent). It is used for both oil and pickling. A significant portion of the annual yield is processed as table olives, producing large or average-sized salted black olives that are of excellent quality, delicious, and have a good aroma.

7.2.4 French

Picholine

Picholine originated on the Mediterranean coast of France, and is a classic dual-purpose variety. The oil is light in color, and has a fine, delicate taste. It is fruity and aromatic, mildly bitter and pungent. This variety can be adapted to various climatic conditions. Its production under irrigation is high. It begins production relatively early, and in intensive orchards a commercial yield (60 percent of the mean potential) can be achieved in the fourth year. With good annual pruning, fruiting is quite regular unless limited by extreme environmental conditions. The trees are of medium vigor, rather wide, and require larger planting distances than many other varieties, both within the rows as well as between them, requiring annual pruning. The leaf's inflorescence is rather long, with 19–25 largish flowers. The fruit is medium in size (4 g), and uniform in both size and color. It is elliptical in shape, elongated, and slightly asymmetrical. For optimal oil extraction, reddish-black fruits are used; these will also render the best oil aroma. Maturation is relatively late, and therefore the harvesting period is relatively long. The oil content is medium to low, but the oil is of very high quality. Picholine is susceptible to peacock spot, olive fly and excessive chilling.

Aglandau

This is a dual-purpose variety, widely planted in the Alpes de Haute Provence and the Valcluse. It has medium productivity and tends towards alternate bearing, which can be controlled through pruning. It has vigorous trees with a spreading habit and dense canopy. The leaves are lanceolate, medium in length and narrow in width. The fruit is medium in weight, ovoid, and slightly asymmetric. It has good rooting ability. Its oil is considered to be stable, with a highly fruity, bitter and pungent taste, and is of the finest quality; it is therefore often blended with other varieties to impart fruitiness and stability. The oil content is medium. Aglandau is resistant to *Verticillium* wilt, cold and drought.

Tanche

Tanche is considered to be the best French black-olive variety, suitable both for oil and for pickling. It is a good producer if actively farmed with good cultural practices. It prefers light, deep soils. The tree is very vigorous, with an upward growth habit and an expansive global crown. It has broad leaves, which are light green in color and rather large. The fruit is large

(5.5 g) with 80 percent pulp, and ellipsoid in shape with a rounded tip. Maturation is late (November–December), when drupes are of a violet-black color. It is strongly resistant to cold and olive knot.

Bouteillan

Bouteillan is widely planted in the Var and Languedoc regions, and is used exclusively for oil because of its high oil yield. It is a hardy variety that requires light but frequent pruning and irrigation. It is highly resistant to cold. The tree grows quickly, and its productivity is high and constant. The ripening time is intermediate. The tree is vigorous, with erect, dense growth. The leaves are elliptic-lanceolate in shape, of medium length and width. The fruit is clingstone, and generally large, ovoid, and slightly asymmetric in shape. Its oil content is high, and the oil has a light golden color. Its flavor is of citrus and fresh flowers, and it possesses faint aromas of almonds and green herbs. The oil is often used for blending with other olive oils. It is susceptible to olive fly, olive scale and olive moth.

Salonenque

Salonenque is the main variety grown in Bouches du Rhône, occupying over 60 percent of the cultivated surface. It is used for both oil and table pickles, and is a rustic variety fairly resistant to cold and drought. It can be grown in poorer soil, although it adapts well to good farming practices. It has a low rooting ability and an intermediate start to bearing. Salonenque flowers early and ripens late, and production is constant and high. The tree is of average or smaller size, with a low, spreading, globular shape. The leaves are elliptic-lanceolate, short and narrow. The fruit is of medium size, ovoid and symmetrical in shape, with a high pulp to pit ratio. The oil is light golden in color, with a flavor of green artichokes and a slight aroma of almonds. The oil content of fruits is high. It tolerates the olive fly and peacock spot.

Grossanne

Grossanne is primarily used for making sweet-tasting black olives. It is resistant to cold and drought. Bearing is intermediate, but can be accelerated by irrigation and fertilization. Flowering and ripening are intermediate. Production is medium and constant. The tree is highly vigorous, with an erect habit and medium canopy density. The leaves are lanceolate, and of medium length and width. Its rooting ability is poor, so it is often grafted. The fruit is spherical in shape and of medium weight. It has a low oil content, and the oil is quite fragrant though not durable. It tolerates *Verticillium* wilt and is susceptible to olive scale, olive moth and olive fly.

Cailletier

Cailletier is the main variety of the Alps Maritime region near Nice. It is a dual-purpose variety; its olives are cured black in the area around Nice, and are hence called Niçoise, and are also used to make delicate-tasting oil. It is highly tolerant to cold, and can easily be propagated from cuttings. Bearing is generally intermediate and constant. The tree is very vigorous, and grows into a large tree with downward drooping branches. The fruit is small in size and ovoid in shape, with a high oil content. Cailletier is susceptible to the olive fly, olive scale and olive knot.

7.2.5 Turkish and Syrian

Memecik

Memecik is the most important variety in Turkey, occupying around 45 percent of cultivated olive groves in the country. It is a durable variety that can resist cold and drought, and roots easily. The drupes are used for both oil and pickling. They are used as table olives because of their size (5.2 g) and high pulp/stone ratio, and also for oil due to their high oil content and good organoleptic characteristics. The olive oil is of an average olive fruitiness, with a hint of apple and other ripe fruits; it is not very bitter although a little pungent. It can become rancid, so its stability is considered medium–low. It is, to some extent, susceptible to the olive fly.

Ayvalik

Ayvalik trees are vigorous and can adapt extremely well to drought. The fruits are used mainly for oil, due to their high content of good-quality oil. The drupes are large (4.5 g), mature earlier than Memecik, and are easier to harvest mechanically. Ayvalik is resistant to the olive fly.

Sourani

Sourani is a durable variety that can resist cold, drought and salinity. It is an autogamy, which flowers late and also begins to produce late. The drupes are medium in size; they mature in mid-season and have a low force of detachment. They are used for both oil and pickling. Their oil content is high and of excellent quality. Sourani is resistant to peacock spot and olive knot, and susceptible to *Verticillium* wilt.

Zaity

Zaity is tolerant, to an extent, to cold and salinity, but susceptible to drought. The trees are autogamous, and flower late; they are characterized by a high percentage of parthenocarpy fruits which are of a much reduced size, low commercial cost and difficult to harvest. The fruits mature early and have a low force of detachment. Zaity is favored for its high oil content and good-quality oil. It is resistant to peacock spot and olive knot.

7.2.6 Lebanese, Israeli, Palestinian and Jordanian

Souri

Souri has been cultivated for many years in Lebanon, Northern Israel and the West Bank. It can be cultivated in arid areas and in shallow soil. The trees are known for their high alternate bearing characteristics when cultivated in a traditional non-irrigated way. Intensive cultivation has led to a reduction in the rate of alternate bearing, but an "off" year still occurs every second or third year. In the Middle East the trees naturally usually develop a small canopy; however, under intensive cultivation the trees are able to grow fast and develop relatively large canopies.

Traditionally, Souri olives produce strong aromatic oil. Their oil content is relatively high (above 20 percent). Cultivation of this variety in the Negev Desert has not changed the basic characteristics of the oil. Souri trees are highly susceptible to leaf spot disease, olive fly and *Verticillium* wilt.

Ayrouni

Ayrouni is a Lebanese variety that has been a true blessing for the olive growers, since it has never needed pesticides because it has never been attacked by any parasites or bugs. The fruits are used for both oil and pickling. Ayrouni is cultivated all over the country. Its fruit is small to medium in size (2.5 g), hardy and succulent, with a relatively high oil content. Fruits are harvested late; olives for pickles are harvested around the end of November, and for oil towards the end of December.

Nabali Muchasan

Nabali Muchasan is the selected variety in the West Bank, and the most common among Arab olive-growers in the Middle East. It responds to improved cultivation conditions and techniques, although trees grown in the Negev area with a full irrigation regimen reached a relatively small size. Due to extensive cultivation practices, Muchasan trees usually develop small canopies. The trees have a good rooting ability, and the productivity and oil content of Muchasan olives is relatively moderate. The quality of the olive oil is well accepted, mainly by the Arab consumers that are used to this oil. The trees are susceptible to leaf spot disease and the olive fly.

Nabali Baladi

Nabali Baladi responds to improved cultivation conditions and techniques. Especially impressive is the Muhsan sub-variety, which originated on the West Bank of the Jordan River; cultivation in Israel began after the Six Day War. The Muhsan olive is an enhancement of the Nabali variety. It is strong, takes root easily, and has larger fruits that are suitable for eating. Nabali Baladi and Muhsan are cultivated extensively using advanced irrigation technology. This variety's share of the olive market is constantly increasing.

Barnea

Barnea was named for the Kadesh Barnea region in which it was found, on the border between the Sinai Desert and Israel. The cultivar was developed for oil extraction, but is also suitable for producing high-quality black pickles. The tree (and particularly the fruit) is sensitive to stress caused by a lack of water in summer. Its fruiting potential is very high. Barnea trees are well known and easily recognized by their tall apical dominant tree characteristics. They are vigorous, erect-growing trees with thin fruiting branches and a loose, open crown. The leaves are medium to large, widening at mid-length, with a light green upper surface. Their vigorous growth can be controlled by increasing the salinity level in the irrigation water, up to 4.2 dS/m, without suffering any yield reduction. Above this level the sustainability of Barnea trees is affected due to high level of salts accumulating in their vegetative tissues. Barnea trees are usually intensively cultivated. In Negev Desert conditions, trees have a very high yield (~15 Mt/y) with a relatively low rate of alternate bearing. The fruit is medium-sized, elongated, bent, wide at the base and narrow at the apex, often ending in a nipple. The green fruit has a rough surface. At green maturation, the fruit is pale yellow-green; it turns smooth with a deep black color at black maturation. The mesocarp is bright and has a smooth texture. Barnea is propagated from cuttings that root easily, and is only grown on its own roots. Flowering is relatively early, as is green maturation. Full black maturation is in mid-season. The oil content in mature Barnea olives is about 18 percent; however, due to high productivity Negev cultivation practices, the Barnea oil yield per land is higher (~2.5 tons/ha). The quality of Barnea olive oil is accepted as being somewhat lower than the strong Souri oil and other more delicate European olive oils. Barnea trees are highly susceptible to Leopard moth larvae. The trees are well suited to mechanical harvesting using trunk shakers.

Maalot

Maalot is a modern, leaf spot disease-resistant Eastern Mediterranean cultivar derived from the North African Chemlali cultivar. It is grown almost exclusively for oil production, and is a vigorously growing tree. Owing to their alternate bearing characteristics, even under intensive Negev cultivation, Maalot trees were found to produce a multi-annual moderate yield of olives. The fruit is medium-sized, round, and its oil is of good quality, with a fruity flavor. Oil content in the olives is relatively high (~25 percent). Maalot's characteristics have not been negatively affected by saline drip irrigation practices in the Negev Desert area.

Rasei

Rasei is an improved cultivar of Nabali, used mainly for oil, even though its fruit is relatively large. Its oil is noted for its intriguing mixture of sweet and pungent flavors, and the hint of an apple taste. This variety is also quite popular in the West Bank areas and Northern Israel.

7.2.7 North African

Chemlali de Sfax

Chemlali de Sfax is a vigorous variety that occupies around 60 percent of cultivated olive groves in Tunisia. It has spread throughout the central, eastern and southern areas of the country, where precipitation is low, and is resistant to drought and salinity. It is an autogamy, with a very small fruit (1.3 g), and matures late. The oil content of the drupes is high, though low in oleic acid. The oil has a strongly aromatic fruitiness with a notable varietal character. Chemlali de Sfax is susceptible to olive knot.

Chetoui

Chetoui trees are durable, and can resist cold and salinity. They put down roots easily. Flowering is early; however fruits mature late. The drupes are small to medium at 2.3 g, as is their oil content;

in addition, the oil contains a large amount of polyphenols that contribute to its stability against becoming rancid. The oil is of an intense fruitiness, with a green leaf aroma; it is bitter, pungent and astringent, with hints of fresh bark and green almond. Chetoui black olives are occasionally used for pickling. It is susceptible to peacock spot.

Meski

Meski is the main pickling variety in Tunisia. It is cultivated especially in the north, and has a low capacity of rooting. It is favored for green pickles due to its size, high pulp/stone ratio, the pulp's high quality, and the ease of separating it from the stone. It is highly susceptible to peacock spot, and tolerates olive knot.

Barouni

Barouni is a dual-purpose variety, though known mainly for its large green table fruit. It can be grown both in warm and cold climates, and is a consistent bearer of good crops. The tree forms a spreading canopy that eases harvesting. The fruit ripens mid- to late-season, and its oil has a peppery taste with a lusty spiciness.

Phicoline di Morocco

Phicoline di Morocco is the dominant variety of Morocco and Algeria. The tree can endure drought and salinity, and it has a medium rooting ability. The drupes are used for both mill and pickles, and are medium to large (3.5 g). The oil is fair, and the oil itself is rich in oleic acid and highly resistant to freezing, with a very fruity and aromatic flavor with dominant flavors of green leaves, bitterness and pungency. Tasters will be quick to recognize the fig and fresh bark flavor of this olive oil. The stability of this olive oil is considered to be average. Phicoline di Morocco is known to be sensitive to peacock spot, and tolerant to olive knot and *Verticillium* wilt.

7.2.8 California

Mission

Mission originated on the California Missions. It is prized as a dual-purpose variety for both green and black pickles, and for oil production. It is highly resistant to cold. Its productivity is medium and alternate, and ripening is late. Flowering occurs in late May, and the beginning of bearing is intermediate. The tree is vigorous, with an erect habit and medium-density canopy. The leaves are of medium length and width, with an elliptic-lanceolate shape. The fruit is medium sized, with a slightly asymmetrical ovoid shape. It has a medium force of detachment, and is freestone. The oil content at maturation is relatively high (around 22 percent), and the oil derived from this variety has a good level of fruitiness, with hints of green apple; it is somewhat bitter and a bit pungent, although the overall sensation when tasting is sweet. The strongest flavor is of almond. Mission is susceptible to olive leaf spot and *Verticillium* wilt, and resistant to olive knot.

Sevillano

Sevillano is the largest California commercial variety of olive, and is used mainly for pickling, due to its low oil content. It is hardy to cold. The tree is a strong grower and regular bearer. To prosper, it requires deep, rich, well-drained soil. The fruit is relatively large, bluish-black when ripe, with a large clinging stone. It ripens early.

7.2.9 South American

Azapa

Azapa is Chile's most highly rated cultivar, and is named after the valley in which it is grown, close to the Peruvian border. It prospers in warmer regions, though it has a medium yield. The fruit is very fleshy, with a thin skin; it is very large in size and elongated in shape, with a somewhat pronounced point on the end. It ripens mid–late season. Researchers believe it is related to Sevillano.

Arauco

Arauco is Argentina's main variety of olive. It has been cultivated in the Arauco Valley in La Rioja, Argentina, since the seventeenth century. Its fruits are used mainly for the preparation of green and black olives in brine. The fruit is big (8 g), asymmetrical and conical in shape, with a large base and fine top. The peduncle is quite elongated (0.4 mm). It is known for its strong flavor as a consequence of the high acidity developed during its fermentation.

7.3 Selection of genetic materials

As new plantations of olive trees are rapidly increasing in many new countries beyond the traditional areas, including Australia, South America, the United States, China, India and other drier areas, selection of the genetic material or variety of the olive is more crucial now than previously.

In modern olive plantations, including those in dryland conditions, cultivars are either derived via traditional or modern technologies in order to achieve genetic improvement, or are selected from the numerous genotypes of the Mediterranean region or elsewhere. Desirable criteria that are taken into consideration when selecting genetic materials should solve both agronomic and commercial problems. The prime criteria that should be considered when selecting genetic materials for olive-growing include (Rugini and Pesce, 2006):

1. Auto-production or auto-fertile plants

2. The regulation of fruit ripening and increase of oil content and quality

3. The production of plants with parthenocarpic fruits

4. Increased tolerance to cold and salt

5. Modification of vegetative habitats

6. Resistance to biotic stress.

Until the past two decades, morphological descriptions of the olive tree and the olives were the most common method used to identify the selected olive trees, which were later described as new varieties. It was common to monitor the trees' shape, size, vigor, growth, leaf and olive visual parameters, together with various pest and disease tolerance and/or susceptibilities in order to identify the variety (Barranco *et al.*, 2000). However, it was very difficult to provide accurate and reproducible data based on morphological criteria in order to identify olive varieties, and even more difficult to differentiate between olive varieties of similar genetic origin. The recent development of molecular biology allows evaluation of the genetic variability in olive trees in an objective and precise manner.

7.3.1 Molecular markers for genetic fingerprinting of olive

Today, the olive industry requires well-characterized cultivars with elite agronomic characteristics, or cultivars adapted to modern intensive mechanized orchards. In recent years, new genetic marker biotechnologies have been developed and further refined in order to identify the genetic relationship between olive varieties, and their link with ancient genetic materials. These technologies allow evaluation of the genetic variability in olive trees in an objective and precise manner that emphasizes that eventual genetic change is both biochemical and molecular. Molecular markers are a readily detectable sequence of DNA or protein whose inheritance can be monitored. To be useful, molecular markers must possess certain characteristics; they must be polymorphic and reproducible, preferably display co-dominant inheritance, be fast, and be inexpensive to detect. Although molecular markers have only been established in recent decades, they have already been able to estimate basic germplasm variability and trace the origin of olives (Hatzopoulos *et al.*, 2002). A brief summary is provided here of the molecular marker biotechnologies available today.

Restriction fragment length polymorphisms (RFLPs)

RFLP is one of the common groups of DNA-based molecular markers. This was the first technology developed that enabled the detection of polymorphisms at the sequence level. The technology is based on electrophoretic comparison of the size of defined restriction fragments derived from genomic DNA. In this technology, the following steps are carried out:

1. Isolation of high-quality DNA

2. Digestion with a combination of restriction enzymes

3. Fractionation of digested samples by electrophoresis

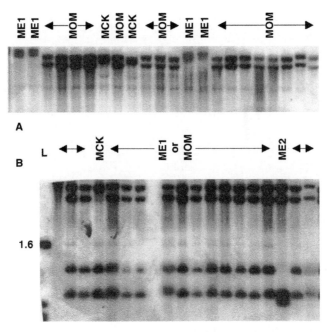

Figure 7.2: RFLP profiles of total DNA from oleasters or cultivars distinguishing the four mitotypes, ME1, ME2, MOM and MCK (Besnard and Berville, 2000).

4. Transfer of fragments to membrane

5. Hybridization with radioactively labeled DNA probe(s), and detection by autoradiography.

The sequence of RFLP may be known (e.g., a cloned gene) or unknown (e.g., from random cDNA or a genomic DNA clone). Specific probe/enzyme combinations give highly reproducible patterns for a given individual, but variation in the restriction patterns between individuals can arise when mutations in the DNA sequence result in changed restriction sites.

In olives, this approach was first applied to differentiate 70 olive cultivars, approximately 90 old trees and 101 oleasters, using chloroplast DNA (Amane *et al.*, 1999), which showed five distinct chlorotypes. Later, Bensard and Berville (2000) were able to distinguish four mitotypes by RFLP – ME1, ME2, MOM and MCK – using mitochondrial DNA (Figure 7.2).

Polymerase chain reaction (PCR)

Another common group of DNA-based molecular markers is PCR-based. This group consists of RAPDs, AFLPs, SCARs and SSRs. In PCR, amplification of tracts of DNA is defined by border sequences that hybridize to selected DNA polymerase primers. The PCR reaction requires:

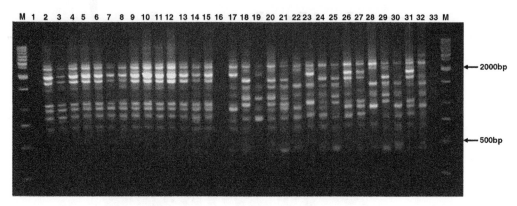

Figure 7.3: RAPD profiles of 33 different olive genotypes by 10 base primer BAO3. Lanes: M, 1 Kb, DNA ladder (Soleimani *et al.*, 2006).

1. Target DNA

2. Thermostable DNA polymerase (Taq)

3. Oligonucleotide primer(s) (7–30 nt)

4. dNTPs + Mg^{++}.

The DNA does not need to be of high quality. Tests are extremely sensitive and easily contaminated, so the reaction conditions must be optimized and controlled. It is a rapid and relatively inexpensive method.

Random amplified polymorphic DNA markers (RAPDs)

In this technology, an arbitrary primer is used to amplify DNA from total genomic DNA. The use of RAPD markers is considered an attractive method for fingerprinting of genotypes. This technique has also been successfully used for cultivar identification of other plants (Burgher *et al.*, 2002; Yamagishi *et al.*, 2005). The randomly amplified polymorphic DNA markers are widely used mainly because this technology is fast, easy and cheap (Figure 7.3).

The first RAPD study on olives was carried out by Fabri *et al.* (1995), who screened 17 oil and table-olive cultivars originating from throughout the Mediterranean region using this method and achieved a high degree of polymorphism in the olive germplasm. Later, Vergari *et al.* (1996), Cresti *et al.* (1997) and Wiesman *et al.* (1998) (Figure 7.4) also simultaneously used this RAPD technology to evaluate the olive germplasms in different countries around the Mediterranean.

OPZ-9   OPZ-11   OPZ-13

M 1 2 3 4 5 6 M 1 2 3 4 5 6 M 1 2 3 4 5 6 M

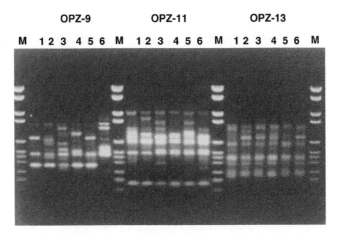

Figure 7.4: A typical RAPD differentiation between six common olive varieties grown in Israel, using three primers (OPZ-9, OPZ-11 and OPZ-13). The olive varieties analyzed were: 1, Barnea; 2, Souri; 3, Muhasan; 4, Maalot; 5, B-69; and 6, Manzanillo (Wiesman *et al.*, 1998).

Amplified fragment length polymorphisms (AFLPs)

AFLP is a combination of PCR and RFLP that provides information fingerprints by amplified fragments. The technology is very sensitive and of good reproducibility, but is technically demanding and relatively expensive. Discrimination of homozygotes from heterozygotes requires band quantitation (gel scanner) and bands are anonymous, so the interpretation of patterns can be challenging.

This methodology can be used to study genetic variation within and among populations of cultivated olives, wild olives, or species belonging to different *Olea* species. In the first of this kind of study, Angiolillo *et al.* (1999) scored 290 bands in two major clusters from samples of 43 different olive varieties, 30 wild olives, and 9 *Olea* species obtained from the Mediterranean and north-western Africa (Figure 7.5).

Sequence-tagged sites as markers

These include microsatellite-based markers (SSR: short sequence repeats), SCARs (sequence-characterized amplified regions) and CAPS (cleaved amplified polymorphic sequences). Among them, microsatellite-based markers are most common.

Microsatellites, short tandem repeats (STR) or simple sequence repeats (SSR) consist of a number of tandemly repeated short DNA sequences (one to six base pairs long). They are distributed throughout the eukaryotic genome. In addition, microsatellites are multiallelic due to their high intraspecies variability, and are easily amenable to polymerase chain reaction (PCR)-based analysis. Both characteristics make them the DNA markers of choice for human DNA profiling analyses. However, microsatellite-based markers have found wider application in

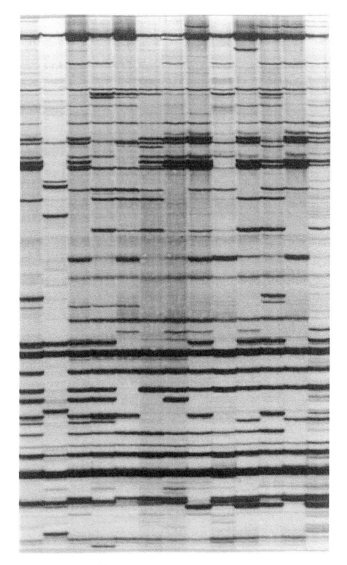

Figure 7.5: Example of AFLP banding patterns in olives using the primer combination E-AGC/ M-CTG (Angiolillo _et al._, 1999).

different branches of animal and plant sciences (Stambuk _et al._, 2007). At present, microsatellite-based DNA sequences provide one of the most appropriate genetic markers used in olive cultivar characterization and classification. Many microsatellites have been isolated from olives, and their respective primer pairs have been developed (Rallo _et al._, 2000).

Recently, Omrani-Sabbaghi _et al._ (2007) used this technology to study the genetic variation in 17 Iranian and 30 foreign olive cultivars, and found most of the Iranian olive accessions

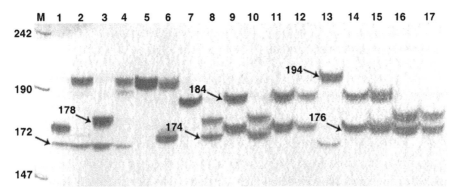

Figure 7.6: Microsatellite polymorphism in 17 olive cultivars at GAPU89 locus. The genotypes are as follows: 1, Dan; 2, Sevillana (R); 3, Kalamata (R); 4, Manzanilla (R); 5, Gorgan no. 1 (D); 6, Gorgan no. 2; 7, Dezful; 8, Shenge (D); 9, Dakal (D); 10, Shenge (A); 11, Dakal (A); 12, Khastavi; 13, Arbequina (R); 14, Barbar; 15, Zahedi; 16, Gorgan no. 1 (A); 17, Khoramabad. M, molecular weight marker (Omri-Sabbaghi *et al.*, 2007).

formed a main distinct group cluster different from the Syrian and introduced accessions (Figure 7.6)

CAPS procedure In this technology, site-specific primers are used to amplify a particular locus; the amplicon is digested with a restriction enzyme; fragments are electrophoretically resolved; and polymorphisms are detected as differences in fragment sizes.

The significant advancement in molecular marker technologies in recent years has been specifically directed toward this aim, and many studies have been reported.

7.4 Evaluation of desert-suitable olive varieties of various origins

For further development of cost-effective olive oil production in the desert environment, along with other technological improvements, the selection of olive varieties suitable for desert cultivation has played a crucial role. In order to select suitable varieties for desert areas, several attempts have been made to cultivate various varieties in the Negev Desert over the past two decades. In these studies, many olive varieties grown in various parts of the world were collected, cultivated in the Negev Desert, and evaluated for their suitability for the desert environment – especially under saline irrigation conditions. The studies included:

1. Evaluation of locally grown varieties

2. Evaluation of European varieties

3. Evaluation of Spanish semi-arid varieties

4. Evaluation of Central Asian varieties.

7.4.1 Evaluation of locally grown varieties

Our initial research approach to studying the suitability of olive varieties for cultivation under Negev Desert conditions was directed towards locally grown olive varieties that had been selected and commonly used in the Middle East for many years. Olive varieties that originated from Lebanon, Israel, the West Bank and Jordan but had been grown in the Negev region for a long time were considered to be local varieties; among them, Souri, Nabali Muchasan, Barnea and Maalot were used in this study.

Souri is one of the selected olive varieties in northern Israel, the West Bank and Lebanon. It has been cultivated in this region for many years. Traditionally, Souri olives produce strong and aromatic oil which is well recognized in the Middle East. The oil percentage in Souri olives is relatively high (above 20 percent). Souri trees developed in Middle Eastern conditions have small canopy, and are known for their high alternate bearing characteristics when cultivated in a traditional, non-irrigated way. However, under intensive cultivation in the Negev Desert, Souri trees were able to grow rapidly and develop relatively large canopies. Due to the intensive cultivation, the rate of alternate bearing was somewhat reduced; however, every second or third year an "off" year still occurred. Negev Desert cultivation of Souri oil did not change the basic oil characteristics. Souri trees were also found still to be susceptible to leaf spot, *Verticillium* wilt and olive fly under Negev Desert conditions.

Nabali Muchasan is another of the selected varieties of the West Bank, and the most common among Arab olive-growers in the Middle East. Under extensive cultivation practices, Muchasan trees usually develop small canopies; in the Negev area, too, even with the full irrigation regimen, these trees are relatively small in size. The productivity and oil content of Muchasan olives is relatively moderate even when the trees are irrigated. Muchasan trees were still found to be susceptible to leaf spot disease and olive fly in the Negev Desert.

Barnea was selected from the Negev Desert area in the 1970s. Barnea trees are well known and easily recognizable by their tall apical dominant characteristics. Our study in the Negev Desert has shown that Barnea's vigorous growth can be controlled by increasing the salinity in the irrigation water, to a level of 4.2 dS/m, without any yield reduction. Above this level, the sustainability of Barnea trees is affected due to the accumulation of too high a level of salts in the vegetative tissues of the tree. Usually Barnea trees intensively cultivated in Negev Desert conditions are very high-yielding ($\sim$15 Mt/y), with a relatively low rate of alternate bearing. The average oil percentage in mature Barnea olives is about 18%, but due to the high productivity in the Negev cultivation practices, the Barnea oil yield per land is high ($\sim$2.5 Mt/y). The quality of

Barnea olive oil is accepted as being somewhat lower than that of the strong Souri oil and other, more delicate, European olive oils. Barnea trees are highly susceptible to Leopard moth larvae. They are well suited to mechanical harvesting using trunk shakers, under intensive cultivation in the Negev (Wiesman *et al.*, 2004).

Maalot has been selected in recent years in north Israel, mainly for its high tolerance of leaf spot disease. It is a vigorously growing tree. Owing to its alternate bearing nature, even under intensive Negev cultivation, Maalot trees produce multi-annual moderate olive yields. The oil percentage accumulated in the olives is relatively high ($\sim$25 percent), and the quality of the oil is good. These characteristics were not found negatively affected by saline drip irrigation practices in the Negev Desert area.

General observations of the above varieties in various locations and environments of the Negev area, from the north to the south, have proved the feasibility of their intensive cultivation in desert areas. While dealing with these olive cultivation trials, we exposed the uncertainty and confusion in the Israeli olive industry regarding the identity of the local olive varieties available to Israeli growers. From the 1960s to the 1990s, it was accepted among growers that there were several lines of Souri olive varieties. To take better marketing positions in Israel, each olive nursery used to claim to produce a unique and improved Souri clone. Olive-growers then faced a very confusing situation when they tried to choose plant material for planting in new olive plantations. To address this topic, which is highly important in desert olive cultivation, and to try to introduce more systematic order to the field of olive plant material in Israel, our research group carried out one of the first studies using the technology of random amplified polymorphic DNA fragments for molecular markers to identify Middle East olive native genetic material (Wiesman *et al.*, 1998). However, our study concluded that, because of the long period of selection and vegetative propagation, all the Souri plant materials produced by different olive nurseries in Israel are identical. This study also established:

1. Molecular fingerprintings of the most common olive varieties in Israel and the West Bank like Souri, Nabali Baladi, Nabali Muchasan, Barnea and Maalot

2. The determination of the relationships between the genetic material of local Middle Eastern and selected European olive varieties

3. Characterization of the genetic markers of Maalot which might be responsible for the resistance of Maalot trees to leaf spot disease, which is a highly limiting factor for cultivation of olives in many environments.

Several similar studies using RAPD molecular technology were also reported in the 1990s, by research groups in Europe and elsewhere (Wiesman *et al.*, 1998). Following these positive initial basic horticultural feasibility tests of the suitability of local olive genetic materials, and the basic molecular markers identifying these varieties, a list of criteria has been made for further and more intensive selection of varieties to develop desert olive oil intensive cultivation. The following are the most important of these parameters:

1.   Compatibility with the Negev soil type

2.   Vegetative growth rate in Negev conditions

3.   Genetic and Negev cultural effects on tree size

4.   Genetic and Negev cultural flowering and ripening dates (early, mid or late season)

5.   Pollination needs of the evaluated variety in Negev conditions

6.   Productivity, alternate bearing and variety yield rate in Negev conditions

7.   Oil percentage and oil yield rate per unit of land in the Negev

8.   Susceptibility to pests and diseases in Negev conditions

9.   Ease of harvesting (fruits' force of detachment) in Negev cultural conditions

10.   Suitability for mechanical harvesting in Negev cultural conditions

11.   Harvest timing (early, mid or late season).

These selection criteria were utilized in the next selection studies, analyzing a wider range of olive genetic material for its suitability for cultivation under Negev Desert conditions.

7.4.2 Evaluation of European varieties

In a second set of studies, 12 selected olive varieties of various Mediterranean origins, including Italian, Spanish, Greek, French and Moroccan, and locally grown varieties, were planted in the Central Negev Desert area in a special experimental plot at Ramat Negev in 1997. Systematic evaluation has continued from that date. The olive varieties used in this study were Barnea, Souri and Maalot, of local origin; Leccino and Frantoio, of Italian origin; Arbequina, Picual and Picudo, of Spainsh origin; Kalamata and Koroneiki, of Greek origin; Phicolin, of French origin; and Phicholin de Morocco, of Moroccan origin. The name of the variety, its origin and the number of plants in one sub-plot are presented in Table 7.2.

In order to support the rapidly growing olive industry in the Negev Desert and similar areas in the world, a special saline-irrigation controlled experimental plot was established with 12 selected superior local olive varieties and varieties from various Mediterranean countries. A similar plot

Table 7.2: Olive varieties, their origin, and the number of trees planted in each sub-plot for the study

Variety	Origin	Number of trees
Barnea	Local	10
Souri	Local	10
Maalot	Local	6
Frantoio	Italy	10
Leccino	Italy	10
Arbequina	Spain	10
Picual	Spain	10
Picudo	Spain	10
Kalamata	Greece	6
Koroneiki	Greece	10
Picholin	France	10
Picholin di Morocco	Morocco	6
Total		**108**

was irrigated with fresh water obtained from the National Water Carrier of Israel (tap water; EC 1.2 dS/m), while the second was irrigated with moderately saline water (EC 4.2 dS/m) with enough leaching provision. This study aimed to evaluate and compare the vegetative and reproductive multi-annual response of mature, yielding trees of these tested olive varieties under fresh water and moderate saline water irrigation regimes, using the drip system, in a commercial orchard simulation study in a semi-arid area.

The trees were planted in an area 4×6 m, representing a density of 240 plants per hectare. The olive trees were drip-irrigated, using a straightforward monthly irrigation formula that was developed based on multi-year local pan evaporation data (described in Chapter 5). The formula was rechecked daily, and corrected accordingly. The average quantity of water applied annually was 6560 m^3 per hectare. For the first three years after planting, the practice was to irrigate immediately following any rainfall, to prevent salinization of the root zone area. Twice a year, in March and November, a supplement of 1000 m^3 was added to the monthly water to leach the soil and remove the salt excess from the root system zone (Wiesman *et al.*, 2004). The electrical conductivity (EC) of the water for the saline treatment was adjusted by mixing the two types of water or adding NaCl. The rate of NPK fertilization was determined based on the results of annual nutrient leaf analysis.

The trunk circumference of each tree was measured twice a year, 30 cm above the ground, and individual trees were harvested separately. The olive yield of each variety was assessed for every tree in each sub-plot. The fruit, yield and oil analyses were carried out after that.

In order to assess the effect of salinity and nutritional imbalance, leaf analysis was periodically carried out in three pools. Furthermore, the soil was analyzed at different depths to assess the salinity effect on soils.

Effect on trunk development

The changes over five years in the trunk circumference of mature olive varieties irrigated with moderately saline water (4.2 dS/m) in comparison those irrigated with tap water (1.2 dS/m) are presented in Table 7.3. The trunk growth during the earlier two years was found to be significantly lower in saline irrigation; however, no differences in growth rate were observed after two years. The variety of the olive tree, though, consistently showed differences; the greatest growth in trunk circumference was observed in P. Morocco under saline irrigation treatment. Maloot, a local variety, showed the second greatest growth in circumference, but this was under fresh-water irrigation conditions.

The fact that no effect was found in terms of trunk development of all 12 olive varieties irrigated with saline or fresh water in the present study is both surprising and interesting. Moreover, in terms of foliage development, no visual differences between trees irrigated with the two types of water could be observed in the two sub-plots in advanced years of growth (data not shown). The results obtained in this study raised the question: "What is the main reason for the similar vegetative response of olive trees irrigated with saline (4.2 dS/m) or with tap water (1.2 dS/m)?" Addressing this question may enable significant improvement in olive cultivation in semi-arid areas using saline water for irrigation; however, before trying to do this directly, some additional horticultural parameters characterizing the response of mature olive varieties to saline irrigation water might be useful.

Effect on fruit yield

General fluctuation in tree productivity was observed among all the olive varieties across the various years analyzed in this study. Indeed, fluctuation in the yield of olive trees is a common phenomenon because of alternate bearing (Barranco *et al.*, 1998). However, even after allowing for the alternate bearing effect, a high level of variation was still observed among the varieties tested over the four-year study period (Table 7.4). Fresh-water irrigated Barnea (2003), Frantoio (2002), Picudo (2002), Koroneiki (2002), Picholin (2002) and Picholin di Morocco (2002) trees yielded significantly more olives than the saline-irrigated trees of the same variety. Usually, but not significantly, in the following year the yields from the saline-irrigated trees tended to be greater than those of the control trees. Picholin trees irrigated with saline water did not follow this pattern, and were found to produce a significantly higher yield than the control trees in 2003 and 2004. Looking at the average data over four years, Picholin produced the highest yield with both forms of irrigation, followed by Barnea, while Souri showed the lowest yield. However, no significant differences could be found between trees of all the varieties irrigated with the two water treatments (Table 7.4). Analysis of variance clearly showed, on one hand, a significant

Table 7.3: Annual trunk circumference of olive trees of different origin, planted in 1997 and cultivated in the Ramat Negev Station under two different salinity conditions

Variety	Salinity (dS/m)	Trunk circumference (cm)					Average growth (cm/y)
		2001	2002	2003	2004	2005	
Local							
Barnea	1.2	37.0	42.4	48.5	50.c8	57.4	5.1
Barnea	4.2	34.5	38.2	45.9	51.2	57.6	5.8
Souri	1.2	33.4	42.8	50.5	55.3	60.9	6.9
Souri	4.2	32.6	38.8	46.5	51.5	55.6	5.7
Maalot	1.2	37.8	47.3	54.7	59.7	65.0	6.8
Maalot	4.2	34.2	43.6	53.0	60.7	69.2	8.7
Italian							
Frantoio	1.2	37.2	41.8	47.7	51.4	56.3	4.8
Frantoio	4.2	31.1	37.0	44.8	48.7	55.9	6.2
Leccino	1.2	38.7	46.2	53.2	56.6	63.0	6.1
Leccino	4.2	37.7	44.4	52.5	57.5	63.7	6.5
Spanish							
Arbequina	1.2	27.1	35.2	41.0	44.1	50.8	5.9
Arbequina	4.2	27.0	33.2	40.3	43.2	48.1	5.3
Picual	1.2	37.8	43.2	50.0	54.1	62.9	6.3
Picual	4.2	34.0	38.2	46.0	51.0	59.7	6.4
Picudo	1.2	31.5	35.1	43.8	46.3	50.2	4.7
Picudo	4.2	26.4	32.6	40.3	46.9	53.3	6.7
Greek							
Kalamata	1.2	25.9	33.5	41.7	46.5	51.2	6.3
Kalamata	4.2	22.2	30.7	39.7	45.0	50.5	7.0
Koroneiki	1.2	29.9	35.6	43.5	47.9	50.6	5.2
Koroneiki	4.2	26.4	30.4	42.2	43.8	48.1	5.4
French							
Picholin	1.2	33.8	39.4	48.5	53.6	59.2	6.3
Picholin	4.2	35.3	37.8	48.1	52.5	60.0	6.2
Moroccan							
P. Morocco	1.2	28.9	38.0	46.8	52.0	57.0	7.0
P. Morocco	4.2	23.5	38.7	45.3	52.3	60.7	9.3
Significance							
Variety		***	***	***	***	***	***
Treatment		*	*	NS	NS	NS	NS
Variety & treatment		NS	NS	NS	NS	NS	NS

*NS, not significant; ***, highly significant (at 0.001); *, of low significance (at 0.05), by Tukey-Kramer.*

Table 7.4: Olive fruit yield per tree of different origins, planted in 1997 and cultivated in the Ramat Negev Station under two different salinity conditions

Variety	Salinity (dS/m)	Olive fruit yield (kg/tree)				
		2001	2002	2003	2004	Four-year average
Local						
Barnea	1.2	38.3	55.1	41.3	76.6	54.8
Barnea	4.2	33.7	52.8	23.0	84.5	52.2
Souri	1.2	5.0	20.3	33.1	51.1	31.8
Souri	4.2	4.2	22.6	36.9	52.8	31.5
Maalot	1.2	no	32.0	39.6	40.7	37.5
Maalot	4.2	no	21.0	27.5	52.8	33.8
Italian						
Frantoio	1.2	10.5	54.5	0.0	66.7	38.6
Frantoio	4.2	23.8	36.6	0.0	66.0	31.6
Leccino	1.2	9.2	36.0	38.4	71.2	46.1
Leccino	4.2	12.2	37.2	27.7	78.7	45.7
Spanish						
Arbequina	1.2	12.3	38.2	43.7	51.6	42.5
Arbequina	4.2	25.5	31.9	50.0	57.9	45.3
Picual	1.2	30.5	41.4	11.0	77.4	42.5
Picual	4.2	16.5	44.3	13.0	77.9	43.3
Picudo	1.2	2.0	42.6	8.3	50.6	31.9
Picudo	4.2	5.5	25.0	17.0	55.0	30.7
Greek						
Kalamata	1.2	no	16.2	37.5	40.4	30.3
Kalamata	4.2	no	15.1	41.1	51.9	38.5
Koroneiki	1.2	11.0	53.5	24.1	58.0	44.4
Koroneiki	4.2	15.8	28.6	21.7	55.7	33.1
French						
Picholin	1.2	19.1	67.6	11.5	88.4	53.6
Picholin	4.2	26.6	57.0	23.4	102.9	59.0
Moroccan						
P. Morocco	1.2	no	34.0	21.7	54.0	36.6
P. Morocco	4.2	no	7.33	28.6	69.1	35.0
Significance						
Variety		***	***	***	***	***
Treatment		NS	NS	NS	NS	NS
Variety & treatment		NS	NS	NS	NS	NS

*NS, not significant; ***, significant at 0.001 level of significance, by Tukey-Kramer.*

effect of olive variety on the rate of olive yield and, on the other hand, no effect of the irrigation treatments or of the interaction between the varieties and irrigation treatments.

It is clear that almost all the varieties tested can be grown under both fresh-water and moderately saline irrigation, without any significant losses in the fruit yield, in desert conditions. Though this study provides indications for the future, the data are from very young trees in early years of production; therefore, in order to confirm the result more data are required and the study is continuing.

Effect on percentage oil recovery, oil yield and quality

A comparison of the average data regarding four-year oil recovery and fatty acid profiles in saline and fresh-water irrigated trees is presented in Table 7.5. Both the oil percentage and fatty acid profiles were found to vary significantly among the varieties of different origin, although no variation was observed between the treatments (i.e., fresh water or moderately saline water irrigation). However, both the fatty acids profile and oil percentage varied from year to year (data not shown). Four-year data revealed that the highest percentage of oil was found in the Maalot variety, followed by Koroneiki, Souri and Barnea, under either of the irrigation conditions. Interestingly, in most of the varieties tested a higher oil percentage was recorded in the fruit obtained from trees treated with the moderately saline irrigation.

Though no variation between fresh-water and saline irrigated trees with regard to the amount of oleic acid was shown among the tested varieties, in most cases a higher level of oleic acid was found in trees undergoing saline irrigation rather than in those undergoing than fresh-water irrigation. The variety-specific genetic effect was again clearly observed in terms of its effect on oleic acid level. Most other fatty acids in all the tested varieties were found to be within the IOOC limit, as described in Chapter 11.

Effect on leaf sodium and potassium concentration and soil EC level

In order to discover the response of the different varieties under the two different types of irrigation in desert conditions, leaf and soil analysis was performed during the study period. Specific soil EC measurements in various fractions of the soil column (0–30, 30–60 and 60–90 cm) in the root zone and leaf sodium and chloride analysis were carried out routinely in all 12 tested olive varieties cultivated in both fresh-water and saline irrigated experimental plots.

In line with the previous studies, the levels of Na^+ and Cl^- were found to be higher in leaves of saline irrigated trees compared to fresh-water irrigated trees (Al-Saket and Aesheh, 1987; Benlloch *et al.*, 1991; Klein *et al.*, 1994; Gucci and Tattini, 1997; Aragues *et al.*, 2005). The highest level of Na^+ was found in the saline irrigated leaves of Picual, Arbequina, Koroneiki and Picholin trees (1.455, 1.291, 1.268, and 1.125 mg/g, respectively). The highest level of Cl^- was found in the saline irrigated leaves of Picual, Picholin di Morroco, Kalamata and Souri

Table 7.5: Average oil percentage and fatty acid profile of the different olive varieties planted in 1997 and cultivated in the Ramat Negev Station under two different salinity conditions, 2001–2004

Varieties	Salinity (dS/m)	Oil (%)	Fatty acid profile				
			Palmitic 16:0	Stearic 18:0	Oleic 18:1	Linoleic 18:2	Others
Israeli							
Barnea	1.2	20.1	13.18	2.66	68	14.52	0.75
Barnea	4.2	19.6	14.52	2.79	65.3	16.3	0.91
Souri	1.2	20.3	11.65	1.4	70.2	15.6	1.02
Souri	4.2	19.4	6.8	2	71.5	18.2	1.1
Maalot	1.2	24.3	11.3	3.5	68.2	14.5	2.1
Maalot	4.2	22.6	11.5	1.3	70.1	15.6	1.5
Italian							
Frantioio	1.2	12.5	10.1	3.2	67.2	17.2	1.3
Frantioio	4.2	15.9	13	2.1	68.9	15	0.98
Leccino	1.2	17.5	16.22	3	65.14	14	1.9
Leccino	4.2	16.6	16.2	1.01	65.62	14.06	1.33
Spanish							
Arbequina	1.2	18.7	13.8	1.8	70.1	13.2	1.8
Arbequina	4.2	19.3	9.85	2.87	72.5	12.1	2.1
Picual	1.2	14.7	10.01	1.84	70.17	10.45	3.64
Picual	4.2	14.7	15.35	3.18	70.55	8.92	1.27
Picudo	1.2	16.5	13.6	2.4	69.8	12.1	2.5
Picudo	4.2	15.5	14	2.7	70.2	11.2	1.8
Greek							
Kalamata	1.2	16.6	15.5	2.3	68.2	12.3	1.2
Kalamata	4.2	15.4	14.2	2.1	70.2	11.5	1.8
Koroneiki	1.2	19.6	15.17	2.2	67.15	10.73	4
Koroneiki	4.2	20.9	14.82	2.52	71.53	10.12	1.11
French							
Picholin	1.2	16.8	14.43	1.18	73.6	7.05	2.26
Picholin	4.2	14.4	14.82	1.48	71.3	9.09	2.26
Moroccan							
P. Morocco	1.2	13.5	11.94	3.7	65.9	16.83	0.9
P. Morocco	4.2	13.8	13.87	0.95	65.92	17.44	1.15
Significance							
Variety		***	***	***	***	***	***
Treatment		NS	NS	NS	NS	NS	NS
Variety & treatment		NS	NS	NS	NS	NS	NS

*NS, not significant; ***, significant at 0.001 level of significance, by Tukey-Kramer.*

(1.480, 1.466, 1.333, and 1.275 mg/g, respectively). Aragues and colleagues (2005) reported that the growth of the Arbequina olive might be reduced when the leaf chloride and sodium levels are higher than 2.3 and 1.5 mg/g, respectively; however, this finding was not observed in our study, although this might be because there were lower concentrations of sodium and chloride in the leaves we analyzed, even from the saline irrigated trees. Analysis of variance showed a relatively low significant effect of the olive varieties on sodium accumulation in the leaves, but no significant effect of the varieties was obtained for chloride (data not shown).

Study of the EC of three fractions of the soil column showed a peak of about 4.5 dS/m in the middle soil fraction (30–60 cm) of the fresh-water irrigated trees, but the pattern of the soil fractions' EC level in the saline irrigated trees was different, showing a linear curve along the soil depth. However, the middle soil fraction, which contained the main root mass and activity in this experimental plot, showed no significant difference between the two irrigation treatments; the EC level of the saline irrigated soil was 5.3 dS/m. In the other two soil fractions (0–30 and 60–90 cm), a significant difference between the two irrigation treatments was obtained, but even in the lower soil fraction the EC was no more than 6 dS/m (Figure 7.7). Analysis of variance clearly indicated a significant effect of the irrigation treatments and a lower rate of soil depth effect. The data obtained from the EC of the soil fraction study are in good agreement with previous reports suggesting that most olive varieties may develop well, with no significant reduction of growth, development or yield, within a range of 3–6 dS/m (Bernstein, 1964; Maas and Hoffman, 1977; FAO, 1985; Aragues *et al.*, 2005).

Based on the data obtained in the present study, it seems that the key factor for cultivation of various olive varieties using moderately saline water for irrigation lies in the ability to

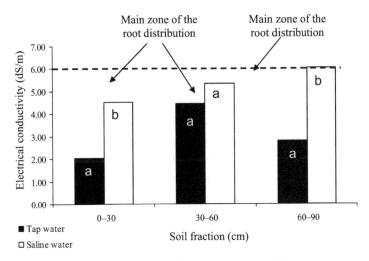

Figure 7.7: Average analytical results of the soil analysis from different fractions at two different degrees of salinity.

maintain the EC of the soil in the root zone growth area at a level lower than 6 dS/m, as suggested by the FAO (1985) recommendation for olive cultivation and supported by many other reports (Bernstein, 1964; Maas and Hoffman, 1977; Aragues *et al.*, 2005). In semi-arid regions, where most of the crop water requirements are supplied via irrigation, and water often contains large quantities of dissolved salts, salinity control is frequently a major objective of irrigation management (Shalhevet, 1994). Leaching is the process of applying more water to the crop than can be held by the root zone so that the excess water drains below the root system, carrying salts with it (Shalhevet, 1994; Grattan and Oster, 2002). When saline water is supplied to crops, leaching becomes indispensable in order to exclude or reduce the excess salt from the root zone (Beltran, 1999). Therefore, a proper leaching methodology throughout the entire year has been developed specifically for olive cultivation in semi-arid areas, and had been used in previous studies carried out in the Ramat Negev area (Wiesman *et al.*, 2002, 2004). This leaching process, together with accurate weekly drip irrigation, kept the soil conditions well set for most of the tested olive varieties to develop very similarly to the fresh-water irrigated trees of the same variety. Interestingly, using a minirhizotron system based on a video camera, it was found that the roots of both saline and tap-water irrigated olive trees seem to be highly distributed in the two upper soil fractions (Weissbien, 2006). The EC in these soil fractions is lower than the upper level of 6 dS/m, which is well recognized as limiting the regular development of olive varieties (FAO, 1985).

The results of this study clearly indicate that although variation is shown among the varieties in terms of growth, yield and oil quality when grown under desert conditions, there are no significant differences in these factors between fresh-water and moderately saline irrigated trees. This is a very positive result for olive oil cultivation in desert environments, because most desert areas lack fresh water and the only available source of irrigation may be saline. However, successful irrigation with moderately saline water is closely linked to a proper soil leaching methodology and maintenance of the soil EC level in the root zone at less than 6 dS/m.

7.4.3 Evaluation of Spanish semi-arid varieties

As a follow-up to the successful second study, and in order to extend the spectrum of desert-suitable olive varieties, we carried out a survey of olive genotypes well adapted to desert conditions in the Spanish semi-arid area of Catalonia. Twelve selected olive varieties were obtained from Catalonia and planted at the Ramat Negev experimental station in the spring of 2000. Directly after being planted, these olives were treated with the 4.2-dS/m irrigation methodology described in Chapter 6. The varieties used in this study are presented in Table 7.6.

In this study, we aimed to evaluate different genetic material from that tested already. The Spanish dryland-adapted varieties chosen have demonstrated a different pattern of response to desert conditions. While local olive varieties, such as Souri and Nabali Muchasan, are slow-growing and produce a small canopy when cultivated in extensive non-irrigated desert conditions,

Table 7.6: Spanish olive varieties used for evaluation of genotypes in Negev Desert

Varieties	Remarks
Babilonga	Small tree, relative of Arbequina
Balais	Small tree, relative of Arbequina
Blanqueta	Small tree, relative of Arbequina
Empeltre	Early ripener
Hojiblabca	High quality, from Andalusia
Iumed	Strong tree, may be used as rootstock
Memia	Strong tree, may be used as rootstock
Noccelara de Belicce	Italian (used as control)
Okal	Strong tree, may be used as rootstock
Palomar	Most common
Rozala	Strong tree, may be used as rootstock
Spanish Leccino	Early ripener

varieties such as Rozala, Okal, Memia and Iumed grow vigorously and produce a large canopy in their natural Spanish desert habitat. This might be related to intensive root-system development assisting them to reach water in the soil. It has been interesting to evaluate the response of these olive varieties to intensive cultivation based on controlled root-system development due to drip irrigation of saline water and exclusion of salts from the root zone. If these varieties adapt well to intensive cultivation, they may be useful in the future development of the desert olive industry. Even if they do not produce superior oil quantity and quality, they may be used as rootstock, controlling the vigor of the desert olive trees. Another aspect being dealt with in the present study is a search for olive varieties that mature and ripen earlier or later than most common olive oil varieties. For this reason, Empeltre, Palomar and Spanish Leccino were chosen for evaluation. The small-canopy varieties of Blanqueta, Babilonga and Balais, known to be genetically related to Arbequina, the leading Spanish olive variety, have been evaluated to assess their potential for establishment in dense olive groves in the Negev area. The common Spanish Hojiblabca variety and the Italian Noccelara de Belicce variety were chosen as controls.

Horticultural evaluation

Under drip irrigation with moderately saline water irrigation in the Negev Desert, Spanish Leccino, Palomar, Rozala, and Okal trees developed more rapidly than the other varieties. The characteristic strong and vigorous trees in Spanish conditions, Memia and Iumed, were found to develop moderately in comparison to the other varieties in this plot. It is suggested that there was limited root-system development due to the saline drip irrigation system applied in this plot. Balais developed the smallest canopy, probably as a result of genetic and environmental factors. Empeltre (Figure 7.8), Hojiblanca, Nuccelara de Belicce and Blanqueta produced a medium-sized trunk circumference and canopy (Table 7.7).

Figure 7.8: Empeltre tree grown in the Negev Desert.

Table 7.7: Trunk circumference increase of selected Spanish dryland olive varieties cultivated under drip irrigation with moderately saline water in the Negev Desert

Varieties	Tree circumference (cm)			
	18/05/2003	02/09/2003	02/01/2004	09/12/2004
Babilonga	22.80 ± 2.63	28.1 ± 3.79	28.5 ± 2.40	34.50 ± 3.42
Balais	18.83 ± 3.84	22.6 ± 4.00	25.93 ± 4.33	29.83 ± 4.60
Blanqueta	25.25 ± 1.16	27.6 ± 1.18	30.85 ± 1.31	34.87 ± 1.27
Empeltre	24.12 ± 1.87	28 ± 2.85	31.65 ± 3.04	39.50 ± 3.59
Hojiblanca	21.96 ± 1.43	25.6 ± 1.64	28.90 ± 1.76	34.00 ± 2.51
lumed	22.25 ± 2.40	25.6 ± 2.96	29.85 ± 3.06	36.00 ± 3.74
Memia	23.70 ± 1.04	26.1 ± 0.60	28.16 ± 0.84	33.00 ± 1.32
Noccelara de Belicce	25.12 ± 1.58	28.2 ± 1.06	30.70 ± 0.39	38.75 ± 0.95
Okal	27.77 ± 3.60	33.1 ± 4.19	38.07 ± 4.92	45.30 ± 4.59
Palomar	30.25 ± 1.65	34.3 ± 2.31	38.40 ± 2.69	45.00 ± 4.06
Rozala	26.62 ± 0.56	30.1 ± 0.54	33.40 ± 0.58	45.10 ± 5.55
Spanish Leccino	30.30 ± 1.62	36.13	41.35 ± 1.95	48.30 ± 3.08

Values are the mean of 3 ± SE.

Table 7.8: Olive yield of selected Spanish olive varieties intensively cultivated with saline irrigation in the Negev Desert

Variety	Fruit yield (kg/tree)		
	2003	2004	2005
Babilonga	14.20 ± 1.67	26.3 ± 2.52	48.3 ± 5.9
Balais	nd	20.7 ± 6.11	37.8 ± 4.6
Belicce	11.60 ± 1.85	nd	22.6 ± 3.7
Blanqueta	18.80 ± 2.89	nd	51.4 ± 4.6
Empeltre	13.60 ± 1.99	28.9 ± 4.56	39.7 ± 5.2
Hojiblanca	7.33 ± 2.54	nd	19.6 ± 4.0
Iumed	3.65 ± 1.26	32.1 ± 6.03	54.5 ± 6.9
Memia	10.00 ± 0.14	19.1 ± 5.29	41.5 ± 6.2
Noccelara de Belicce	nd	nd	20.1 ± 5.1
Okal	3.25 ± 0.94	51.5 ± 10.3	64.8 ± 6.7
Palomar	18.50 ± 4.15	34.6 ± 3.76	53.8 ± 7.7
Rozala	6.00 ± 1.08	32.2 ± 3.69	43.5 ± 4.4
Spanish Leccino	nd	58.7 ± 2.64	52.9 ± 6.4

Values are the mean of 3 ± SE; nd, not detected.

Among the tested cultivars, Empeltre, Spanish Leccino and Palomar were found to ripen early – as early as the first week of October. Iumed, Memia and Rozala ripened in mid-October, and Okal, Babilonga, Blanqueta and Noccelara de Belicce ripened in late October. Hojiblanca and Balais were the latest among the tested varieties, ripening in late November. Okal, Palomar, Iumed and Spanish Leccino and Blanqueta produced fruits with the highest olive yield in the fifth year after planting (Table 7.8). Babilonga, Rozala, Memia, Balais and Empeltre produced a medium yield, while Hojiblanca and Nuccelara de Belicce produced the lowest yield at this stage.

Hojiblanca, Nuccelara de Belicce, Spanish Leccino and Blanqueta showed high fluctuation in alternate bearing, and Rozala, Okal, Palomar and Babilonga low fluctuation. In the sixth year after planting, Iumed, Memia, Balais, Blanqueta, Babilonga, Nuccelara de Belicce and Hojiblanca produced an oil percentage in the range of ∼18–20 percent, while Empeltre, Rozala, Okal and Spanish Leccino produced a lower percentage ranging from ∼14.1–16.0 percent (Table 7.9).

Evaluation

- Most of the Spanish semi-arid olive varieties have demonstrated a high rate of compatibility with the Negev soil and climatic conditions.

Table 7.9: Oil percentage of Spanish olive varieties intensively cultivated with saline irrigation in the Negev Desert

Variety	Oil (%)
Babilonga	18.4
Balais	19.7
Blanqueta	18.9
Empeltre	14.1
Hojiblanca	17.7
Iumed	20.0
Memia	19.6
Noccerala di Belicce	18.5
Okal	15.1
Palomar	15.9
Rozala	15.4
Spanish Leccino	16.3

- Genetics and Negev Desert intensive cultivation practices seem to affect the development of the canopy and root system in all tested olive varieties.

- Genetics and Negev Desert cultural practices seem to affect the flowering and subsequent ripening dates of the trees of the various olive varieties, leading to early, mid- or late season maturation.

- The phenomenon of the pollination of these varieties still needs to be evaluated in Negev Desert conditions.

- Relatively high rates of productivity and reduced alternate bearing level were achieved for most olive varieties intensively cultivated in Negev Desert conditions.

- Relatively high oil percentages and yields per area of land were obtained for most olive varieties intensively cultivated in Negev Desert conditions.

- Owing to relatively low air humidity and to isolation, there is relatively low susceptibility of olives to pests and diseases in Negev Desert conditions.

7.4.4 Evaluation of Central Asian varieties

We completed our study of the suitability of olive genetic material for intensive cultivation in the Negev Desert by assessing some unique and to date unknown olive varieties collected in the Central Asian region. Since this region has been out of bounds for many years to Western researchers, plant materials from this area may help to ensure the future of agriculture both in the Central Asian region and elsewhere around the world. The collection was obtained by Ben-Gurion University in 1998 from a wide variety of fruit trees managed by Professor Levin after the collapse of the USSR. The area from which these olive varieties were collected is located in the very remote part of Turkmenistan and along the Afghanistan border to the Caspian Sea. According to the local people, these cultivars are widely distributed not only within the present bounds of Turkmenistan but also generally in the Central Asian region, including the northern part of Iran itself, and in the various dry plateaux located between Iran and the Caucasus. It seems that among these olive varieties are progeny of ancient olive genetic material distributed from the original olives in the Middle East. The present study concentrated on the idea that local selection of superior olive varieties well adapted to the dry environment of Central Asia may enable further selection of unique olive varieties with the potential for successful cultivation in the Negev Desert area, to produce olives and olive oil closely fitting the selection criteria characterized at the beginning of Negev Desert olive cultivation years ago. Of the collected cuttings of 39 genotypes, 35 were successfully planted.

In 1999, these plants were planted in two plots. One plot was established in the Central Negev Experimental Station, near the other olive varieties of local and European origin, while the

Table 7.10: Central Asian olive genotypes evaluated in the Negev Desert

TOG 2	TOG 21
TOG 3	TOG 22
TOG 4	TOG 23
TOG 5	TOG 24
TOG 6	TOG 25
TOG 7	TOG 26
TOG 8	TOG 27
TOG 9	TOG 28
TOG 10	TOG 29
TOG 11	TOG 30
TOG 12	TOG 31
TOG 13	TOG 32
TOG 14	TOG 34
TOG 15	TOG 36
TOG 17	TOG 37
TOG 18	TOG 38
TOG 19	TOG 39
TOG 20	

Figure 7.9: TOG 26 grown in the Negev Desert.

second plot was established in the Mediterranean coastal area in the Northern Negev. These plots were intensively cultivated according to the drip irrigation technology developed for the desert environment, as described previously, the first with moderately saline water (4.2 dS/m) and the second with fresh water (1.2 dS/m). The name of the olive germplasm and the number of each line planted in the each site in 1999 are presented in Table 7.10; a Turkmenistan olive genotype (TOG 26) is shown in Figure 7.9 .

Evaluation

- Generally, plants grew better in the Central Negev plot in comparison to the coastal plot. This might be related to their better compatibility with the light sandy washed soils of the Central Negev plot.

- A wide range of variation was observed in regard to the vegetative growth and development of the olive varieties during the 7 years of the study. Based on the trunk circumference and tree height, the most vigorous genotypes were found to be TOG 3, TOG 4, TOG 6, TOG 10, TOG 8, TOG 20, TOG 22 and TOG 38, while the slowest growing were TOG 12, TOG 14, TOG 32 and TOG 37.

- The highest yield was found in TOG 22. TOG 8, TOG 11, TOG 16, TOG 20, TOG 24 and TOG 38 also produced high yields.

- In the Negev Desert plot irrigated with saline water, big variations were found in the time of blooming and ripening of the fruit. Some of the genotypes were ready for harvesting as early as September, and some not until very late December or January. Some produced small (1.5 g) and some produced large (6.5 g) fruits.

- No particular susceptibility to pests or disease was observed. Some genotypes were found to suffer lower damage from olive fly in comparison to Barnea and Manzanillo. Three trees were dwarfed due to viral infection. Some mature olives were white when mature, and some were spotted.

- Some of the genotypes were found to be unusually easy to harvest, with the olives dropping very easily, while others were difficult to pick.

Table 7.11: Oil percentage and fatty acid profiles of some of the Turkmenistan olive genotypes (TOG) cultivated in the Negev Desert with moderately saline irrigation (4.2 dS/m) as determined by GC (2006)

Genotypes	Oil (%)	Fatty acid profiles (% total)				
		16:00	18:00	18:01	18:02	Others
TOG 8	17.32	16.5	1.5	69.2	11.2	1.1
TOG 11	16.19	17.2	1.1	67.2	13.2	0.75
TOG 16	17.10	11.6	2.3	70.2	14.2	1.6
TOG 17	13.67	11.5	1.8	69.1	15.2	2.3
TOG 20	15.63	13.4	1.9	72.5	10.5	1.5
TOG 22	22.87	10.5	0.9	75.2	12.5	0.8
TOG 24	19.49	7.8	2.4	70.2	17.2	2.4
TOG 25	13.63	14.8	2.7	62.8	16.1	3.5
TOG 26	16.09	13.5	1.1	68.5	14.8	2.1
TOG 29	12.19	13.5	2.5	64.2	18.2	1.3
TOG 38	14.36	9.2	0.7	75.9	13.0	1.1
Average	16.23	12.68	1.72	69.55	14.19	1.68
Maximum	22.87	17.2	2.7	75.9	18.2	3.5
Minimum	12.19	7.8	0.7	62.8	10.5	0.75

- In terms of mechanical harvesting, some genotypes had a strong trunk and a low fruit retention force (FRF). Some trees with very small canopies may be suitable for dense groves.

- The nature of the pollination behavior of the genotypes has not been yet evaluated, and this needs to be done.

- Of the 35 genotypes planted, just one, TOG 19, died after 2 years; the remainder are continuing to grow successfully.

Oil quality evaluation

The data regarding the percentage oil recovery (the percentage of oil obtained from the fresh fruit) from 11 randomly selected genotypes of TOGs in the 2006 season showed an average level of 16.23 percent (range 12.87 percent to 12.19 percent). Similarly, the fatty acid profiles of the same genotypes also showed strong variation among the genotypes. The highest oleic acid composition was found in TOG 22, and the lowest percentage was observed in TOG 25 (Table 7.11). Study of the composition of the TAG profile among the 11 selected genotypes also

Table 7.12: Major triacylglycerols (TAGs) of the Turkmenistan olive genotypes (TOGs) cultivated in the Negev Desert with moderately saline irrigation (4.2 dS/m) as determined by MALDI-TOF/MS (2006)

Genotypes	Triacyl glyceride profiles (% total)								
	PPL	PPO	LLP	POL	OOP	LLO	OOL	OOO	OOS
TOG 8	3.48	5.40	4.30	9.00	12.67	5.77	14.90	38.00	6.40
TOG 11	3.50	5.90	4.20	6.80	14.90	3.80	10.50	41.70	8.80
TOG 16	nd	5.20	nd	4.20	14.10	nd	12.80	52.80	11.60
TOG17	2.50	3.90	4.10	7.70	9.60	7.50	18.10	39.50	7.50
TOG 20	3.22	5.14	3.96	8.94	12.39	4.91	15.41	38.44	7.59
TOG 22	2.30	4.90	1.70	2.10	13.70	2.50	8.30	51.80	12.40
TOG 24	1.90	3.30	2.50	5.20	9.50	5.20	17.20	46.40	8.60
TOG 25	5.20	60	7.00	12.00	10.50	7.30	14.80	32.20	4.70
TOG 26	2.90	5.66	2.90	6.29	16.35	3.20	10.90	42.86	8.86
TOG 29	3.20	4.40	5.40	8.90	9.20	8.20	17.70	36.50	6.17
TOG 38	nd	4.81	nd	1.60	15.50	2.36	8.06	52.03	15.50
Average	3.13	4.96	4.01	6.61	12.58	5.07	13.52	42.93	8.92
Maximum	5.20	6.00	7.00	12.00	16.35	8.20	18.10	52.80	15.50
Minimum	1.90	3.30	1.70	1.60	9.20	2.36	8.06	32.20	4.70

nd, not detected.

showed a variation among the genotypes (Table 7.12). Except for TOG 16 and TOG 38, in all the tested genotypes nine TAGs (PPL, PPO, LLP, POL, OOP, LLO, OOL, OOO and OOS) were found. The highest TAG was OOO in all TOGs (range 32.2 percent to 52.8 percent; average 42.93 percent of the total).

7.5 Summary

The data from the present study suggest that the main differences found among all tested olive varieties are mainly due to the genetic factors built into each variety. Environmental factors are far less dominant in these differences. Based on this conclusion, the possibility of selecting and isolating superior olive genotypes that may provide optimal quantitative and qualitative characteristics in terms of their yield and oil quality in desert conditions is only a matter of time and determination.

The present study shows that all the 62 tested olive varieties (4 local, 12 European, 12 Spanish and 34 from Turkmenistan) can grow and produce a relatively high olive yield in the Israeli Negev Desert using intensive cultivation with specifically engineered and developed technologies providing proper saline drip irrigation. Based on the data obtained, Barnea trees seem to produce the highest olive yield in comparison to the other local varieties tested (Souri and Maalot). The oil quality aspects will be described and discussed in detail in Chapter 11. Regarding the European varieties, Picholine was found to be the most productive, followed by Picual and Koroneiki, and many other that were found to be similar in the Negev Desert environment. Of the varieties originating in Spanish dryland regions, Okal, followed by Spanish Leccino, Palomar, Iumed and Blanqueta, were found to be better than the other varieties when tested in Negev conditions. Regarding olive genotypes from Central Asia, based on both horticultural and oil quality evaluation TOG 22 seems the most suitable for Negev cultivation, followed by TOG 24, TOG 8 and TOG 16; however, the pollination of these varieties has still to be studied.

References

Amane, A., Lumaret, R., Hany, V., Quazzani, N., Debain, C., Vivier, G., & Deguilloux, M. F. (1999). Chloroplast-DNA variation in cultivated and wild olive (*Olea europaea* L). *Theor Appl Genet*, 99, 133–139.

Angiolillo, A., Mencuccini, M., & Baldoni, L. (1999). Olive genetic diversity assessed using amplified fragment length polymorphisms. *Theor Appl Genet*, 98, 411–421.

Aragues, R. J., Puy, A., Royo, A., & Spada, J. L. (2005). Three-year field response of young olive trees (*Olea europea* L., cv. Arbequina) to soil salinity: trunk growth and leaf ion accumulation. *Plant Soil*, 271, 265–273.

Barranco, D., Cimato, A., Fiorino, P., Rallo, L., Touzani, A., Castaneda, X., Serafin, F., & Trujillo, I. (2000). *World Catalogue of Olive Varieties*. Consejo Oleicola Internacional, Madrid.

Beltran, J. M. (1999). Irrigation with saline water: benefits and environmental impact. *Agric Water Manag*, 40, 183–194.

Bernstein, L. (1964). Effects of salinity on mineral composition and growth of plants. *Plant Anal Fertil Prob*, 4, 25–45.

Besnard, G., & Berville, A. (2000). Multiple origins for Mediterranean olive (*Olea europaea* L. ssp *europaea*) based upon mitochondrial DNA polymorphisms. *CR Acad Sci Paris Sci de la Vie*, 323, 173–181.

Besnard, G., Baradat, P., Chevalier, D., Tagmount, A., & Berville, A. (2001). Genetic differentiation in the olive complex (*Olea europaea*) revealed by RAPDs and RFLPs in the rRNA genes. *Genet Res Crop Evol*, 48, 165–182.

Besnard, G., Khadari, B., Baradat, P., & Berville, A. (2002). *Olea europaea* (Oleaceae) phylogeography base on chloroplast DNA polymorphism. *Theor Appl Venet*, 104, 1353–1361.

Burgher, K. L., Jamieson, A. R., & Lu, X. (2002). Genetic relationship among lowbush Blueberry genotypes as determined by randomly amplified polymorphism DNA analysis. *J Am Soc Hort Sci*, 127, 98–103.

Cresti, M., Linskens, H. F., Mulcahy, D. L., Bush, S., Di Stilio Xu, M. Y., Vignani, R., & Cimato, A. (1997). Preliminary communication about the identification of DNA in leaves and in olive oil of *Olea europeaae*. *Olivae*, 69, 36–37.

Fabri, A., Hormaza, J. I., & Polito, V. S. (1995). Random amplified polymorphic DNA analysis of olive (*Olea europaea* L.) cultivars. *J Am Soc Hort Sci*, 120, 538–542.

Food and Agriculture Organization (FAO) (1985). Water quality for agriculture. Rome: FAO Irrigation and Drainage Paper 29, 174.

Green, P. S. (2002). A revision of Olea L. (Oleaceae). *Kew Bull*, 57, 91–140.

Green, P. S., & Wickens, G. E. (1989). The Olea europaea complex. In: K. Tan (Ed.), *The Davis and Hedge Festschrift* (pp. 287–299). Edinburgh University Press, Edinburgh.

Gucci, R., & Tattini, M. (1997). Salinity tolerance in olives. *Hort Rev*, 21, 177–214.

Hatzopoulos, P., Banilas, G., Glannoulia, K., Gazis, F., Nikoloudakis, N., Milionin, D., & Haralampidis, K. (2002). Breeding, molecular markers and molecular biology of the olive tree. *Eur J Lipid Technol*, 104, 574–586.

Herrera, C. (1995). Plant-vertebrate seed dispersal system in the Mediterranean: ecological, evolutionary and historical determinants. *Annu Rev Ecol Syst*, 26, 705–727.

Klein I, Ben Tal Y, Lavee S, David I (1994). *Olive Irrigation with Saline Water* [in Hebrew]. Beg-Dagan: Volcani Center Report.

Lipschitz, N., Gophna, R., Hartman, M., & Biger, G. (1991). The beginning of the olive (*Olea europaea* L.) Cultivation in the old world: a reassessment. *J Archaeol Sci*, 18, 441–453.

Maas, E. V., & Hoffman, G. J. (1977). Crop salt tolerance – current assessment. *J Irrig Drainage*, 103, 115–134.

Omrani-Sabbaghi, A., Shahriari, M., Falahati-Anbaran, M., Mohammadi, S. A., Nankali, A., Mardi, M., & Ghareyazie, B. (2007). Microsatellite makers based assessment of genetic diversity in Iranian olive (Olea europaea L.). *Sci Hort*, 112, 439–447.

Rallo, P., Dorado, G., & Martin, A. (2000). Development of simple sequence repeats (SSRs) in olive tree (*Olea europaea* L.). *Theor Appl Genet*, 101, 984–989.

Rugini, E., & Pesce, P. G. (2006). Genetic improvement of olive. *Pomologia Croatica*, 12, 43–74.

Shalhevet, J. (1994). Using water of marginal water quality for crop production. *Agric Water Manag*, 25, 233–269.

Stambuk, S., Sutlovic, D., Bakaric, P., Petricevic, S., & Simun, A. (2007). Forensic botany: potential usefulness of microsatellite-based genotyping of Croatian olive (*Olea europaea* L.) in forensic casework. *Croatian Med J*, 48, 556–562.

Vargas, P., Munzoz, G. F., Hess, J., & Kadereit, J. (2001). Olea europaea subsp. Guanchica and subsp maroccana (oleaceae), two new names for olive tree relatives. *Ann Jard Bot Madrid*, 58, 360–361.

Vergari, G., Patumi, M., & Gontanazza, G. (1996). Use of RAPD markers in the characterization of olive germplasm. *Olivae*, 60, 19–22.

Weissbein, S. (2006). Characterization of new olive (*Olea europae* L) varieties' response to irrigation with saline water in the Ramat Negev area. MSc Thesis, Alber Katz International School for Desert Studies, Ben-Gurion University of the Negev, Israel.

Wiesman, Z., Avidan, N., & Lavee, S. (1998). Quebedeaux: molecular characterization of common olive varieties in Israel and the West Bank using random amplified polymorphic DNA (RAPD) markers. *J Am Soc Hort Sci*, 123, 837–841.

Wiesman, Z., De Malach, Y., & David, I. (2002). Olive and saline water – story of success. *Intl Water Irrig*, 22, 18–21.

Wiesman, Z., Itzhak, D., & Ben Dom, N. (2004). Optimization of saline water level for sustainable Barnea olive and oil production in desert conditions. *Sci Hort*, 100, 257–266.

Yamagishi, M., Matsumoto, S., Nakatsuka, A., & Itamura, H. (2005). Identification of persimmon (*Diospyros kaki*) cultivars and phonetic relationships between *Diospyros* species by more effective RAPD analysis. *Sci Hort*, 125, 283–290.

Zohary, D., & Hopf, M. (1993). *Domestication of Plants in the Old World*. Oxford University Press, Oxford.

Zohary, D., & Spiegel-Roy, P. (1975). Beginnings of fruit growing in the Old World. *Science*, 187, 319–327.

Further reading

Burtolucci, P., Dhakal, B.R. (1999). *Prospects for Olive Growing in Nepal*. Kathmandu: Field Report-1. FAO TCP/NEP/6713.

Lavee, S., Avidan, N., & Wiesman, Z. (1999). Genetic variation within the Nabali Baladi cultivar of the West Bank. *Acta Hort*, 474, 129–132.

Soleimani, A., Zamani, Z., Talaei, A. R., & Naghavi, M. R. (2006). Molecular characterization of unknown potentially salt tolerant olive genotypes using RAPD markers. *J Sciences (Islamic Republic of Iran)*, 17, 107–112.

Olive-harvesting biotechnologies

One of the main limiting factors for cost-effective olive oil cultivation is the efficiency of olive harvesting. Although the cost of harvesting depends on the cost of labor, the type of grove and its density, the yield per hectare and, above all, the degree of mechanization, for most olive growers harvesting accounts for 50 percent or more of the annual production costs (Mili, 2004). This cost is even more significant in large-scale desert olive oil cultivation. The efficiency of the harvest – the percentage of fruit removed from the total crop on the tree – is the first component contributing to total processed product value, while the quality of the fruit, which is partially a function of maturity regarding oil, size and condition when delivered to the processing facility, is the second. Harvesting is the final step in field production of the olive crop, but if it is performed at the wrong time or in the wrong way it can have a marked effect on the net return to the grower (Ferguson, 2006). Therefore, since the 1990s our group has tried to address this topic using the most advanced biotechnological means available. Some of the most relevant issues that we have addressed include:

- development of advanced biotechnologies for the prediction and determination of the optimal harvest time for olives;

- development of an advanced control delivery system for reducing the fruit retention force (FRF) to a level suitable for mechanical olive-harvesting systems.

Since olive fruits, their oil content and its quality, and the consequences for the trees are closely interconnected with each other, the following criteria should always be taken consideration before starting olive harvesting:

- Fruits used for oil extraction must contain the maximum oil percentage and must be of the best quality

- Fruits must be at the right stage of ripening required for their use as green or as black olives; this has a significant effect on the quality of the olive oil

- The olive tree must be subject to as little damage as possible

- The overall cost of harvesting must meet the economic parameters for olive oil production.

8.1 Olive fruit maturation

8.1.1 Olive oil yield, quality and fruit maturation

As olive ripening proceeds, a number of changes occur in the fruit – changes in weight, pulp/stone ratio and color, as well as in chemical composition, oil accumulation and enzyme activity, which influence the fruit's firmness and the olive oil chemical composition and sensory characteristics (Beltran *et al.*, 2004). In general, the oil content of olives increases as the fruit grows and matures, then plateaus when the fruit has reached its maximum size, slows as full ripeness approaches, and declines slightly as the fruit becomes over-ripe (Salvador *et al.*, 2001; Beltran *et al.*, 2004). Oil quality improvement in olive fruits is initially concurrent with increasing oil content, but peaks and begins to decline before the maximum oil yield is reached. The fruit detachment force declines steadily as the fruit ripens and drops sharply when the fruit reaches full ripeness and fruit drop increases. The color of the olive changes from lime green to pale 'straw' green as the fruit matures, then from purple to black as it becomes fully ripe (Salvador *et al.*, 2004) (Figure 8.1).

Figure 8.1: The wide variation in shades of olive fruits.

The main determinants of olive oil flavor are maturity, variety, cleanliness, and time to milling. Olives picked a week apart from the same tree can have strikingly different flavor profiles. Like any tree fruit, more mature olives will have sweeter and fruitier notes through the development of alcohols, esters and aldehydes. Oil from mature fruit can often be consumed immediately without racking. Some varieties can produce bland oil if picked too late.

Greener fruits may produce oil with more antioxidants in the form of polyphenols, and thus have a longer shelf-life, but green fruit tends to make more bitter, peppery oil that needs several months of storage before it becomes palatable. Green olives also drastically lower the oil yield compared with mature fruit. Thus fruit maturity influences not only the organoleptic characteristics of the oil and the ultimate oil stability, but also the oil yield (Caponio and Gomes, 2001). However, fruit maturation depends on temperature, sunlight and irrigation. A hot fall can cause fruit to ripen quickly, resulting in a narrow window for optimum picking, while a cool fall may result in green fruit hanging on the tree well into winter. Some farmers are forced to pick green fruit to hedge against frost damage (Mailer *et al.*, 2005).

8.1.2 *Olive fruit pigmentation and maturation*

The olive fruit reaches maximum weight at the same time as it achieves total pigmentation; however, the oil extraction maturation index has to consider the oil percentage. Determination of the oil percentage is necessary to establish when the oil content has peaked. Although in the field it is very difficult to decide when to harvest olives, there is general agreement that high-quality oil requires precise timing of harvest of good-quality fruit (Ayton *et al.*, 2001). It is important to keep in mind that too long a delay in the harvest date may reduce the oil yield and its quality due to a greater proportion of fallen olives and inferior-quality oil, and lead to a further decrease of production the following year due to interference with the accumulation of nutrient reserves and with flower induction phenomena.

Since the oil content changes with environmental conditions, such as climate and irrigation, and the olive fruit undergoes changes during the ripening process, a fast and reliable method must be utilized to determine the best time to harvest olives for oil production. As a non-optimal harvesting date may reduce the yield of oil by as much as 5–10 percent, huge financial losses may accrue from an incorrect decision as to when to harvest. Growers have no simple analytical method to measure oil content, and must rely on experience, tradition, and subjective visual assessment. In this section we will discuss some common popular techniques for achieving optimal harvesting of olives, as well as an innovative, non-destructive method.

As the olive fruits ripen:

- the water content of the fruit decreases;

- the dry weight increases;

- the pulp/stone ratio increases;

- the epicarp's colors evolve from dark green to black, and a decrease in chlorophylls and carotenoids is noted;

- the chemical composition of the fruit changes – there is a decrease in sugars, decomposition of oleorupein, accumulation of aromatic substances, and changes in fatty acid composition;

- oil accumulation increases to a maximum and then plateaus;

- enzyme activity increases as the number of proteolytic enzymes and proteolytic activity rise, influencing fruit firmness (Esti *et al.*, 1998; Ranalli *et al.*, 1998; Ryan *et al.*, 1999; Motilva *et al.*, 2000; Cossignani *et al.*, 2001; Garcia and Yousfi, 2005).

In order to achieve the highest profit from a given crop, it is essential to harvest the fruit while it has the highest amount of oil of the best quality (in terms of aroma and body). Different approaches have been implemented in order to establish the best time to harvest, all of which are subjective and dependent upon intuitive criteria (Ranalli *et al.*, 1998; Beltran *et al.*, 2004) – for example:

- the exact number of days after the first rain;

- a fixed date on the calendar;

- spontaneous fruit dropping;

- fruit detachment force;

- external signs of maturation.

These parameters are not reliable, since the development and ripening process of olive fruit changes with the cultivar and the environmental conditions (see Salvador, 2001; Beltran *et al.*, 2004), and thus differs for each growing area and yield.

The most popular method today to determine the optimal harvesting date, developed and used by the International Olive Oil Council (IOOC), is the maturity index (MI), which is based on the pigmentation of the olive fruit. In this method, 100 olives chosen randomly from 1 kg of newly harvested fruit are used to calculate the MI, giving a value ranging from 0 to 7. For example:

$$\text{Maturation index} = [(0 \times n_0) + (1 \times n_1) + (2 \times n_2) + (3 \times n_3) \\ + (4 \times n_4) + (5 \times n_5) + (6 \times n_6) + (7 \times n_7)]/100$$

where n_0, n_1, n_2,......n_7 are the number of olives belonging to each of the following eight categories:

0 = deep or dark green skin

1 = yellow or yellowish-green skin

2 = yellowish-green skin with reddish spots

3 = reddish or light violet skin

4 = black skin, and completely green flesh

5 = black skin, and flesh completely violet halfway to the pit (stone)

6 = black skin, and flesh violet almost to the pit (stone)

7 = black skin, and dark flesh throughout.

The optimum time of harvest for olive oil for most varieties is when the maturity index is 5. Olives at various stages of ripeness, based on their maturation index are illustrated in Figure 8.2. However, again, this method is not completely reliable, since it relies on subjective visual assessment. The maturity index can be used only as a guideline, and cannot determine the precise harvesting period for each cultivar.

Numerous researchers have tried alternative methods of estimating an accurate harvesting index, all utilizing differences evoked in drupes during maturation, assuming a correlation with the oil content exists:

- Solinas *et al.* (1975) studied the phenolic fraction of the olive pulp by a spectrometric method

- Vlahov (1976) proposed a ripening index based on the malic/citric acid ratio of drupes

- Esti *et al.* (1998) suggested a method based on surveillance of minor components using HPLC

- Ranalli *et al.* (1998) suggested a ripening index in accordance with the respiration rate curve of olives that coincides with the climacteric phase of drupes

- Mickelbart and James (2003) developed a method to determine optimal harvest time for mill olives by calculating their dry matter

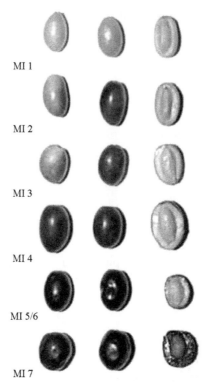

MI 1

MI 2

MI 3

MI 4

MI 5/6

MI 7

Figure 8.2: Olive fruits at various stage of ripeness based on the maturity index (MI). Adapted from Mailer *et al.* (2005).

- Garcia and Yousfi (2005) suggested a non-destructive method based on the firmness of the fruit, assessed using a hand densitometer, and the skin color, using a portable colorimeter.

8.2 New technologies for olive-harvesting automation and management

8.2.1 Prediction of optimal olive-harvest timing
Factors affecting olive-harvest timing

The olive oil industry is one of the fastest growing in the world, and annual revenue is estimated to be over €10 billion worldwide (Luchetti, 2002). This is leading to a need for more efficient production. A major potential area for improvement is in the average amount of oil extracted per unit of olives. The amount of oil that can be extracted from olives is dependent on many characteristics, including the cultivar, irrigation, soil and climate (Vossen, 2004). Most of these parameters are hard to change once the trees have been planted; however, a significant factor that influences the quantity of oil that can be extracted is the timing of harvesting (Salvador *et al.*, 2001). Today, agriculturists decide on the harvesting time mostly based on their experience (Salvador *et al.*, 2001).

Various methods have been proposed for expressing the stage of maturity of olives. Among them, the International Olive Oil Council has suggested a simple technique based on the assessment of the color of skins of 100 olives randomly drawn from 1 kg of the sample, as described above (the maturity index). The first stage of ripening is known as the "green stage," corresponding to green mature fruits that have reached their final size. Following this, chlorophyll pigments in the olive skin are progressively replaced by anthocyanines during ripening, which makes it possible to identify a "spotted stage," a "purple stage" and a "black stage," according to the skin color of the fruits (Salvador *et al.*, 2001). As the olive fruits grow throughout the season, the polyphenol content gradually increases and reaches a maximum just as the skin begins to change color (Vossen, 2004). As the fruit matures and the color penetrates all the way to the pit, the polyphenol content and most of the other flavor components of the fruit decline very rapidly over a period of about 2–5 weeks. Oil quality, therefore, is very strongly tied to fruit maturity (Vossen, 2004).

Oil synthesis and accumulation in the olive fruit occurs over about 34 weeks, beginning about 10 weeks into the season, and increases fairly rapidly until fruit maturity (color change and softening); the rate of accumulation then gradually tapers off (Morello *et al.*, 2006). It appears that there is a much larger increase than is actually the case late in the fruit-ripening season due to the loss of moisture in the fruit. When the fruit becomes very over-ripe, oil synthesis stops completely (Morello *et al.*, 2006).

The ultimate flavor of any variety can be completely changed by harvesting the fruit green (unripe) or mature (ripe) (Vossen, 2004). The subtleties in between those two extremes can have a significant influence on the style of the oil produced (Vossen, 2004). Some producers believe that maturity can have an even greater influence on quality than the variety itself (Salvador *et al.*, 2001).

The climate, elevation, irrigation, soil water-holding capacity, soil mineral content, sunlight intensity, rainfall, etc., undoubtedly have an effect on the ultimate quality of the oil (Mailer *et al.*, 2005), but these things are very difficult to pin down. It is suggested that the influence of territory is minimal compared to that of variety and fruit maturity (Vossen, 2004). Many experiments have shown a strong connection between the color of the fruit and the amount and quality (Figure 8.3) of the oil that can be extracted from it (Salvador *et al.*, 2001); this is because the color of the fruit is an indicator of fruit ripeness, and the oil content and quality are directly related to its maturity (Figure 8.4).

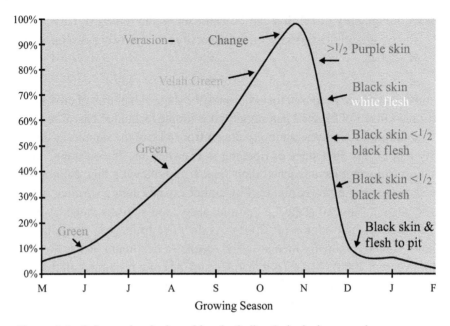

Figure 8.3: Color and polyphenol level of olive fruit during growing season.
Source: Vossen (2004).

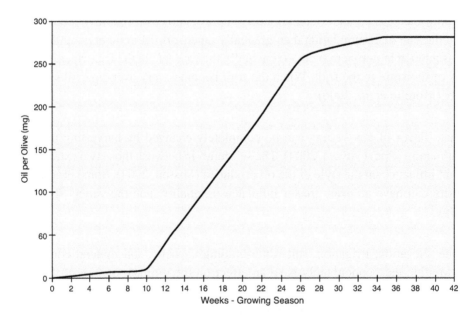

Figure 8.4: Olive oil percentage vs. number of growing weeks.
Source: Vossen (2004).

Computer vision technology for the prediction of optimum olive-harvest timing

Computer vision technology is a well-known and useful tool for tasks such as determining colors, sizes, shapes (Diaz *et al.*, 2000), and has been trialed to assess its ability to determine accurately and repetitively the physical characteristics of the olive fruit. The imaging features serve as input for a model that predicts the amount of oil that can be extracted by harvesting the orchard on a certain day. However, experience has shown a complex connection between the physical characteristics of the olive and the percentage of oil it contains (Carvajal and Nebot, 1998). Simple linear regression failed to explain the majority of the variance; therefore, the model is not valid (Carvajal and Nebot, 1998). As has been determined in many biological-related areas, only intelligent models can explain such complex connections (Carvajal and Nebot, 1998).

Machine vision

Machine vision technology has been widely used in agriculture to define many characteristics of fruits, vegetables, meat and fish (Zheng *et al.*, 2006). The information that can be gained by image-processing techniques can be categorized into color, size, shape and texture (Zheng *et al.*, 2006). Thygesen and colleagues (2001) used such techniques to predict the chemical compounds of potatoes, while Laykin and colleagues (2002) used machine vision technology to classify tomatoes by their color, defects and color homogeneity.

Diaz *et al.* (2004) have shown that machine vision technology has a significant advantage over assessment with the human eye when classifying table olives. Olive experts have made huge errors when separating the best-quality olives by eye because, in cases of doubt, they prefer to assign them to a lower category (Diaz *et al.*, 2000). The results using machine vision technology are a little better when separating the first and fourth classes, but much worse in the intermediate categories due to the great number of olives per image (Diaz *et al.*, 2000).

The big advantage of computer vision technology is its objectivity and consistency over long periods, while human concentration is known to be limited to 400–700 runs, and the dynamic range of human perception is limited to less than 100 gray levels (Saito *et al.*, 2003).

Artificial intelligence

Agricultural systems are known to be complex, and in most cases it is impossible to model their behavior by simple statistical means. Complex behaviors require models based on artificial intelligence. Carvajal and Nebot (1998) built a growth model of white shrimps, based on fuzzy inductive reasoning (FIR), which showed major advantage over linear regression. Salehi and colleagues (1998) improved the prediction of dairy yield by using record classifier combined with an artificial neural network. The main advantage of artificial neural networks is that their flexible structure allows adaptation to the conditions of a classification problem with certain accuracy (Diaz *et al.*, 2004). With suitable structuring of the neural network and an optimum training process, an olive classification system can be obtained (Diaz *et al.*, 2004).

Chemo-optic system: a new and innovative technology for prediction of optimum olive-harvest timing

A team at the Ben-Gurion University of the Negev, Israel, has designed an innovative, non-destructive technique that can be implemented in an automatic classification system by utilizing a fast and efficient algorithm. This is done using machine vision technology to obtain images of olives, and correlating these with oil content measured by low resolution nuclear magnetic resonance (NMR) in order to decide when to harvest. In this system, a camera runs through the field and obtains images of olives, then downloads them to a computer program which, by using statistic analysis, decides on the best time for harvesting. Preliminary results have clearly indicated that, by using digital image processing and statistical tools, it is possible to quantify various parameters (including shape, size and color) and correlate them with oil content, and thus predict the best time for harvesting. This technique has been implemented successfully using images taken in a designated closed compartment. The team is now concentrating on a beta-site phase of the development, in which efforts are focused on demonstration of the system in the field. This innovative model can be used as a decision-making tool for agriculturists all over the world, and has major economic potential.

The technology is divided into three parts: image development (image processing, image acquisition and image analysis), oil content determination, and statistical analysis.

Image processing

Image processing or machine vision deals with the theory and technology of building artificial systems that obtain information from images or multidimensional data. This technology can replace human vision in a variety of industrial operations. It plays a key role in the development of intelligent systems. The principal goal when researching intelligent systems is to develop machines that have the ability to perceive, reason, move on, and learn from experiences.

Digital image processing is the use of computer algorithms to perform image processing on digital images. With the fast computers and signal processors now available, digital image processing has become the most common form of image processing, and is generally used because it is not only the most versatile method but also the cheapest.

Digital cameras generally include dedicated digital image-processing chips to convert the raw data from the image sensor into a color-corrected image in a standard image file format. Images from digital cameras often receive further processing to improve their quality. The digital image processing is typically done by special software programs that can manipulate the images in many ways.

A complete pattern-recognition system consists of a sensor that gathers the observations to be classified or described; a feature extraction mechanism that computes numeric or symbolic information from the observations; and a classification or description scheme that does the actual job of classifying or describing observations, relying on the extracted features. When a beam of light is reflected from or transmitted through an object, it registers into a sensor and the image is formed (Ballard and Brown, 1982). When a sensor collects an image, the image is broken up into discrete pieces of data known as pixels, each of which is identified by coordinates within the image of the sample and the brightness value associated with it. Each pixel on the screen can be represented in the computer or interface hardware (for example, a graphics card) as values of red, green and blue (the RGB color model). These values are converted into intensities, which are then used for display or for analyzing and processing the image.

A color in the RGB color model can be described by indicating how much of each of the red, green and blue color is included. Each can vary between the minimum (no color) and maximum (full intensity). If all the colors are at the minimum level, the result is black; if all the colors are at the maximum level, the result is white. By using an appropriate combination of red, green and blue intensities, many colors can be represented. While color output by a computer is based on these three primary colors and their admixture, the human visual sense of color relies on hue (H), intensity (I) and color purity or saturation (S), and thus a conversion is often made between the two. Limitations of the digitization process are defined by the resolution of the image (Baxes, 1988). Image resolution comprises spatial and brightness resolution. Spatial resolution is the number of pixels representing the entire image. As spatial resolution increases, an image usually gains more detail or looks more natural. Spatial resolution controls the minimum size of a feature, which can be detected using algorithms. Higher resolutions, however, increase the amount of data that must be processed, which slows image processing, segmentation, and classification.

Image acquisition

Images are taken inside a designated cell located in a dark room. A digital camera is located inside the cell, which enables quality photographs in unified conditions (Figure 8.5a). The image acquisition is controlled by the camera's parameters: top and bottom, lighting, and uniform distance between the lens and the olives. Images are acquired for each group of olives according to the following features:

- Resolution unit inch, Exposure time 1\1000 s, F-number f/12

- ISO speed ratings 100, Flash did not fire

- Pixel X dimension 4368, Pixel Y dimension 2912.

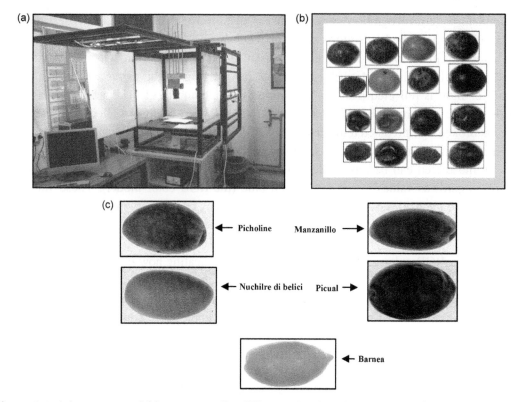

Figure 8.5: (a) Image acquisition system; (b) different classification of olive fruits; (c) olive variety identification based on image analysis in the chemo-optic system.

- Camera features:

  12.8 MP full frame CMOS sensor

  3 fps 60 JPEG image burst

  9-point AF with 6 Assist AF points

  2.5" LCD

  picture style image processing

  DIGIC II

  recording of RAW/JPEG images

  digital Photo Professional software

compact magnesium alloy body

connectivity options.

Image analysis

Images are analyzed using an algorithm based on 28 different features, including color, shape, symmetry, defects and others. After all the raw data have been analyzed and collected by the computer, the team examine the olives themselves and process their observations in order to verify the computer's analysis (Figure 8.5b).

8.2.2 Determination of oil content in olive fruits

Various methods have been developed to estimate the oil content in olive fruits, each with its own advantages and disadvantages. The oil content is defined as the whole of the substance extracted under the operational conditions specified, and is expressed as a percentage by mass of the product as received. In this section, we first discuss the standard techniques widely used around the world to measure oil content, before moving on to a recently developed, non-destructive technique.

Standard techniques

Mechanical method

In the mechanical method, oil is extracted from fruits by subjecting them to high mechanical pressure from a mill. An example is shown in Figure 8.6. This is a simple method that doesn't require much preparation, or highly trained personnel, and no solvents or dangerous materials

Figure 8.6: Olive oil content determination by mechanical method.
Source: http://www.albury.net.au/~jos/tandem%20press.jpg.

are used. On the other hand, it is not an efficient method; around 8–12 percent of the oil is left in the pomace, the machinery is expensive, and the method cannot be used for small amounts of fruit. Different applications have been developed to increase the yield, but this method is not directly used to determine oil content (Karleskind and Wolff, 1996).

Chemical method

In the chemical method, organic solvents such as hexane or petroleum ether are used to extract oil from a solid material in a continuous process. It is based on the fact that oil is soluble in a particular solvent while impurities are not. Two techniques based on this principle are discussed here:

- *Soxhlet apparatus*. This is a continuous process, where a dry ground sample is suspended in the apparatus and constantly washed by solvent. The solvent evaporates, and then condenses on meeting the cooling system. There it makes contact with the sample and extracts the oil. This is repeated for a few hours until the solvent has evaporated from its flask, and oil content can be evaluated (Figure 8.7a). The process is very accurate and easy to handle, but has a few disadvantages: there is poor extraction of polar lipids; it takes a long time; large volumes of solvents are required; and boiling solvents are hazardous.

- *Pressurized solvent extraction (PSE)*. This application describes a technique for the rapid extraction of oil from oilseeds using pressurized solvent extraction for significant time savings and reduced solvent consumption. Solvent is pumped into an extraction vessel containing the sample, and is heated and pressurized. The pressurized solvent at high temperature accelerates the extraction process by increasing the solubility of the analyte in the solvent and also increasing the kinetic rate of desorption of the analyte

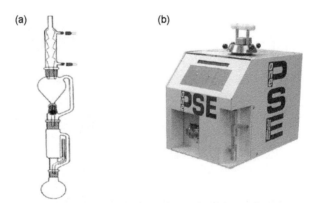

Figure 8.7: Solvent extraction system: (a) Soxhlet apparatus; (b) pressurized solvent extraction.

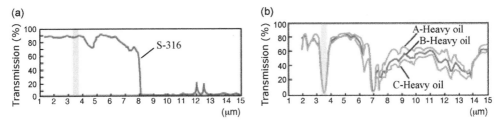

Figure 8.8: Solvent absorption spectrum (a) and hydrocarbons absorption spectrum (b) measured with HORIBA's OCMA-350 IR spectrometer.

from the sample matrix. The method has the same disadvantages as the soxhlet apparatus regarding solvents, but is less time-consuming and more efficient (Fig 8.7b).

IR spectroscopy

In this method, a sample is crushed and dissolved in a powerful solvent and then analyzed using IR absorbance – a non-dispersive infrared spectrophotometric technique that is specific to hydrocarbons such as oil. HORIBA's OCMA-350 IR spectrophotometer is popular and easy to handle for olive oil determination (HORIBA, 1995). All hydrocarbons, including oils, absorb infrared radiation at about 3.4–3.5 micrometers, while the solvent in use has no absorption band within this wavelength range. The OCMA-350 measures the position and strength of the absorption bands in the 3.4–3.5 micrometer range, therefore enabling qualitative and quantitative analysis of the substance, with no distortion of values due to the presence of the solvent (Figure 8.8). Because there is only a very narrow band that absorbs just oil, this method is very accurate; on the other hand, this accuracy is achieved only with advanced technical skills, since the amount of the sample measured is minute. In addition, the solvent used in this method is very powerful (has very high affinity to lipids) and efficient for the extraction process, but is considered to be harmful to humans and to the environment (which is true for tetrachloroethylene, but not for the original solvent).

Other methods

Other standard methods for the determination of oil content are also available (Matthaus and Bruhl, 2001), including supercritical fluid extraction (SFE), near infrared spectroscopy (NIR), accelerated solvent extraction (ASE), microwave-assisted extraction (MAE), solid fluid vortex extraction (fexIKA), Soxtherm 2000, and DGF standard method B-I.

Summary of standard methods

Most of the methods mentioned above have one feature in common – they are destructive, meaning that the fruits go through a number of processes before the oil content can be determined.

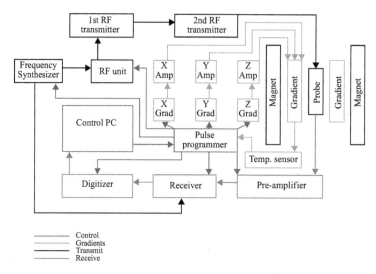

Figure 8.9: Linkage of the low resolution ^{1}H NMR.

In addition, most techniques are time-consuming, rely on the use of hazardous solvents, and depend on the skill and experience of laboratory personnel.

A novel, non-destructive technique

Low resolution nuclear magnetic resonance (NMR)

Low resolution NMR is based on the measurement of resonant radiofrequency absorption by non-zero nuclear spins in the presence of an external static magnetic field. NMR has none of the shortcomings of the other techniques described above. Our team is currently working to develop an innovative method for non-destructive determination of the oil content in olive fruit using low resolution NMR (MARAN ULTRA, 2006).

Low resolution NMR spectrometry is being used extensively in the food, polymer, petroleum and pharmaceuticals industries, to measure the oil and water content of samples that contain less than 10 percent moisture. The instrument employs a permanent magnet, with an operating frequency of between 1 and 65 MHz. It includes a case containing the magnet and the electronics, and is operated using a personal computer. The system's devices and linkage are shown in Figure 8.9. Samples are analyzed in standard NMR tubes, the diameter of which can range from 5 to 60 mm to facilitate the analysis of different materials.

Scientific principle, as explained by Nordon et al. (2001)

NMR spectra are generated by immersing chemicals in a constant homogenous magnetic field generated by a permanent magnet. Application of a variable RF pulse applied by a RF transmitter generates the signal, which is then sensed by means of a detector, recorded, and translated to solid fat content (SFC). The resonance signal of a proton in a chemical compound is shifted in the magnetic field. The external magnetic field is modulated by the secondary magnetic fields induced in the electrons near the proton. The amount of chemical shift is determined by the "electronic environment." Protons with identical electronic environments will show identical chemical shifts – i.e., signals at identical resonance frequencies. The electronic environment is determined by the chemical structure.

In NMR experiments, the sample is placed in a constant magnetic field along the Z-axis, and excited by an RF frequency in an orthogonal direction to the external field. Here, the alignment of the processing nuclei magnetization is tipped from the Z-axis to the XY-plane while absorbing energy. The moment the RF frequency desists, the energy absorbed is emitted until particles return to equilibrium in their ground state. This phenomenon can be monitored by the voltage induced across an RF receiver coil by the alternating magnetic flux through it.

Measurement of solid fat content by low resolution NMR (according to MARAN ULTRA, 2006)
The solid fat content method by NMR involves measuring the FID signal following a radiofrequency (RF) pulse. This signal decay has two components; a rapid decay due to the solid fat phase and a slow decay due to the liquid phase, as shown in Figure 8.10. By measuring the signal soon after the pulse, a signal is obtained due to both the solid and liquid phases; by measuring the signal at a later time, a signal is obtained solely from the liquid fat. These values are then used to determine the solid fat content of the sample.

For the SFC method, two data points are acquired on the FID; one centered at 11 μs following the RF pulse (solid signal, S + liquid signal, L), and the other centered at 70 μs following the

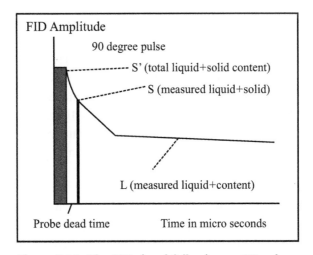

Figure 8.10: The FID signal following an RF pulse.

RF pulse (liquid signal only, L). The SFC is then given by the basic equation:

$$SFC = 1 - 100 \times \frac{L}{S + L}$$

where S is the signal amplitude due to the solid component and L is the signal amplitude from the liquid component.

As the solid signal decays rapidly during the 11 μs following the RF pulse, the total signal obtained is reduced from its original value (S + L). Any measurement of SFC would yield an artificially low value. To overcome this problem, a fixed ratio known as the *f* factor is determined by measuring a set of pre-defined synthetic standards. Once the *f* factor has been calibrated on the instrument, the corrected calculation of the SFC content is given by the equation:

$$SFC = 100 \times \frac{fS}{fS + L}.$$

NMR is not usually available as an absolute measurement technique – i.e., a sample cannot simply be placed in an NMR analyzer and a result for oil content obtained. To perform an NMR analysis, a set of calibration standards, which have a known concentration of the property of interest, must first be measured. A record of calibration is generated from the calibration standards. Users may then measure unknown samples and compare them to the calibration produced from the calibration standards. The procedures are illustrated in Figure 8.11.

The use of low resolution NMR to determine oil and water content has been applied by different researchers, and has proven to be an accurate, fast and simple technique. An International Standard (ISO 10565, 1998) has been established for the simultaneous determination of oil and water content in rapeseeds, soya beans, linseeds and sunflower seeds using pulsed NMR spectroscopy. Kuo *et al.* (2001) carried out an NMR study of water mobility in Pasta Filata and Non-Pasta Filata Mozzarella, while Pathaveerat *et al.* (2001) developed a method to evaluate avocado fruit quality using NMR. Gallo *et al.* (2003) used pulsed NMR for sugar crystallization studies. Lundby *et al.* (2006) used low resolution NMR techniques to determine fat, moisture and protein in fish, and Torbica *et al.* (2006) utilized pulsed NMR to characterize the advantages of solid fat content determination in cocoa butter and cocoa butter equivalents. These are only a few of the studies performed in the low resolution NMR arena.

Low resolution NMR and oil determination in olive fruits
Since NMR works better with a sample that contains less than 10 percent moisture, olives should be dried at a certain temperature in preparation for assessment of their oil content. A valid calibration curve should first be determined before measuring the oil content in olives (Figure 8.12). In order to ensure one calibration curve is sufficient for the entire season, samples

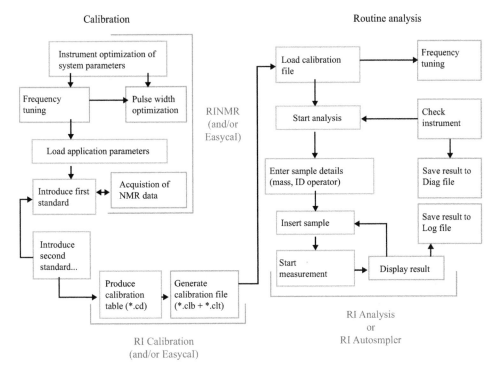

Calibration   Routine analysis

Figure 8.11: Calibration and SFF measurement procedures.

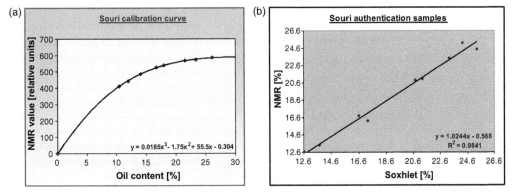

Figure 8.12: Calibration curve of the Souri olive fruit (a), and authentication (b) for the NMR.

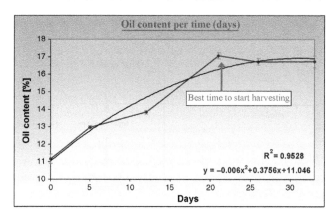

Figure 8.13: Development of oil content of Souri olives throughout 2006–2007 season.

extracted by soxhlet or other extraction methods can be run on GC to inspect the fatty acid profile.

Oil development in Souri olives throughout the season In order to measure oil content by NMR, our team had to produce a calibration curve that converts values obtained by NMR to oil content as measured by a different method. The study indicated that it is possible to determine oil content in olives using low resolution NMR, as long as samples are dried prior to measurement, and a temperature controlled environment is implemented at all times.

As part of the research, the team inspected the development of oil content in Souri olives throughout the entire season. The results are shown in Figure 8.13. Since NMR testing takes only 16 seconds per sample, the team was able to sample a large number of olives and draw an accurate ($R = 0.9528$) profile of oil development throughout the season, producing very similar findings to those of other researchers using other techniques. This shows what an extremely powerful tool the NMR is in terms of oil content evaluation for olives, allowing a great deal of sampling while still maintaining true results.

The feasibility of using Souri's calibration curve for the Barnea cultivar was also tested. Contrary to Souri's results, in the Barnea case a very low correlation factor ($R = 0.537$) was observed when matching soxhlet and NMR oil content. We also documented the fatty acid profile of Barnea, in order to compare it to that of Souri, and found no distinct differences between the two profiles.

The analysis and determination of optimal time After the picture has been analyzed, the information gathered can be interpreted in various ways in order to give every single object a "grade" and categorize it. Shahin and Symons (2001) also analyzed lentils for their color using the following three methods, and compared the three:

- *linear discriminant analysis* (LDA), which is a statistical method to find the linear combination of features that best separate two or more classes of object or event;

- *non-parametric statistics* (NPAR), which is a statistic method to deal with data without taking any assumption on its distribution;

- *Multilayered neural network* (MNN), which is an artificial neural network (ANN) (an information-processing paradigm that is inspired by the way biological nervous systems, such as the brain, process information). The key element of this paradigm is the novel structure of the information-processing system, which is composed of a large number of highly interconnected processing elements (neurons) working in unison to solve specific problems. ANNs, like people, learn by example. An ANN is configured for a specific application, such as pattern recognition or data classification, through a learning process.

According to the profile in Figure 8.13, the best time to harvest is around the twenty-second day after the beginning of sampling, at which stage the oil content is at its maximum value. This correlates with the color of the olives on the trees as inspected during the same season. As shown in Figure 8.14, on the same date where oil begins to peak (the twenty-second day after the beginning of sampling) the olives are mostly black (the fiftieth day from the beginning of the season). Statistical examination of the results regarding Souri cultivar shows a clear correlation between the color of the olives and their oil content. It was found, in accordance with previous research, that the darker the color of the olive, the higher its oil content.

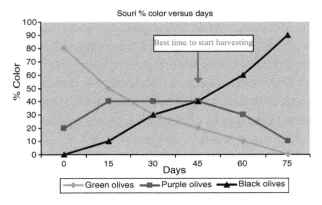

Figure 8.14: Determination of the optimum time for harvest of Souri olives, based on the color and oil content analysis.

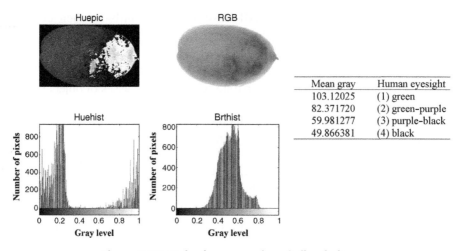

Figure 8.15: Color homogeneity of olive fruits.

The study also found a distinct connection between the mean gray parameter, which reflects the mean gray color of the olive according to the image taken, and the color grade given by researchers. Notice that the darker the color of the olive, the lower the mean gray (Figure 8.15).

A lot of work still has to be done; nevertheless, this research supports the possibility of a tool to assist in determining the ideal date for harvesting olives using computer vision and pattern recognition. Moreover, a clear connection can be seen between the visual properties and the oil content in the Souri cultivar. The results for the Souri cultivar were more satisfying than those for the Barnea, and this might be due to errors in the calibration curve for this cultivar.

8.3 Efficient mechanical harvesting

In many regions of the Mediterranean and most parts of the world, olives are still picked by hand, using wooden tools, or beaten from the trees with poles and caught in large nets (Figure 8.16). In some areas, collection of olive fruits that have fallen naturally to the ground or been blown down by wind and rain is also common. This method is preferred only when the trees are of great height and there is little labor available. The fruits are harvested gradually (at least once every two weeks), otherwise the quality of olive oil is greatly reduced. Another disadvantage is the prolonged harvest period (3–5 months). To obtain the best oil, some growers harvest olives just as they are changing color, indicating that almost all the oil has formed and that they are at peak flavor. Once collected, the olives are rushed to mills for same-day or, at most, next-day pressing – olives not pressed immediately begin to oxidize and ferment.

Damage to the tree is dependent on the method used to pick olives. In many regions, the traditional method is manual collection of the olives using wood sticks. As a result the yield is

Figure 8.16: Traditional methods of olive harvesting: **(a)** hand picking; **(b)** fruit weighing; **(c)** olive fruits ready to take to olive press; **(d)** olive collecting; **(e, f)** stick harvesting.

low, the trees are severely damaged, and the cost of the harvesting operation is high. Economic analysis of an average olive grove has shown that the traditional harvesting system costs as much as 80–90 percent of annual expenses.

In recent years, mechanical harvesting, using different types of tree shakers, has been intensively developed. Such technological developments have improved the efficacy of olive harvesting, and in some varieties up to ~80 percent of the olives can be removed from the tree by a mechanical shaker. However, in many varieties the efficacy of mechanical systems and the general cost-effectiveness is very low because of the severe damage caused to the harvested trees, and the slow rate of harvesting per tree. Most recently, new vine-type harvesters have been introduced to the olive industry. However, this new olive harvesting technology requires that high-density olive orchards be planted using dwarf varieties that are not yet readily available, and the problem of severe damage to the trees also needs to be addressed. It is suggested that the combination of engineering solutions together with chemical means of loosening the olive fruits' force of detachment may provide the optimal solution for mechanical harvesting of olives.

**Figure 8.17: (a) Hand-held pneumatic comb; (b) tractor-mounted shaker.
Source: TDC-Olive (2004).**

8.3.1 Available mechanical systems

Hand-held pneumatic comb

Using this system, a single operator can harvest 300–400 kg of olives in a day. A well-trained operator can pick 150–170 kg in about an hour if working in an intensive manner. The system uses a tool with raked teeth of two sizes to facilitate penetration into the crown of the tree. The combing action of the teeth harvests without damaging the fruit or trees. The tool consists of a pneumatic comb assembled on a variable-length 2.5 to 3 m telescopic rod (Figure 8.17a).

Tractor-mounted shaker

This is a vibrating-type olive picker. There are several models of vibrating-type harvesters. The most common device can be attached to an 85 hp tractor, and utilizes a hydraulic pump to transfer power to the vibrating head. Harvesting nets are first placed under the trees, and the operator then grasps the trunk with the harvester and vibrates it for 5–15 seconds, depending on the olive variety and tree design. The vibration works its way up the tree and the olives come raining down (Figure 8.17b).

Side-pass comb harvester

This is also called a Korvan harvester, and is a new type of harvester directed mainly towards table olives. It can also be used for harvesting other fruit crops, such as almonds, walnuts and grapes. It consists of a self-propelled platform with multiple rotating heads with long plastic fingers, and a catch frame and conveyor system (Figure 8.18a). It needs one operator to drive and keep the catch-frame skirts in contact with the trunk, and another to operate the rotating picking heads on swing arms. The machine moves at 0.5 mph, and can pick a tree in about 60 seconds (i.e., a rate of 1.5 ha/h).

(a)  (b)

Figure 8.18: (a) Side-pass comb harvestor; (b) straddle harvester.
Source: www.oliveoilsource.com.

Straddle harvesters
Straddle harvesters straddle the row, and are generally used for dwarf olive varieties. For efficient operation of this system, the olive trees must be planted at high density, treated with growth retardants and pruned into hedges (Figure 8.18b).

This long-term research and developmental work, carried out over more than a decade, into the mechanical harvesting of olives has led to significant progress in the field of olive harvesting technologies. However, to increase the cost-effectiveness of mechanical harvesting, the machine still needs to address two common problems:

- substantial fruit damage;

- poor fruit removal efficiency.

8.3.2 Factors affecting mechanical harvesting

The ability to harvest successfully using mechanical harvesting technology is affected by horticultural, environmental and physiological factors.

Horticultural and environmental factors
Tree shape, canopy density and pruning
The tree trunk and branch architecture, and the height, shape and density of the canopy of the tree markedly affect mechanical harvesting efficiency (Ferguson, 2006). For shaker-type olive harvesters, the most important requirement is a trunk that is straight for the first meter from the ground. This section of trunk must be free of branches to allow the harvester's head to grip the trunk securely, without any obstruction, thus allowing the harvester to work more quickly while also avoiding damage to the tree.

The canopy shape, width, length and density also affect mechanical harvesting (Tombesi *et al.*, 2002). Proper design of the olive tree structure during the first two years after planting is essential. Initially, when the tree is only 30–60 cm tall, any branches that begin to grow from the trunk less than 30 cm from the ground must be removed. Branches more than 30 cm from the ground can be left to grow with moderate pruning. This clearing will make it easier for weed spraying, and will also allow the tree to focus its growth into the main leader trunk and some higher lateral branches. In the early years it is important to leave as much growth as possible on the tree, because foliage promotes root growth, which in turn promotes the production of more foliage.

Trees of some olive varieties usually develop a straight central trunk with small side branches. Others may head straight for the sky as a single trunk with no side branches. Both cases are acceptable, but with a single trunk the growing tip must be nipped off at about 1.25 m to encourage side or lateral branches to grow at this point. It is these lateral branches that will form the main structure of the mature tree. Mature olive trees need to be kept reasonably open in the center to allow light penetration, for better tree health and fruit production. This is best achieved through a vase-shaped, sturdy growth habit which also facilitates mechanical harvesting. Olive trees should have quite a number of lateral branches at about 1 m or so from the ground when the tree is in its second growing year. Four evenly spaced lateral branches should then be chosen; these do not all need to come from exactly the same height, but should be no lower than 90 cm from the ground. As these branches form the vase framework of the tree, if possible they should be growing at at least 30 degrees to the horizontal. This will lead to a vase- rather than a flat plate-shaped tree structure.

When the tree is about 1 m tall, and if it has plenty of leafy branches towards its top, any branches growing from the trunk up to 30–60 cm from the ground should be removed. A well-shaped olive tree should have a straight, clean trunk to 60 cm, and a good number of branches above 60 cm from the ground.

For straddle harvesters, dwarf and/or growth-retarded olive varieties are preferred. There is no need for special trunk design, as there is with mechanical shakers. Owing to the fact that today only very limited numbers of naturally dwarf olive varieties that are also productive are available, the second option, of growth control and retardation, is more common. Proper irrigation and fertilization methodologies that are specifically developed for the local environmental conditions are very important in keeping the olive trees as small as possible. This issue is even more complicated when dealing with olive cultivation under desert conditions. As explained in Chapter 6, for olive oil production under intensive cultivation conditions in the desert environment, the high rate of transpiration means that efficient irrigation is essential, and the use of saline water means that it is important to leach the salt accumulated in the soil, which requires large volumes of water. From the large amount of experimental data accumulated in recent years (Weissbein,

2006; Weissbein *et al.*, 2008) it is clear that, generally, after a few years' intensive cultivation in desert conditions, most common olive varieties tend to produce tall trees – including Arbequina, Koroneiki and others that are accepted as producing small canopies under extensive cultivation. This pattern is also stimulated in high-density olive plantations. Based on many trials carried out using various olive varieties cultivated in relatively low-spaced plots, after a few years all the trees develop tall foliage while looking for light. To overcome this problem, it is important to systematically top and side-prune every year. In large olive plantations, this pruning practice should be carried out by mechanically pruning. However, in many cases pruning may not be enough, and recently it has been demonstrated that additional treatment with anti-gibberellin growth-retardants, such as Paclobutrazol and Majic, may significantly contribute to the efforts to keep the olive trees at a moderate size and even increase their productivity. There is an ongoing study regarding the concentrations and long-term effects of these growth retardants, but at this stage results are looking promising.

Orchard density

Plot spacing is one of the most important factors in the planning of olive plantations designed for mechanical harvesting. Generally, based on many experts' experience with olive shakers, gained over a number of years, most of the common recommended plot spacings of less than 8×5 m between the rows and trees, respectively, are far too close for efficient mechanical harvesting, and this is true for any olive mechanical shaking harvester.

Plot spacing for a straddle harvester should be about 7 m between the rows and 3 m between the trees. Even when designing a high-density olive plot, it is important that there is space between the rows of the tree to ensure that the machine can pass through easily. In the earlier years of most olive varieties it is less important, but later on, when the trees are older and larger, the spacing may become a limiting factor for mechanical harvesting. Leaving enough space may extend the lifetime of such high-density plantations.

Cultivar

Olive cultivars vary greatly regarding fruit weight and fruit removal force (FRF) and thus the ratio between the two is one of the factors that affects the mechanical harvesting of olive fruits. Though in all cultivars there is a decline of this ratio towards fruit maturity, the rate varies significantly among cultivars (Fridley, 1971; Kaurabi *et al.*, 2004). The fruit weight and FRF can be measured directly and the ratio calculated; however, the color development of the olive fruits as they mature has the greatest significance. Although many studies have been made in this sector, mechanical harvesting technology is still being developed and has not yet become the major factor in cultivar selection (Ferguson, 2006).

Physiological factors

It is well established that in advanced mature and senescent reproductive olive tissues, such as flowers and fruits, an abscission zone is formed (Weis *et al.*, 1991). This abscission process is regulated by the main plant hormones, including auxin, gibberellins, cytokinin, abscissic acid and ethylene (Goren, 1993; Huberman *et al.*, 1997; Bangerth, 2004). Ethylene is well-recognized as dominating abscission zone formation (Bleecker and Kende, 2000).

Using molecular biology, the biochemical pathways that are involved in this process have been studied intensively in recent years (Lashbrook *et al.*, 1998). It is clear today that ethylene induces the synthesis of cellulose enzymes as well as regulating the activity of a list of additional genes and enzymes involved in cell wall degradation, such as pectinases, extensions, expansins and others (Bleecker and Kende, 2000; Sharova, 2007). All these enzymes that are affected by ethylene are involved in the control of the viability, senescence and degradation of the olive-fruit stem tissue. Interestingly, the stem abscission zone is formed specifically in the stems attached to mature and senescent olives. It is suggested that this specificity is due to the fact that mature tissue is much more sensitive to ethylene than young tissue. The specificity is maintained due to the extremely low levels of ethylene in the plant tissues and a delicate biochemical sensing system. In recent studies, two genes revealed that ethylene sensing involves a protein kinase cascade and have been isolated and characterized (Kieber, 1997). One of these genes encodes a protein similar to the ubiquitous Raf family of Ser/Thr protein kinases; the second shows similarity to the prokaryotic two-component histidine kinases, and most likely encodes an ethylene receptor. Additional elements involved in ethylene signaling have also been identified genetically.

Most ethylene-related and -regulated enzymes are known to respond strongly to a wide range of biotic and aboitic stress events (Bleecker and Kende, 2000). Desert environmental conditions such as heat, low humidity, high irradiance, drought and salinity directly affect the production of ethylene and all other related biochemical chain reactions. Therefore, the rate of olive-stem abscission zone formation may also vary in comparison to the same olive varieties cultivated in more moderate climates. Usually, in desert olive-oil cultivated trees, maturation is somewhat delayed (unpublished data). As a direct result of this delay, olive-stem abscission zone formation takes place later than in more moderate environments.

Weis *et al.* (1991) reported that naturally occurring abscission in olive tissues is preceded by plasmolysis of abscission zone cells, loss of cell-wall materials (as evidenced by changes in stain intensity), and lacunar cell separation. They further demonstrated the phytotoxic symptoms caused by treatment with synthetic ethylene-producing agents, such as ethephon. These symptoms include plasmolysis and cell senescence throughout all explant tissues. All active abscission zone cells exhibit small size, isodiametric shape, dense cytoplasm, and starch accumulation. Cell division does not occur in any abscission zone with any treatment, suggesting the death of this layer of cells, which leads to removal of the olive fruit (Figure 8.19).

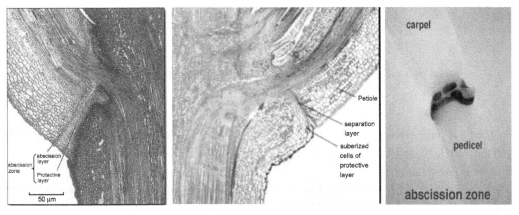

Figure 8.19: Abscission zone induced by ethylene that promotes secretion of cell wall degrading enzymes such as cellulases.
Source: Weis *et al.* (1991).

8.3.3 *Improvements in mechanical harvesting*

Terminology

Fruit retention force (FRF) has been commonly used in the research of natural olive-stem abscission zone formation (Ben Tal, 1992). FRF is also intensively used for evaluation of the effect of various chemical materials on loosening of olives for mechanized harvesting. FRF declines as physiological olive maturity is reached, and continues to decline until the fruit fall naturally.

Tension meters measure the "grip" of the olive to the tree. It is well accepted that the lower the tension, the riper the fruit, and the nearer it is to falling from the tree. Different olive varieties have different release parameters, and appropriate benchmarks need to be established for each variety. Well-designed olive sample tests from various parts of the tree and olive grove may give an average indication of the ripeness of the olive fruit.

Ethephon is the trade name of a synthetic plant growth regulator. Upon metabolism by the plant, it is converted into ethylene – a potent regulator of plant growth and maturity. It is often used on wheat, coffee, tobacco, cotton and rice crops in order to help the plant's fruit to reach maturity more quickly. In cotton, which initiates fruiting over a period of several weeks, ethephon is used to encourage all bolls to open simultaneously in order to enhance harvest efficiency.

Although many environmental groups worry about the toxicity of growth hormones and fertilizers, the toxicity of ethephon is actually very low, and any ethephon applied to plant material is usually converted very quickly to ethylene (Figure 8.20).

Ethephon is stored in very strong acid conditions, and needs to be activated to release ethylene by increasing the pH to a neutral level so that all the ethylene is released at once (Shulman *et al.*, 1982).

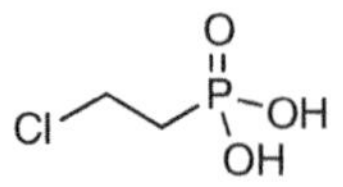

Figure 8.20: Scheme of ethephon (ethylene-releasing agent).

Over the final three decades of the twentieth century, many attempts to develop methodologies based on ethylene and ethylene-releasing agents, such as ethephon and others, to improve mechanical olive-harvesting efficiency were found to fail at field testing (Klein *et al.*, 1978, 1979; Lavee *et al.*, 1982; Shulman *et al.*, 1982; Ben Tal, 1987, 1992). The main reason for this failure relates to the imbalanced and uncontrolled effect of ethylene-releasing agents on the olive trees' foliage and fruits. The ethylene response of the olive tissues is strongly affected by the concentration of application, and more so by environmental conditions. The same concentration of ethylene-releasing agents may cause severe damage if applied in high temperatures and low humidity; on the other hand, it may be completely ineffective if applied in low temperatures and high humidity or rainy conditions. The effect of ethephon is closely related to the water status of the olives and foliage, and the maturation state of the olives. In many cases ethephon was found to cause massive leaf drop and even to have a lethal effect on whole branches and trees. Because of the unpredictable and complicated cascade of processes following regular spray application of ethylene-releasing agents, this practice was not accepted in the olive industry.

It was only in 2002 that this new aid to mechanical harvesting, based on an ethylene-releasing compound developed in the Phyto-Lipid Biotechnology Laboratory of Ben-Gurion University, was patented and internationally registered as an aid to mechanical olive harvesting (Wiesman and Markus, 2004). The biotechnological concept of the newly developed system is based on the following parameters:

1. The active materials are encapsulated in small particles and have a controlled and long-term effect.

2. Both the phosphorus compound and the ethylene-releasing agent are almost inactive outside the sprayed plant tissue; thus too high and toxic a level is avoided, significantly reducing the ethylene risks. The agent is activated only after penetrating the leaf and fruit tissues. Upon penetration, the active material induces endogenous ethylene production by the tissue, so that an almost natural senescence process occurs.

3. The novel adjuvant delivery system assures long-lasting penetration of the active ingredients through the massive cuticle membrane of the olive leaf as well as the olive. This continuous penetration lasts throughout both day and night for at least 4–7 days, leading to a greater chance of success in various environmental climate conditions.

Table 8.1: Effect of controlled delivery harvest-aiding treatment in mechanical shaker harvesting of various olive varieties in different locations

Variety	Location	Treatment	FRF		Harvesting efficacy (%)	Speed (s/tree)
			Day 0	Day 7		
Barnea	Israel (Negev)	Control	490	460	55	20–25
Barnea	Israel (Negev)	Aided		190	95	5–10
Picual	Israel (Negev)	Control	710	630	45	20–30
Picual	Israel (Negev)	Aided		220	90	5–15
Picholine	Israel (Negev)	Control	360	280	80	15–25
Picholine	Israel (Negev)	Aided		200	95	5–10
Manzanillo	Israel (Negev)	Control	620	600	55	25–30
Manzanillo	Israel (Negev)	Aided		260	85	< 5
Koroneiki	Israel (Negev)	Control	370	310	75	20–25
Koroneiki	Israel (Negev)	Aided		180	95	5–10
Frantoio	Israel (Negev)	Control	290	270	75	20–25
Frantoio	Israel (Negev)	Aided		145	95	< 5
Arbequina	Spain (Catalunia)	Control	460	410	65	20–25
Arbequina	Spain (Catalunia)	Aided		190	95	< 5
Hojiblanca	Spain (Andulasia)	Control	790	720	45	25–30
Ojiblanca	Spain (Catalunia)	Aided		320	75	< 5
Coratina	Italy (Bari)	Control	430	360	70	20–25
Coratina	Italy (Bari)	Aided		160	95	5–10
Leccino	Italy (Bari)	Control	520	440	55	20–25
Leccino	Italy (Bari)	Aided		270	85	< 5
Memicik	Turkey (Nizip)	Control	650	610	–	–
Memicik	Turkey (Nizip)	Aided		240	–	–
Ivolek	Turkey (Nizip)	Control	680	580	–	–
Ivolek	Turkey (Nizip)	Aided		200	–	–

4. The overall effect of the developed technology stimulates the maturation of almost all the olives in the tree. The green olives are advanced to purple, and the more mature olives are advanced to black coloring. This long-lasting effect even increases slightly the potential oil content of the treated trees.

5. The date of spray application of this mechanical olive-harvesting system is determined by systematic assay of the fruit retention force (FRF) according to the maturity index of the olives in every plot.

6. The long-lasting treatment significantly reduces the FRF to 200 or less, depending on the olive variety and the length of the time between the spray application and the beginning of mechanical harvesting.

7. The plan is for this new aid to mechanical harvesting systems to be further developed for vine-type harvesters (to reduce the number of hammer hits needed for efficient harvesting), and that this technology be combined with the recently developed chemo-optic system to predict the optimal date for olive harvesting.

The system has been tested in various locations, using a long list of olive varieties (Barnea, Picual, Arbequina, Hojiblanca, Blanqueta, Coratina, Leccino, Frantoio, Moraiolo, Mission, Kalamata, Koroneiki and others) in Asia, Europe (Turkey, Italy, Spain, Greece), the USA, South Africa, Australia and New Zealand. Some of the data demonstrating the efficiency of the new aid to mechanical harvesting, with various olive varieties and in various locations are shown in Table 8.1.

References

Ayton, J., Mailer, R. J., & Robards, K. (2001). Changes in oil content and composition of developing olives in a selection of Australian cultivars. *Aust J Exp Agric*, 41, 815–821.

Ballard, D. H., & Brown, C. M. (1982). *Computer Vision*. Prentice Hall, Englewood Cliffs, NJ.

Bangerth, J. (2004). Abscission and thinning of young fruit and their regulation by plant hormones and bioregulators. *Plant Growth Reg*, 31, 43–59.

Baxes, G. A. (1988). *Digital Image Processing*. Cascade Press, Denver, CO.

Beltran, G., Del Rio, C., Sanchez, S., & Martinez, L. (2004). Seasonal changes in olive fruit characteristics and oil accumulation during ripening process. *J Sci Food Agric*, 84, 1783–1790.

Ben-Tal, Y. (1992). Quantification of ethephon requirements for abscission in olive fruits. *Plant Growth Reg*, 11, 397–403.

Bleecker, A. B., & Kende, H. (2000). Ethylene: a gaseous signal molecule in plants. *Annu Rev Cell Dev Biol*, 16, 1–18.

Caponio, F., & Gomes, T. (2001). Phenolic compounds in virgin olive oils: influence of the degree of ripeness on organoleptic characteristics and shelf life. *Eur J Food Res Technol*, 212, 329–333.

Carvajal, R., & Nebot, A. (1998). Growth model for white shrimp in semi-intensive farming using inductive reasoning methodology. *Comput Electr Agric*, 19, 187–210.

Cossignani, L., Simonetti, M. S., & Damiani, P. (2001). Structural changes of triacylglycerols and diacyglycerol fractions during olive drupe ripening. *Eur Food Res Technol*, 212, 160–164.

Diaz, R., Faus, G., Blasco, M., & Molto, E. (2000). The application of a fast algorithm for the classification of olives by machine vision. *J Food Eng Food Res Intl*, 33, 305–309.

Diaz, R., Gil, L., Serrano, C., Blasco, M., Molto, E., & Blasco, J. (2004). Comparison of three algorithms in the classification of table olives by means of computer vision. *J Food Eng*, 61, 101–107.

Esti, M., Cinquanta, L., & La Notte, E. (1998). Phenolic compounds in different olive varieties. *J Agric Food Chem*, 46(1), 32–35.

Ferguson, L. (2006). Trends in olive fruit handling previous to its industrial transformation. *Grasas y Aceites*, 57, 1–7.

Fridley, R. B., Harmann, H. T., Melschau, J. J., Chen, P., & Whisler, J. (1971). Olive harvest mechanization in California. *Calif Agric Exp Station Bull*, 855.

Gallo, A., Mazzobre, M. F., Buera, M. P., & Herrera, M. L. (2003). Low resolution pulsed NMR for sugar crystallization studies. *Latin Am Appl Res*, 33, 97–102.

Garcia, M. J., & Yousfi, K. (2005). Non-destructive and objective methods for the evaluation of the maturation level of olive fruit. *Eur Food Res Technol*, 221, 538–541.

Goren, R. (1993). Anatomical, physiological and hormonal aspects of abscission in Citrus. *Hort Rev*, 15, 33–46.

HORIBA (1995). *Oil Content Analyzer OCMA-350 Instruction Manual.* Kyoto: HORIBA, Chapter 10, pp. 1–2.

Huberman, M., Riov, J., Aloni, B., & Goren, R. (1997). Role of ethylene biosynthesis and auxin content and transport in high temperature-induced abscission of pepper reproductive organs. *J Plant Growth Reg*, 16, 129–135.

ISO (International Standard) 10565 (1998). Oilseeds – simultaneous determination of oil and water contents– method using pulsed nuclear magnetic resonance spectrometry. Geneva: ISO.

Karleskind, A., & Wolff, J.-P. (1996). *Oils and Fats Manual.* Intercept Limited, Andover, 1, 783–786.

Kauraba, K. J., Gil Ribes, J., Blanco Roban, G. L., de Jaime Reyuelta, M. A., & Navero, D. B. (2004). Suitability of olive varieties for mechanical harvester shaking. *Olivae*, 101, 39–43.

Kieber, J. J. (1997). The ethylene response pathway in Arabidopsis. *Annu Rev Plant Physiol Plant Mol Biol*, 48, 277–296.

Klein, I., Epstein, E., Lavee, S., & Ben-Tal, Y. (1978). Environmental factors effecting ethephon in olive (*Olea europeae* L.). *Sci Hort*, 9, 21–30.

Klein, I., Lavee, S., & Ben Tal, Y. (1979). The effect of vapor pressure on the thermal decomposition of 2-chloroethylphosphonic acid. *Plant Physiol*, 63, 474–477.

Kuo, M.-I., Gunasekaran, S., Johnson, M., & Chen, C. (2001). Nuclear magnetic resonance study of water mobility in Pasta Filata and Non-Pasta Filata Mozzarella. *J Dairy Sci*, 84, 1950–1958.

Lashbrook, C. C., Tieman, D. M., & Klee, H. J. (1998). Differential regulation of the tomato ETR gene family throughout plant development. *Plant J*, 15, 243–252.

Lavee, S., Avidan, B., & Ben-Tal, Y. (1982). Effect of fruit size and yield on the fruit removal force within and between olive cultivars. *Sci Hort*, 17, 27–32.

Laykin, S., Alchanatis, V., & Edan, Y. (2002). Image processing algorithms for tomato classifications. *Am Soc Agric Eng*, 45(3), 851–858.

Luchetti, F. (2002). Importance and future of olive oil in the world market – an introduction to olive oil. *Eur J Lipid Sci Technol*, 104, 559–563.

Lundby, F., Sorland, G.H., & Eilertsen, S. (2006). Determination of fat, moisture and protein in fish powder within 30 minutes, by combining low resolution NMR techniques and microwave technology. *The 8th International Conference on the Applications of Magnetic Resonance in Food Science, Nottingham.*

Mailer R, Conlan D, Ayton J, (2005). *Olive Harvest – Harvest Timing for Optimal Olive Oil Quality.* Kingston: RIRDC Publication No 05/013, RIRDC Project No DAN-197A.

MARAN-ULTRA (2006). *MARAN-ULTRA SFC User Manual* V6.2. Oxford: Oxford Instruments, Molecular Biotools Ltd.

Matthaus, B., & Bruhl, L. (2001). Comparison of different methods for the determination of the oil content in oilseeds. *J Am Oil Chem Soc*, 78(1), 95–102.

Mickelbart, V. M., & James, D. (2003). Development of a dry matter maturity index for olive (*Olea europaea*). *NZ J Crop Hort Sci*, 31, 269–276.

Mili S (2004). Market dynamics and policy reforms in the EU olive oil industry: an exploratory assessment. Paper presented at the 98th EAAE Seminar on Marketing Dynamics within the Global Trading System: New Perspectives, 29 June–2 July 2006, Chainia, Crete, Greece.

Morello, J. R., Romero, M. P., & Motilva, M. J. (2006). Influence of seasonal conditions on the composition and quality parameters of monovarietal virgin olive oils. *J Am Oil Chem Soc*, 83, 683–690.

Motilva, M. J., Tovar, M. J., Romero, M. P., Alegre, S., & Girona, J. (2000). Influence of regulated deficit irrigation strategies applied to olive trees (Arbequina cultivar) on oil yield and oil composition during the fruit ripening period. *J Sci Food Agric*, 80, 2037–2043.

Nordon, A., Macgill, C. A., & Littlejohn, D. (2001). Process NMR spectrometry. *Analyst*, 126, 260–272.

Pathaveerat, S., Chen, P., & McCarthy, M.J. (2001). On-line evaluation of avocado fruit quality. St Joseph, MI: ASAE Paper No. 013003.

Ranalli, A., Tombesi, A., Ferrante, M. L., & De Mattia, G. (1998). Respiratory rate of olive drupes during their ripening cycle and quality of oil extracted. *J Sci Food Agric*, 77, 359–367.

Ryan, D., Robards, K., & Lavee, S. (1999). Changes in phenolic content of olive during maturation. *Intl J Food Sci Technol*, 34, 265–274.

Saito, M., Sato, Y., Ikeuchi, K., & Kashiwagi, H. (1999). Measurement of surface orientations of transparent objects by use of polarization in highlight. *J Opt Soc Am A*, 16, 2286–2293.

Salehi, F., Lacroix, R., & Wade, K. M. (1998). Improving dairy yield predictions through combined record classifiers and specialized artificial neural networks. *Comput Electr Agric*, 20, 199–213.

Salvador, M., Aranda, F., & Fregapane, G. (2001). Influence of fruit ripening on 'Cornicabra' virgin olive oil quality, A study of four successive crop seasons. *Food Chem*, 73, 45–53.

Shahin, M. A., & Symons, S. J. (2001). A machine vision system for grading lentils. *Can Biosyst Eng*, 43(7), 7–14.

Sharova, E. I. (2007). Expansins: proteins involved in cell wall softening during plant growth and morphogenesis altered expression of expansin modulates leaf growth and pedicel abscission in *Arabidopsis thaliana*. *Russ J Plant Physiol*, 54(6), 713–727.

Shulman, Y., Avidan, B., Ben-Tal, Y., & Lavee, S. (1982). Sodium bicarbonate, a useful agent for pH adjustment of ethephon controlling grapevine shoot growth and loosening olive fruits. *Riv Della Ortoflorofruitt Ital*, 66, 181–187.

Solinas, M., Di Giovacchino, L., & Cucurachi, A. (1975). Variations of some polyphenols of olive oil during the ripening cycle of olives. Note 1. *Ann Ist Sper Elaiot*, 5, 105–126.

Thygesen, L. G., Thybo, A. K., & Engelsen, S. B. (2001). Prediction of sensory texture quality of boiled potatoes from low-field 1H NMR of raw potatoes. The role of chemical constituents. *Lebensm Wissensch Technol*, 34, 469–477.

Tombesi, A. M., Boco, M., Pill, M., & Farinelli, D. (2002). Influence of canopy density on efficiency of trunk shaker olive mechanical harvesting. *Acta Hort*, 586, 291–294.

Torbica, A., Jovanovic, O., & Biljana, P. (2006). The advantages of solid fat content determination in cocoa butter and cocoa butter equivalents by the Karlshamns method. *Eur Food Res Technol*, 222, 385–391.

Vlahov, G. (1976). Organic acids of olives: malic acid/citric acid ratio as a maturation index. *Ann Ist Sper Elaiot*, 6, 93–110.

Vossen, P., Diggs, L., & Mendes, L. (2004). Santa Rosa Junior College's super high density orchard. *Olint Oct*, 6–8.

Weis, K. G., Webster, B. D., Goren, R., & Martin, G. C. (1991). Inflorescence abscission in olive: anatomy and histochemistry in response to ethylene and ethephon. *Botanical Gazette*, 52, 51–58.

Weissbein, S. (2006). Characterization of new olive (*Olea europae* L.) varieties' response to irrigation with saline water in the Ramat Nege area. MSc thesis, Ben-Gurion University of the Negev, Beersheva, Israel.

Weissbein, S., Wiesman, Z., Efrath, Y., & Silverbush, M. (2008). Physiological and reproductive response olive varieties under saline water. *Hort Sci*, accepted.

Wiesman Z, Markus A (2004). Multi-layer adjuvants for controlled delivery of agro-materials into plant tissues. US Patent No. 10/488,388, registered by BGU.

Zheng, C., Sun, D. W., & Zheng, L. (2006). Recent developments and applications of image features for food quality evaluation and inspection e a review. *Trends Food Sci Technol*, 17, 642–655.

Further reading

Cho, S. I., Lee, D. S., & Jeong, J. Y. (2002). Weed–plant discrimination by machine vision and artificial neural network. *Biosyst Eng*, 83(3), 275–280.

Friedl, M. A., & Brodley, C. E. (1997). Decision tree classification of land cover from remotely sensed data. *Remote Sens Environ*, 61(3), 399–409.

Friedl, M. A., Brodley, C. E., & Strahler, A. H. (1999). Maximizing land cover classification accuracies produced by decision trees at continental to global scales. *IEEE Trans Geosci Remote Sens*, 37(2), 969–977.

Goel, P. K., Prasher, S. O., Patel, R. M., Landry, J. A., Bonnell, R. B., & Viau, A. A. (2003). Classification of hyperspectral data by decision trees and artificial neural networks to identify weed stress and nitrogen status of corn. *Comput Electr Agric*, 39(2), 67–93.

Heinemann, P. H., Pathare, N. P., & Morrow, C. T. (1996). An automated inspection station for machine-vision grading of potatoes. *Machine Vis Appl*, 9, 14–19.

Karimi, Y., Prasher, S. O., McNairn, H., Bonnell, R. B., Dutilleul, P., & Goel, P. K. (2005). Classification accuracy of discriminant analysis, artificial neural networks and decision trees for weed and nitrogen stress detection in corn. *Trans ASAE*, 48(3), 1261–1268.

Shahin, M. A., Tollner, E. W., McClendon, R. W., & Arabnia, H. R. (2002). Apple classification based on surface bruises using image processing and neural networks. *Trans ASAE*, 45(5), 1619–1627.

Vossen, P. (2004). Variety and maturity: the two largest influences on olive oil quality. (available at http://ucce.ucdavis.edu/files/filelibrary/2161/17338.pdf).

CHAPTER 9

Oil extraction and processing biotechnologies

Olive oil has a privileged position among edible vegetable oils because it is the juice of a fruit, and as such is consumed almost entirely in its natural state. Therefore, it contains (among other important constituents) phenolic compounds, which are usually removed from seed oils during the refining process. The olive oil extraction process refers to extracting the olive oil that is present in the olive fruits, botanically termed "drupes," for food use. In olive fruits the oil is produced in the mesocarp cells, and stored in a particular type of vacuole called a lipovacuole (i.e., every cell contains an olive oil droplet). Hence, olive oil extraction can be said to be a process of separating the oil from the other fruit contents (vegetation tissue, water and solid material), and this separation is done only by physical means. This is totally different from other vegetable oils, which are extracted with chemical solvents (generally hexane).

Since olive oil is present in the form of minute drops in the vacuoles of mesocarp cells in the olive fruit, as well as being scattered to a lesser extent in the colloidal system of the cell's cytoplasm and, to a still lesser extent, in the epicarp and endosperm, it is possible for all the oil in the vacuoles to be released during processing. It is difficult to obtain oil dispersed in the cytoplasm, so this oil generally remains in the olive pomace – a by-product of processing (Kiritsakis, 1998). The olive oil extraction process is normally divided into crushing, malaxation (mixing), and separation of the oil. Below, we describe the main tasks that need to be performed, after harvesting the olives, in order to obtain good-quality olive oil.

Delivery of the olive fruits

After the olives have been picked in the proper manner, they are delivered to the mill and stored there in the yard in collection boxes. This offers the opportunity to determine the quality and rate of yield, and serves as the basis for the settlement of accounts between the oil mill and olive farmer. At harvest, there is generally 18–32 percent oil, 40–55 percent water and 23–35 percent seeds and vegetable tissue in olive fruits.

Washing and cleaning of the olive fruits

Stems, twigs and leaves should be removed, and the olive fruits may or may not be cleaned with water to remove pesticides, dirt, etc. Rocks and sand will quickly wear out a centrifugal decanter or oil separator, reducing its lifespan.

Grinding, mixing or malaxation of olive pulp

After cleaning, the olive fruits are ground up together with their stones. For grinding, either stone rollers or a metal tooth grinder or hammer mill can be used. Each has its own advantages and disadvantages. After this process, mixing or malaxation should be performed. Malaxation is the action of slowly churning or mixing milled olives in a specially designed mixer for 20–40 minutes. This mixing allows the smaller droplets of oil that were released by the milling process to combine into larger ones that can be more easily separated. The paste is normally heated to around 27 °C during this process. It is now possible, with newer equipment, to use a blanket of inert gas (such as nitrogen or carbon dioxide) over the olive paste, which greatly reduces oxidation; this allows for an increased yield without compromising the quality of the oil.

After malaxation is complete, the paste is then sent to a phase separator. Nearly all producers use a decanter centrifuge for this phase.

The most common mixer is a horizontal trough with spiral mixing blades. Malaxation of the pulp is carried out, dependent on the oil extraction procedure – i.e., warm water may be added. To aid in further breakdown of the olive cells, and to create large oil droplets, the pulp is beaten. Salt is often added at this stage, to aid the osmotic breakdown of cells in the olives and thus ease the separation of the oil from the water. The olives are beaten several times.

Separation of the oil and water

In this step, separation of oil and water from the olive paste is carried out. Many procedures can be used to achieve this, and will be discussed later in the chapter; centrifuge/decanters, modern di- and tri-phase centrifuge, and percolation-Sinolea methods are common. In small oil mills, the olive oil is extracted in a batch operation using the traditional press method. The oil extracted is collected in containers and clarified by sedimentation. In a big oil extraction system, large amounts of olive fruits are used for extraction by more modern systems. To improve the separation of oil and pulp, biological or chemical aids that attack the cell walls can be added.

Separation of oil from the water

This is a very important process, where the olive oil is separated from the olive's water. As in cream separation in a dairy, the liquid is spun to separate the heavier water from the oil. Gravity decanter or vertical centrifuges with perforated conical discs are commonly used in this process.

Extraction of residual oil

After extraction of the first oil – generally virgin olive oil – the residual waste or pomace derived after first pressing or centrifugation undergoes further extraction by different means. The process of extraction of the residual oil differs according to the system originally used. The solid waste from oil extraction by pressing still contains about 6 percent oil or, using the continuous decanter, 4 percent olive oil. The oil content in the solid–liquid mixture from the dual-phase decanting process contains 2.5–3.5 percent oil. In specialized extraction plants, a solvent extraction process is used for this purpose. First, the waste is completely dried; hexane is then used as a solvent to extract further oil, known as olive-pomace oil.

The utilization of by-products of olive oil extraction

The dry residues can be used as concentrated fodder. In some extraction plants, the stones are separated from the pulp after extraction and used as fuel for heating the driers.

The pulp is sold as fertilizer or fodder. In some oil mills, the solid waste from the press is directly used as fuel for the heating of water.

Vegetable water, the liquid phase obtained as a result of centrifugation, is abundant in the three-phase extraction method due to water injection into the paste before centrifugation. As vegetable water still contains oil, it is treated a second time in order to remove the maximum amount of oil. However, since the result is a combination of water and fat, it is difficult to recycle it. Vegetable water is highly polluting and has a negative effect on underground water. The most serious ecological problem in olive oil production is the recycling of vegetable water.

9.1 Olive oil extraction technologies

9.1.1 Traditional method

The traditional method of olive oil extraction is by using an olive press. The olive press works by applying pressure to olive paste to separate the liquid oil and vegetation water from the solid materials. In this method, after separating the oil and water from the fruit paste, the water and the oil themselves are then separated from one another by a standard decantation procedure. This method has been in use since the Greeks first began pressing olives over 5000 years ago,

Figure 9.1: Traditional olive oil press.
Source: Wikipedia (www.wikipedia.org), 2007.

and is still widely used today, with some modification. Even now, it is considered a valid way of producing high-quality olive oil if adequate care is taken (Figure 9.1). However, this type of method is suitable only for small growers.

In this method, after cleaning the olive fruits, the olives are ground into an olive paste using large millstones. The olive paste generally remains between the stones for 30–40 minutes in order to guarantee that the olives are well ground; this allows enough time for the olive droplets to join to form large drops of oil, and for the fruit enzymes to produce some of the oil aromas and taste. Some traditional oil presses also use modern crushing methods. After grinding, the ground olive paste is spread on fiber disks, which are stacked on top of each other, then placed into the press. Traditionally the disks were made of hemp or coconut fiber, but nowadays they are made from synthetic fibers, which are easier to clean and maintain. These disks are then threaded onto a hydraulic piston, forming a pile. Pressure is applied on the disks, thus compacting the solid phase of olive paste and percolating the liquid phase (oil and vegetation water). In this system, the applied hydraulic pressure can be as high as 400 atm. To facilitate the liquid phase, water is run down the sides of the disks to increase the speed of percolation. The liquid is then separated either by a standard process of decantation or by the faster vertical centrifuge.

The main advantages of the traditional method are that it produces good-quality oil by more efficient grinding of the olives, reducing the release of oil oxidation enzymes, and that it also reduces added water, which minimize the washing of polyphenols. The pomace obtained from this method has a low water content, so it is easier to manage. The disadvantages of this method include the difficulty of cleaning the disks. Also, since this method is not a continuous process, it means that there are periods when the olive paste is exposed to oxygen and light, which can cause deterioration of oil quality. Finally, the traditional method requires more manual labor, and also needs a longer time period between harvest and pressing.

This traditional method, also called discontinuous extraction, is an ancient procedure that only distinguishes two phases by pressing or centrifugation. The liquid phase is filtered later in order to obtain oil. In this case the by-product is a plastic paste, which has the advantage of avoiding the production of vegetation water. However, although it is better ecologically, this technique provides a lower yield, which is not always seen as an advantage by the main producing countries. Nearly all olive oil producers in the traditional olive-growing regions, such as Spain, changed from the traditional presses to modern systems in the mid-1980s.

9.1.2 Modern methods

The modern method of olive oil extraction uses an industrial decanter to separate all the phases by centrifugation. In this method, olives are crushed to a fine paste by a hammer crusher, a disk crusher, a depitting machine or a knife crusher. The paste is then malaxed for 30–40 minutes to allow the small olive droplets to agglomerate and the fruit enzymes to create aromas. After this the paste is pumped into an industrial decanter, where the phases will be separated. Depending upon the system, water might be added with the paste to facilitate the extraction process. The decanter is a large-capacity horizontal centrifuge rotating at around 3000 rpm; the high centrifugal force created allows the phases to be readily separated according to their different densities (solids > vegetation water > oil).

The most common industrial processing method is a continuous extraction system with two centrifuges (first horizontal and then vertical). Over the years, a few technological variations have been introduced to the industrial decanter. In the three-phase decanter (oil, pomace and vegetable water), as shown in Figure 9.2, some of the oil polyphenols are washed out due to the higher quantity of water added, which produces a large quantity of vegetable water that needs processing.

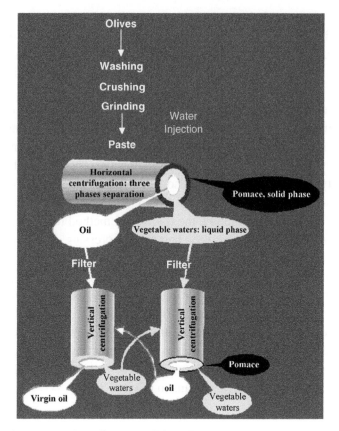

Figure 9.2: Flow diagram of the olive oil extraction process.
Source: Unctad (www.unctad.org (2004).

The procedure shown in Figure 9.2 is the most widely used in intensive production areas, and dates back to the 1970s and 1980s. The main disadvantages of this process are the huge amount of water needed, which leads to production of large quantities of waste and the resultant pollution. In an attempt to solve this problem, a two-phase oil decanter has been created; here, no water injection is used and just the oil and a plastic paste are produced. Since no water is added in this two-phase system, there is less phenol washing and oil and hard pomace are obtained. Instead of three exit points, for oil, water and solids, this type of decanter has only two (Figure 9.3). The water is expelled by the decanter coil together with the pomace; resulting in a wetter pomace; this is much harder and needs to be processed industrially, leading to high energy costs so that the pomace can be dried in order to extract further oil from it. In practice, therefore, the two-phase decanter solves the phenol washing problem but increases the residue management problem. This has led to the creation of a new two-and-a-half phase oil decanter (sometimes still referred to as a two-phase decanter system). This separates the olive paste into the standard three phases, but has a lesser need for added water and thus a smaller

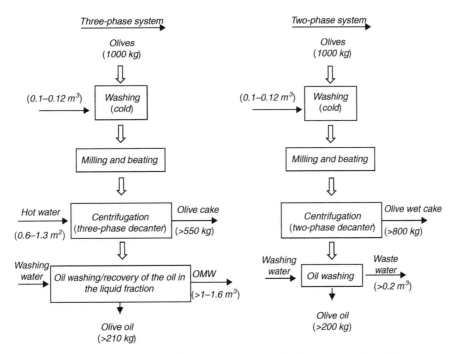

Figure 9.3: Three- and two-phase centrifugation system of olive extraction (Alburquerque *et al.*, 2004).

vegetable-water output. This system is considered to be more economically and environmentally viable than the others.

This new industrial technique of continuous extraction is commonly practiced these days and, because it needs less water, is being used more and more widely. A detailed comparison of results obtained from a two-phase extraction and a three-phase extraction is represented in Table 9.1. The study involved different varieties of olives, composed of 48–51 percent water and 19–23 percent oil.

Olive oil obtained from the two-phase decanter system also contained higher amounts of polyphenols, artho-diphenols, hydroxytryrosol, tocopherols, trans-2-hexenal (the most important aroma component, with a herbaceous aroma) and total aroma volatile compounds than that obtained from the three-phase decanter. Moreover, the oil obtained from the two-phase decanter showed higher oxidative stability, lower turbidity, and lower contents of pigments, steroid hydrocarbons, waxes, and aliphatic and triterpenic alcohols than did the oil from the three-phase decanter. Some of the differences in the quality of oil obtained from two- and three-phase decanter systems, as found by Di Giovacchino (1996) are shown in Table 9.2. This shows that the two-phase decanter system produced better-quality oil than the three-phase system.

Table 9.1: A detailed comparison of results obtained from two- and three-phase olive oil extraction systems

Parameters	Two-phase extraction	Three-phase extraction
Oil extraction capacity	86%	85%
Pomace		
Quantity (kg/100 kg of olives)	72.5 a	50.7 b
Moisture (%)	57.5 a	52.7 b
Oil (%)	3.16 a	3.18 b
Oil (% dry matter)	7.44 a	6.68 a
Oil (kg/100 kg of olives)	2.28 a	1.60 b
Dry pomace (kg/100 kg of olives)	30.7 a	23.9 b
Vegetable water		
Quantity (l/100 kg of olives)	8.30 a	97.2 b
Oil (g/l)	13.4 a	12.6 a
Oil (kg/100 kg of olives)	0.14 a	1.20 b
Dry residual (kg/100 kg of olives)	1.20 b	8.3 b
Oil in by-products (kg/100 kg of olives)	2.42 a	2.80 a

Values followed by identical letters are not statistically different at $P < 0.05$.
From Amirante et al. (1993).

Table 9.2: Quality characteristics of olive oils obtained from the two- and three-phase decanter system (Di Giovacchino, 1996)

Quality characteristic	Two-phase decanter	Three-phase decanter
Acidity (%)	0.35	0.34
Peroxide value (meq/kg)	3.8	4.3
Total polyphenols (mg/l gallic acid)	333	220
O-diphenols (mg/l caffeic acid)	342	165
Rancimat stability (h) (induction time at 120 °C 20 l/h air-flow rate)	15.3	11.6
Chlorophyll pigments (ppm)	6.3	6.6
K_{232}	1.548	1.438
K_{270}	0.105	0.091
Sensory evaluation	7.1	7.2

However, apart from the yield and the level of polyphenols, other basic oil-quality parameters, such as the acidity level, fatty acid composition, triacylglycerol molecular species and stability, were found to be similar whether the oil was extracted using the discontinuous (pressing) or continuous (centrifuging) procedure (Ben Miled *et al.*, 2000). That said, the quality and the quantity of oil obtained from the different varieties did show differences.

Figure 9.4: Percolation of the olive oil from the olive paste in a Sinolea unit.

9.1.3 Extraction by selective filtration (Sinolea)

This is considered to be the most up-to-date method of extracting oil from olives. In this method, selective filtration combined with centrifugation is used for the separation of olive oil from the olive paste. The most common selective filtration system is the Sinolea process (Figure 9.4). This process is the opposite of pressing, since no pressure is applied to the paste. In Sinolea, rows of metal plates are dipped into the olive paste; the oil preferentially wets and sticks to the metal, and is removed with scrapers in a continuous process. This process is based on the different interferential tensions of oil and water coming into contact with steel plates. A steel plate will be coated with oil when plunged into the olive paste, since the interfacial tension of the oil is less than that of the water. These different physical behaviors allow the olive oil to adhere to the steel plate while the other two phases remain behind. In this system, 350–750 kg of olive paste can be handled at one time. Sinolea works by continuously introducing several hundred steel plates into the paste, thus extracting the olive oil; the moving plates slot through the slits in the grating units, and slowly penetrate the paste in a reciprocating motion.

In this process, the oil does not suffer any physical damage, so its quality is very high. Furthermore, the temperature can be kept very low, which can help to protect many oil-quality parameters – especially the polyphenols, aroma and flavor. The process is automated, so labor costs are low. However, it is not completely efficient, and a large quantity of oil remains in the

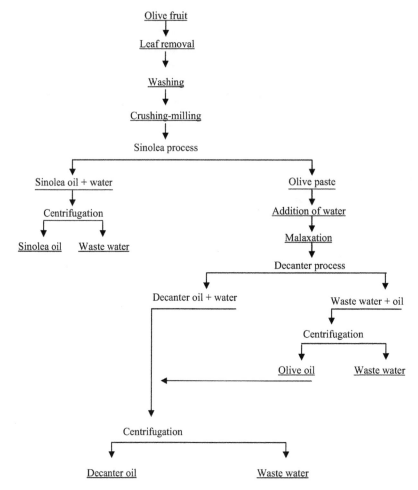

Figure 9.5: Flow diagram of combined processes of olive oil extraction (selective filtration and centrifugation).

paste. Therefore, the remaining paste has to be processed by the standard modern method (industrial decanter), and generally many commercial oil-extracting companies use Sinolea processing combined with the decanter method, as shown in Figure 9.5. Furthermore, the Sinolea equipment is complicated and requires frequent cleaning, maintenance of the stainless steel blade mechanism, and a constant heat source to keep the paste at an even temperature. Extraction is stopped when vegetable water begins to appear in the oil.

9.1.4 Summary

It is widely accepted that, apart from polyphenols, most of the other olive oil-quality parameters (such as free fatty acids, peroxide values, ultraviolet absorption and sensory character) are not

Table 9.3: Characteristics of virgin olive oil obtained from good-quality olives by three different processing systems

Characteristic	Pressure	Sinolea	Centrifugation/decanter
Free fatty acid (%)	0.23	0.23	0.22
Peroxide value (meq/kg)	4.0	4.6	4.9
Total polyphenols (mg/l gallic acid)	158	157	121
o-diphenols (mg/l caffeic acid)	100	99	61
Induction time (h)	11.7	11.2	8.9
Chlorophyll pigments (ppm)	5	8.9	9.1
K_{232}	1.93	2.03	2.01
K_{270}	0.120	0.129	0.127
Sensory evaluation (panel taste)	6.9	7.0	7.0

generally affected by the different modes of extraction, provided high-quality olive fruits are used and harvesting is carried out at the optimum time period (Di Giovacchino *et al.*, 1994) so that there is minimal enzymatic alteration in the olives before they are processed (Kiritsakis, 1991; Di Giovacchino, 1996). The oil-quality parameters of these three systems of oil extraction – pressure, Sinolea and centrifugation/decanter – are presented in Table 9.3.

9.2 Bio-enzymatic technology for improved olive oil extraction

Modern industrial olive oil extraction systems, either discontinuous (pressing) or continuous (centrifugation or decanter), both provide a better yield compared to traditional processing; however, these systems are still not optimal regarding either the yield or the quality of the olive oil produced. In fact, these mechanical systems are capable of extracting no more than 80–90 percent of the oil contained in the olive fruit (Ranalli *et al.*, 2003). Furthermore, the overall residual oil content in the byproducts (olive pomace and vegetable water) can be equivalent to the percent of the olives pressed. Thus, there is still a great economic and environmental loss to the olive oil sector. Although the modern double-phase system has been found to be much more effective than the three-phase system, this method remains less than optimal and in fact, despite the increased production costs and the positive effect only on the yield, there is little improvement in the quality of the oil produced.

In this context, efforts have been made to use natural enzymes or enzyme preparations (vegetable extracts) to enhance the mechanical extraction system during olive fruit processing. Enzymes are biocatalysts that facilitate the breakdown of cellulose and pectin, thereby helping to release the droplets of oil by reducing the stability and resistance of the cell walls. The addition of enzymes during grinding provides better results; however, it has become more common to add enzymes during the malaxation step.

Studies by Montedoro and Petruccioli (1972) and Petruccioli *et al.* (1988) have reported an increase in olive oil yield and a decrease in processing time with the use of the enzymes pectine depolymerase, papain, cellulose, hemicellulase and acid proferase during olive oil extraction. Use of an endo-polygalacturonase has also been found to increase the oil yield and lead to a better aroma. Olivex – an enzyme preparation produced by the fungus *Aspergillus acculeatus* (Novo Nordisk Ferment Ltd., Dittiingen, Switzerland) – has been found to increase olive oil yield, and has subsequently been recommended for use during the malaxation process. Olivex (Olivex Ltd., Turkey) has also been found to increase the concentration of phenolic compounds in olive oil when used during the mechanical extraction process of virgin olive oil production (Vierhuis *et al.*, 2001). Although these endogenous enzymes are natural products and increase both the quality and the yield of olive, studies have shown that the enzymes are largely inactivated during the critical crushing step, which could be due to oxidized phenols binding to their prosthetic group (Ranalli *et al.*, 2003). In order to overcome these problems, a new complex enzyme preparation called *Rapidase addax* D (Gist-Brocades, Seclin City, France) that degrades the uncrushed vegetable cell walls and promotes the release of functional components has been developed (Ranalli *et al.*, 2003).

9.2.1.1 Features of the complex enzyme preparation Rapidase Adex D

This enzyme preparation essentially comprises pectolytic, cellulolytic and hemicellulolytic enzyme species. The activity is not less than 200 units/ml, where the activity of the enzyme is defined as the amount of enzyme complex that liberates 1 μmol of reducing sugars per unit from pectins. The Rapidase Adex D degrades the vegetable colloids (pectins, hemicellulose, proteins, etc.) that emulsify the minute oil droplets. This enzyme preparation is water soluble and comes out in the liquid effluent (waste water) during the final step (oily must centrifugation) of the extraction cycle.

When Ranalli and colleagues (2003) used this enzyme preparation in extraction of virgin olive oil from three typical Italian olive species, namely Dritta, Leccino and Coratina, cultivated organically, they obtained the results shown in Table 9.4. These clearly show that although values were quite different among the three varieties tested, most of the oil-quality parameters were meaningfully and positively increased when the enzyme was used during oil extraction. The enzyme-treated oils showed an increase in the content of total phenols, o-diphenols and secoiridoid derivatives, such as major free phenols. They also had higher phenol/polyunsaturated fatty acid ratios, which are considered to be of great importance in predicting shelf-life (Ranalli and Angerosa, 1996).

Enzyme treatment also led to substantial increases in the oil yield, which ranged from 11.3 to 16.7 kg/ton of olives, regardless of the olive cultivars. The evidence shows that, with the use of the enzyme, greater amounts of oil can be freed from the vegetable tissue and in addition the coalescence phenomenon occurred regarding the minute oil droplets. Furthermore,

Table 9.4: Values of the major analytical and quality parameters in three enzyme-treated virgin olive oils as compared to the controls

Characteristic	Cv. Dritta		Cv. Leccino		Cv. Coratina	
	Enzyme	Control	Enzyme	Control	Enzyme	Control
Acidity (as oleic acid, g/kg)	8.0	7.1	16.1	19.2	8.1	7.2
Peroxide value (mequiv of O_2/kg)	14.3	14.1	12.5	14.0	19.2	19.2
Chlorophyll pigments (mg/kg)	18.3	15.4	21.5	18.7	20.4	16.0
Carotenes (mg/kg)	30.7	29.6	27.7	23.6	38.3	34.4
K_{232}	1.80	1.71	1.70	1.82	1.90	2.01
K_{270}	0.10	0.11	0.11	0.10	0.20	0.21
Sensory scoring (panel taste)	7.5	7.0	7.0	6.5	7.6	7.2
Pleasant volatiles[a] (as nonan-1-ol, mg/kg)	305	218	440	283	969	858
Phenols (as caffeic acid, mg/kg)	97	64	89	60	128	93
O-diphenols (as caffeic acid, mg/kg)	49	36	52	31	73	37
Secoiridoid derivatives[b] (as resorcinol, mg/kg)	36.3	24	33	20	66	46
Phenolic antioxidants/polyunsaturated fatty acids	8.4	5.8	8.8	6.1	13.9	10.3
Tocopherols ($\alpha + \gamma$, mg/kg)	100	86	242	185	210	185

Source: Ranalli and Angerosa (1996).

[a] *Includes trans-2-hexenal, trans-2-hexen-1-ol, cis-3-hexenyl acetate, cis-3 hexenyl acetate, cis-3-hexen-1-ol, penta-1-en-3-one, cis-2-pentenal, trnas-2-pentenal, and other.*

[b] *Free tyrosol and hydroxytyrosol and their aglycons.*

the enzyme preparation was found to aid in producing a more environmentally friendly liquid waste.

9.3 Negev olive oil extraction studies

Until the past decade, most of the olive production in Israel and, consequently, the oil presses were located in the northern and central parts of the country. However, following the development of advanced saline water irrigation technology, specifically engineered and approved for olive trees (Wiesman *et al.*, 2004; Weissbein, 2006; Weissbein *et al.*, 2008), a significant increase in olive cultivation has taken place in the southern Negev region. The rapidly increasing trend for olive oil cultivation in the Negev Desert area may soon lead to the area producing approximately 30 percent of Israel's total olive oil output. As the cultivation of olives has become more intensive in the Negev Desert area (Figure 9.6), the oil extraction process has also advanced. A study

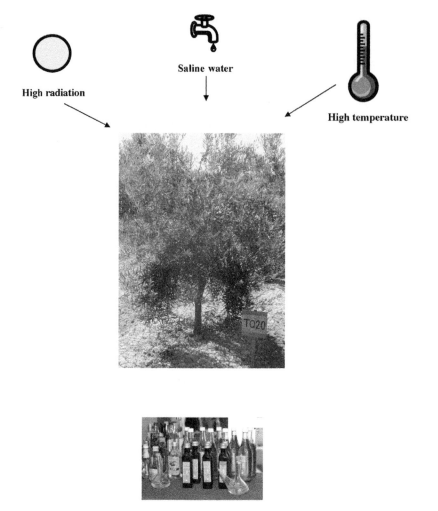

Figure 9.6: The main environmental factors affecting olive oil production in the Negev Desert.

on olive oil extraction was initiated and has been further developed in several units located in various parts of the Negev. An experimental research station, the Ramat Negev Desert Agro-Research Center (RNDARC), established by pioneers such as Yoel DeMalach and others many years ago, provides basic horticultural and cultivation facilities, mainly using saline irrigation and drip irrigation technologies for efficient development of the olive industry. Ben-Gurion University, which is located in the capital of the Negev area at Beer-Sheva, has initiated a long-term research and development study of desert olive oil extraction technologies and oil analyses.

Until recent years, the main olive variety cultivated in the Negev region was Barnea. However, other varieties from different sources of origin have been introduced and were used in these olive oil studies. Among these are varieties originating in Israel (Barnea, Souri, Maalot), Italy (Frantio, Leccino), Spain (Arbeqina, Pecual, Picudo), Greece (Kalamata, Koroneiki), France (Picholine), Morocco (Picholin di Morocco) and many others, as described in Chapter 7.

9.3.1 Oil potential of Negev-grown olives

In 1997, the Phyto-Lipid Biotechnology Laboratory of the Ben-Gurion University of the Negev initiated a study to try to gain an indication regarding the potential oil production from olives cultivated in the Negev Desert region, and to develop an efficient method of oil extraction. The study continued until 2007.

Initially, the basic systems for oil extraction and determination of oil content were organized (a brief description is given below; a more detailed discussion is provided in Chapter 8).

- *Soxhlet extraction.* This is quite an old method, which detects the oil content using solvents. In this method, all traces of oil are removed from the olive material. Soxhlet apparatus is generally used, and hexane is the most common solvent. The extracted oil does not meet the requirements for extra-virgin olive oil, so this method is employed only in the production of refined pomace oil and for analyzing the amount of oil accumulated in the olive fruits. Soxhlet extraction is generally agreed to provide very accurate results in comparison to most of the other common chemical oil analysis methods; however, the method is time-consuming, which limits the number of olive samples that can be analyzed.

- *Near infrared (NIR) analyzer.* NIR technology can measure the oil and moisture in milled olives and fatty acids, and the moisture in oil. This instrument works by passing near-infrared light through the oil sample and measuring the emerging energy, which relates to the concentration of oil, moisture and fatty acids. This method only determines the percentage oil content.

- *IR spectroscopy.* In this method, a small sample of olives is crushed and dissolved in strong solvent, and then analyzed using infrared absorbance – a non-dispersive infrared spectrophotometric technique which is specific to hydrocarbons such as oil. The HORIBA's OCMA-350 oil content analyzer is commonly used for this purpose (Weissbein, 2006). IR spectroscopy is also used to detect the percentage of oil in a particular olive fruit. Once a good oil calibration curve has been achieved, this technique gives very similar results to those from the soxhlet extraction method.

- *Low resolution nuclear magnetic resonance (NMR).* Low resolution NMR is a rapid, non-destructive and highly reliable technique for detecting the oil content in olive fruits (Nordon *et al.*, 2001; MARAN ULTRA, 2006). This is a new technique, based on a 23-MHz low resolution NMR system that has recently (2006) been introduced

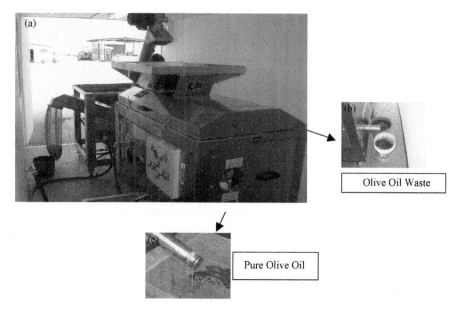

Olive Oil Waste

Pure Olive Oil

Figure 9.7: Oliomio mini-olive mill system: (a) main system; (b) olive oil waste material; (c) pure olive oil drainage.

and used in the Phyto-Lipid Biotechnology Laboratory at Ben-Gurion University of the Negev. Following two years' experience of its use, the system appears to be highly reliable and as accurate as the soxhlet system, but the oil content can be determined in just 16 seconds.

- *Oliomio mini-olive oil mill.* The Oliomio mill is a small mill suitable for olive oil extraction (Figure 9.7). This mini olive-oil system provides high-quality olive oil in a similar way to the large-scale industrial olive oil mill systems. The proximity of this system to the experimental plots in the Negev Desert enabled production of the best olive oil, soon after harvesting, from trees of the various olive varieties being cultivated in the region using the advanced cultivation technologies developed over the past decade (Wiesman *et al.*, 2004). Subsequently, two main long-term studies were carried out simultaneously. One was dedicated to evaluating the olive oil potential of various olive varieties tested in the Negev Desert area, and the second was focused on developing proper and efficient oil extraction in the same region. These studies were based on advanced laboratory olive oil analysis and used the Oliomio mini olive oil mill system for extraction.

The olive oil content of Souri olives as determined by various systems is shown in Table 9.5. The level is relatively similar in the three systems tested, although the percentage obtained using the Horiba system is slightly lower than with the soxhlet system or low resolution NMR.

Table 9.5: Determination of percentage olive oil content of Souri olives cultivated in the Negev Desert region in the 2006 season, measured by three different methods

Analysis date	Olive oil percentage		
	Soxhlet	Horiba 350	LR NMR
Nov 21	12.9	11.9	12.2
Nov 28	13.8	12.7	13.7
Dec 7	14.5	13.9	14.9
Dec 12	18.3	17.6	18.1
Dec 19	18.8	18.1	18.7
Dec 29	18.6	18.0	18.7

These values are the means of five samples for the Soxhlet and Horiba systems, and of 50 samples for the LR NMR system.

Table 9.6: Olive oil extraction from Negev Desert cultivated olive varieties using the Oliomio system and laboratory Horiba 350 analysis, 2005 season

Olive variety	Olive oil percentage	
	Oliomio	Horiba 350
Barnea	14.3	17.2
Souri	16.6	19.8
Leccino	14.1	16.4
Frantoio	12.9	15.2
Picual	14.0	16.5
Arbequina	15.0	16.9
Kalamata	10.3	16.7
Koreneiki	15.8	18.6
Phicolin	18.0	20.9

These values are the means of three separate Horiba analyses and one Oliomio extraction of each olive variety.

The data obtained in Table 9.6 indicate an average difference in oil extraction of about 2–3 percent between the Oliomio mini-olive oil system and the Horiba system. Usually, such a difference between an olive oil mill and laboratory testing is well accepted, as laboratory analysis is supposed to provide information regarding the oil potential of olives. Furthermore, the difference is explained by the fact that for laboratory analysis only the flesh of the olive is crushed and assessed for oil content, whereas in the actual mill system the crushed paste consists of both flesh and stone, and of course a certain amount of oil is lost in separating the oil fraction from the organic waste material. Only in the case of Kalamata was a large difference in oil extraction shown; here, the Oliomio result was more than 6 percent less than that obtained by the laboratory system (10.3 versus 16.7 percent). Laboratory analysis therefore provides a

Table 9.7: Olive oil extraction from Negev Desert cultivated olive varieties by the Oliomio system: comparison of trees cultivated with saline and with fresh water, 2004 season

Olive variety	Olive oil	Percentage
	Saline water (4.2 dS/m)	Fresh water (1.2 dS/m)
Barnea	15.3	16.2
Souri	19.6	17.8
Leccino	12.6	13.4
Frantoio	12.9	10.2
Picual	12.8	10.5
Arbequina	14.4	12.6
Kalamata	14.9	12.7
Koreneiki	19.8	16.6
Phicolin	21.4	18.3

relatively good estimate of the oil potential of olives; however, in an actual milling system it can be difficult to separate the oil and the waste material. The relatively high levels of emulsification agents such as phospholipids mean that it is necessary to add a significant volume of water to precipitate the heavier organic material from the lighter oil fraction. In Kalamata and some other olives, this seems to be the cause of the unusually low level of oil extraction. It is not clear whether this phenomenon is directly related to desert environmental conditions or to cultural practices, or whether it is mainly related to genetics. A study to address these issues is already underway.

The percentage oil content in the waste material obtained in the final Oliomio process is shown in Table 9.7. The results presented in this table show that some olive varieties produce more oil when irrigated with saline water, while others produce more oil when irrigated with fresh water. This suggests that the use of saline water does not directly affect the rate of olive oil accumulation. As discussed in Chapter 7, it appears that the main factor dominating the oil yield of each variety is genetic. However, saline water may affect the rate of maturation of each olive variety, and thus indirectly affect the oil content of olives at an early or later date. Based on experience gained over recent years, it seems that saline water irrigation combined with Negev Desert conditions delays somewhat the rate of olive maturation in comparison to other areas in the northern part of Israel that are not considered to be desert. This issue needs to be systematically studied. In any case, as discussed in Chapter 8, it is very important to monitor the level of olive oil content regularly.

Another study carried out in the Negev Desert area regarding the effect of intensive irrigation versus dry regulated irrigation on Muchasan olive oil extraction clearly showed that the water content of the intensively irrigated olives (58 percent) was significantly higher than in olives that were cultivated under supplemental irrigation only (42 percent). Obviously, the oil content was higher in the less intensively irrigated olives (26 percent per fresh weight) than in the intensively irrigated olives (21 percent). Furthermore, the efficiency of olive oil extraction in a commercial tri-phase olive oil mill was greater in the less irrigated olives than those that had been more intensively irrigated. This was concluded from the oil percentage obtained from the solid waste material (6.5 versus 9 percent, based on dry weight, respectively). These results suggest that it is important to gradually decrease the rate of irrigation in the final stages of olive maturation. This practice is indeed risky in desert environments, but may pay for itself in easier mechanical harvesting (as discussed in Chapter 8) and also the higher operational efficiency of the olive oil mill in obtaining a greater amount of oil from the olives.

9.4 Decantation, bottling and shelf-life, with an emphasis on Negev Desert conditions

Premium-quality oils should be stored in stainless steel containers and maintained at a constant temperature between 8° and 18 °C. After processing, the oil should be stored in bulk for 1–3 months to allow any particulate matter and vegetable water to settle out. Bulk storage and decantation eliminate the problems of sediment in bottles and oil contact with processing water residues that could lead to off flavors in the oil.

When olive oil is too old and has oxidized, it usually becomes rancid. Rancidity is most commonly detected by taste, but can be checked chemically. The "rancimat" chemical method is mostly used for large industrial frying operations. Oil doesn't suddenly go rancid; it slowly becomes more oxidized and, as it does, the flavor suffers.

Different oils age at different rates. Some olive varieties make oil with more natural antioxidants, which resist ageing. These oils may be good for up to three to four years if properly stored in unopened containers. Other oils, particularly unfiltered, may be unpalatable in a year even if stored well.

A two-year-old olive oil may taste rancid to some while not to others. Most people would be put off by the taste of any vegetable oil more than four to five years old. Rancid oil has fewer antioxidants, but is not poisonous. A good percentage of the world's population routinely eats rancid oil because of lack of proper storage conditions, and some actually prefer the taste. In historical times, olives that had dropped to the ground or may have spoiled were made into olive oil which was stored in open-mouthed earthenware vats. Practices like these encouraged rancidity. People have come to expect non-rancid oil in the past 50 years because of chemical refining and better production and storage methods.

Heat, exposure to light, and exposure to air are the three major factors that affect the oxidation and rancidity of the oil in bottles. Since desert areas are prone to all these conditions, extra precautions need to be taken, especially regarding processing and bottling, and storage periods. A tightly capped bottle that is full will oxidize less than a large tin that is only half-full. The longer the oil sits, the more rancid it will become. The following factors also affect the quality of the oil and its shelf-life.

- *Olive variety*. Polyphenols and other natural antioxidants in the oil help to keep the oil from going rancid, and some varieties have more antioxidants than others. The Tuscan varieties – Frantoio, Coratina, Pendolino and Leccino – tend to be higher in polyphenols than others.

- *Time of picking*. Olives picked earlier in the year may have more polyphenols and a longer shelf-life. Low-polyphenol olive oils often needed to blend with high-polyphenol oils to give a longer shelf-life.

- *Picking method*. In some parts of the world, nets are placed under the trees and the olives are allowed to drop for weeks as they ripen. Any remaining olives are then beaten from the tree and the nets are gathered. Some olives may therefore have been off the tree for weeks or months, spoiling in the interim. Olives that are badly bruised during picking will become more rancid if not pressed within hours. However, in intensive cultivation olive fruits are harvested only by machines, so there it is important that great care is taken during picking time, especially in desert areas.

- *Time to milling*. Live olives start to die once they have been picked, and the longer it takes to get to the mill, the more oxidized the oil will be. Producers who want to produce extra-virgin oil must get the olives to the mill within 48 hours of picking. Refrigeration and good ventilation of the orchard bins extends the olives' life.

- *Milling method*. During the milling process, the olive paste may be exposed to air. The old-fashioned stone wheel and hydraulic press with jute mats or Rapanelli discs may be very picturesque, but expose the paste to far too much air. In modern techniques, where malaxation is carried out in specially made tanks, it is important that the paste is always enclosed with minimal contact with the air. Some mills bathe the paste with an inert gas during the mixing or malaxation step, or perform it in a vacuum. Heating the paste during malaxation will extract more oil, but will hasten oxidation. Centrifugal presses expose the paste to zero air.

Extra-virgin olive oil is not filtered, because it can lose some aromatic and taste properties during the procedure, so it is decanted. This process takes about 40 days; the olive oil is left to rest at a temperature of about 18 °C, and the residues in suspension begin to sediment on the bottom of the tank. It is then only necessary to remove the oil via an outlet situated above the level of the solid residues.

References

Alburquerque, J. A., Gonzálvez, J., García, D., & Cegarra, J. (2004). Agrochemical characterisation of "alperujo," a solid by-product of the two phase centrifugation method for olive oil extraction. *Bioresource Technol*, 92, 195–200.

Amirante, P., Di Renzo, G. C., Di Giovacchino, L., Bianchi, B., & Catalano, P. (1993). Evolución tecnológica de las instalaciones de extracción del aceite de oliva. *Olivae*, 48, 43–53.

Ben-Miled, D. D., Smaoul, A., Zarrouk, M., & Cherif, A (2000). Do extraction procedures affect olive oil quality and stability? *Biochem Soc Trans*, 28, 929–933.

Di Giovacchino, L. (1996). Influence of extraction systems on olive oil quality. *Olivae*, 63, 52–55.

Di Giovacchino, L., Solinas, M., & Miccoli M, M. (1994). Effect of extraction systems on the quality of virgin olive oil. *J Am Oil Chem Soc*, 71, 1189–1194.

Kiritsakis, A. (1991). Extraction of olive oil. In: A. Kiritsakis (Ed.), *Olive Oil* (pp. 61–69). American Oil Chemists Society, Champaign, IL.

Kiritsakis, A. K. (1998). *Olive Oil from the Tree to the Table* (2nd edn.). Food & Nutrition Press, Inc., Trumball, CT, 26.

MARAN-ULTRA (2006). *MARAN-ULTRA SFC User Manual* V6.2. Oxford: Oxford Instruments, Molecular Biotools Ltd.

Montedoro, G., & Petruccioli, G. (1972). Enzymatic treatments in the mechanical extraction of olive oil. *Acta Vitam et Enzym*, 26, 171.

Nordon, A., Macgill, C. A., & Littlejohn, D. (2001). Process NMR spectrometry. *Analyst*, 126, 260–272.

Petruccioli, M., Servili, M., Montedoro, F., & Federici, F (1988). Development of recycle procedure for the utilization of vegetation waters in the olive oil extraction process. *Biotechnol Letts*, 10, 55–59.

Ranalli, A., & Angerosa, F. (1996). Integral centrifuges for olive oil extraction. The qualitative characteristics of products. *J Am Oil Chem Soc*, 73, 417–422.

Ranalli, A., Gomes, T., Dicuratolo, D., Contento, S., & Lucer, L. (2003). Improving virgin olive oil quality by means of innovative extracting biotechnologies. *J Agric Food Chem*, 51, 2597–2602.

Vierhuis, E., Servilli, M., Baldioli, M., Schols, H. A., Voragen, A. G. J., & Montedoro, G. F. (2001). Effect of enzyme treatment during mechanical extraction of olive oil on phenolic compound and polysaccharides. *J Agric Food Chem*, 49, 1218–1223.

Weissbein, S. (2006). Characterization of new olive (*Olea europae* L.) varieties' response to irrigation with saline water in the Ramat Nege area. MSc Thesis, Ben-Gurion University of the Negev, Beersheva, Israel.

Weissbein, S., Wiesman, Z., Efrath, Y., & Silverbush, M. (2008). Physiological and reproductive response olive varieties under saline water. *Hort Sci*, 43, 320–327.

Wiesman, Z., Itzhak, D., & Ben Dom, N. (2004). Optimization of saline water level for sustainable Barnea olive and oil production in desert conditions. *Sci Hort*, 100, 257–266.

Further reading

Ranalli, A., & Martinelli, N. (1995). Integral centrifuges for olive oil extraction at the third millennium threshold. Transformation yield. *Grasas y Aceites*, 46, 255–263.

Suarez, M. J. M. (1975). Preliminary operations. In: J. M. Moreno Martinez (Ed.), *Olive Technology*. FAO, Rome.

CHAPTER 10

Olive-mill wastes: treatment and product biotechnologies

10.1 Olive-mill wastes and their characteristics

Two kinds of waste are produced by olive mills. The first is pomace, which is the pulpy material that remains after removing most of the oil from the olive paste. Olive pomace constituents generally include water, nitrogenous compounds, non-nitrogenous compounds, cellulose, ash and oil. The second waste product is vegetation water, or olive-mill wastewater (OMW). OMW comes mainly from the water content of olive fruits (olives contain almost 50 percent water) and the water added during oil processing in the olive mill, which collectively end up as wastewater. OMW comprises a huge number of organic constituents, such as sugars, lipids, nitrogenous compounds, organic acids, polyalcohols and inorganic constituents, and more than 80 percent water (Table 10.1). The amount of each constituent depends on the type of processing.

The disposal of OMW is presently one of the most serious environmental problems in Mediterranean countries such as Spain, Italy and Greece (Rozzi and Malpei, 1996), as it is highly phytotoxic. Research into OMW treatment and disposal has had little success in finding an environmentally friendly and economically viable solution for general adoption. The

Table 10.1: Composition of olive-mill wastewater (OMW)

Constituent	Content (%)
Water	83.2
Inorganic constituents (ash)	1.8
Organic constituents	
Lipids	1.0
Nitrogenous compounds	2.0
Sugars	7.5
Organic acids	1.5
Polyalcohols	1.5
Pentose, tannins	1.5
Glycosides	Traces

Source: Carola (1975).

difficulties in disposing of solid and liquid waste are mainly related to their high organic matter content. In terms of the pollution effect, $1 \, m^3$ of OMW is equivalent to $100–200 \, m^3$ of domestic sewage (Evagelia et al., 2006). Its high phosphorus content also accelerates the growth of algae due to eutrophication. DellaGreca et al. (2001) have shown that OMW has a negative effect on aquatic ecosystems, while Fiorentino et al. (2004) have found that exposure corresponding to just 1 l of unprocessed OMW mixed in 100,000 liters of circulating water is highly toxic to some aquatic creatures (local river fishes, and some crustaceans).

The typical composition of OMW is: 83–96 percent water, 3.5–15 percent organic material and 0.5–2 percent mineral salts (Evagelia et al., 2006). Contamination-related parameters of OMW are described in Table 10.2, and the chemical characteristics of OMW are listed in Table 10.3.

Olive mills are usually small-scale enterprises that cannot afford the cost of proper wastewater treatment unless the procedure is very simple and cheap. Most treatment technologies require high investment costs. Thus, centralized treatment plants that can treat the OMW produced by several mills are considered more viable. This creates an extra burden in operational costs, as high transportation costs due to geographic dispersal of the mills must be taken into account. The disposal of these materials represents a problem that is further aggravated by legal restrictions.

Therefore, agro-industrial waste is often utilized as feed or fertilizer, although demand for these products varies depending on agricultural production. Valuable nutrients contained in agro-industrial waste are lost when they are disposed of, and therefore new ideas regarding the exploitation of these waste by-products as food additives or supplements with high nutritional value are of increasing interest, since the recovery of such wastes may be economically attractive (Vasso and Constantina, 2006).

Table 10.2: Maximum and minimum values of the main contamination-related parameters of olive-mill wastewater (Rafael *et al.*, 2006)

Parameter	Units	Maximum	Minimum
pH		6.7	4
Redox potential	mV	−330	−80
Conductivity	mS	16	8
Density	g/l	1.100	1.016
Color	U pt-Co	180, 000	52, 270
Turbidity	UNT	62, 000	42, 000
Suspended solids	g/l	9	1
Settleable solids	ml/l h	250	10
Biochemical oxygen demand, BOD_5	mg/l	110, 000	35, 000
Chemical oxygen demand, COD	mg/l	170, 000	45, 000
Oxygen uptake rate, OUR	mg/l h	100	50
Total bacteria	10^6 col/ml	5	–
Total yeasts and fungi		5	–

10.2 Treatment of vegetation wastewater

10.2.1 Chemical treatment

The effect of olive-mill effluents on sewers is quite severe, and is related to the acidity and the suspended solids content of the waste. Extensive damage to sewerage systems, caused by OMW, has been reported in the Apulia region of Italy (Mendia and Procino, 1964). Fermentation also occurs if OMW is stored in open tanks before later being discharged to the surrounding soil (Balice *et al.*, 1986). During the oil production period, offensive smells due to fermentation can be very strong near the mills, and give rise to endless complaints.

Chemical oxidation treatments are often carried out in order to eliminate hazardous organic pollutants, using single oxidants such as chlorine, ozone, UV radiation and hydrogen peroxide. However, the degradation of pollutant content by single chemical treatments may be limited if the organic matter is especially refractory to the oxidants. Advanced oxidation processes (AOPs) are based on the generation of highly reactive and oxidizing free radicals which possess high oxidant power. The production of radicals is achieved by combining hydrogen peroxide with ferrous ions (Fenton's reaction) (Walling, 1975). The single Fenton's reaction and the photochemically enhanced Fenton's reaction are considered the most promising processes for the remediation of highly contaminated waters (Oliveros *et al.*, 1997).

Table 10.3: Main chemical characteristics of two-phase olive-mill wastewater given by several authors (Roig *et al.*, 2006)

Parameter	(a)	(b)	(c)	(d)	(e)	(f)	(g)	(h)
Humidity (%)	61.8	64	57	64.5	65	64	49.6	71.4
pH	4.9	5.32	nd	5.23	5.4	5.5	6.8	5.19
EC (dS/m)	1.78	3.42	nd	5.24	nd	3.47	1.2	2.85
OM (%)	97.4	93.3	98.5	94.3	95.4	91.6	60.3	94.5
C_{OT}/N_T	53	47.8	59.7	49.3	29.3	42	32.2	46.6
N_T (g/kg)	10.5	11.4	100	11.3	18.5	13.5	11.0	9.7
P (g/kg)	nd	1.2	0.5	0.9	nd	1.4	0.3	1.5
K (g/kg)	nd	19.8	6.3	24.3	nd	15.9	29.0	17.1
Ca (g/kg)	nd	4.5	2.6	nd	nd	2.3	12.0	4.0
Mg (g/kg)	nd	1.7	nd	nd	nd	0.9	1.0	0.5
Na (g/kg)	nd	0.8	nd	nd	nd	nd	0.2	1.0
Fe (mg/kg)	nd	614	nd	526	nd	769	2600	1030
Cu (mg/kg)	nd	17	nd	17	nd	21	13	138
Mn (mg/kg)	nd	16	nd	13	nd	20	67	13
Zn (mg/kg)	nd	21	nd	18	nd	27	10	22
Lignine (%)	41.2	42.6	19.8	47.5	nd	46.8	nd	35
Hemicellulose (%)	nd	35.1	15.3	38.7	nd	nd	nd	nd
Cellulose (%)	nd	19.4	33.7	17.3	nd	nd	nd	nd
Lipids (%)	3.76	12.1	10.9	18.0	11.0	12.7	nd	8.6
Protein (%)	nd	7.2	6.7		nd	nd	nd	nd
Carbohydrate (%)	nd	9.6	19.3	9.6	12.7	10.4	nd	nd
Phenols (%)	0.54	1.4	2.4	1.5	2.1	0.5	nd	nd
Pb, Cd, Cr, Hg (mg/kg)	nd	nd	nd	<5	nd	nd	nd	<1

nd, not determined; (a)–(h) are various sources used by the authors.

10.2.2 Biological treatment

Chemical treatment of the olive-mill wastewater can be replaced or combined with biological methods. Of the biological treatments reported for OMW, aerobic processes in a completely mixed and activated sludge reactor have shown high COD (chemical oxygen demand) removal efficiencies. A full-scale activated sludge plant has been in operation in Italy since 1979 for the combined treatment of olive-mill effluents and domestic sewage (Giorgio *et al.*, 1981). Both aerobic and anaerobic digestion of wastewater using fluidized beds and hybrid reactors have been developed.

Aerobic treatment of olive mill effluent

Aerobic processes can operate efficiently only if the feed concentration is relatively low (on the order of 1 g COD/l). Aerobic microorganisms degrade a fraction of the pollutants in the effluent

Table 10.4: Some parameters of wastewater from olive wash prior to oil production (Borja *et al.*, 1995)

Parameter	Value
pH	5.8
COD (mg/l)	3025
Soluble COD (mg/l)	2375
BOD_5 (mg/l)	1920
Total suspended solids (mg/l)	300
Volatile suspended solids (mg/l)	280

Table 10.5: Main parameters of the wash waters for virgin olive oil purification (Borja *et al.*, 1995)

Parameter	Value
pH	5.5
Total solids (mg/l)	1610
Soluble COD (mg/l)	2200
BOD_5 (mg/l)	1400
Total suspended solids (mg/l)	820
Volatile suspended solids (mg/l)	690

by oxidizing them with an external oxygen source (air or pure oxygen). These microorganisms use most of the remaining fraction of the pollutants to produce biomass, which is later removed from the water. Borja and colleagues (1995a) reported an aerobic treatment of wastewater from the washing of olive fruits prior to the oil production; the wastewater characteristics are described in Table 10.4. The aerobic process was carried out in a completely mixed reactor without recycling. The wastewater from the washing of olives was more efficiently treated at an operating temperature of 28 °C (compared to 14 °C), because the kinetic parameters are temperature dependent. Over 93 percent of soluble COD was removed at a retention time of 6 h.

Borja and colleagues (1995b) also reported on aerobic treatment of wash waters derived from the purification of virgin olive oil in the two-phase extraction process. The process used a completely mixed activated sludge reactor, and the main parameters of the wash waters are described in Table 10.5.

The results obtained indicated more than 93 percent removal of COD under various operational conditions (four values of solids retention time, ranging from 4 to 15 days; hydraulic retention times ranging from 8 to 10 hours).

Aerobic treatment is costly for olive mills on a local scale. The main reason is that the aerobic treatment of concentrated wastewaters yields huge volumes of excess secondary sludge that has to be removed from the system. In addition, the treatment cannot easily remove pollutants such as polyphenols and lipids. Aerobic treatment can be used as a pre-treatment or post-treatment step to increase the efficiency of the main treatment process used (Evagelia *et al.*, 2006).

Anaerobic treatments of olive mill effluent

Anaerobic treatment of moderate- and high-strength wastewaters can achieve a high degree of purification. Importantly, a combustible biogas is generated. The production of biogas containing flammable methane could eventually provide a large proportion of the energy needs in the plant (Wheatley, 1990).

Anaerobic digestion is carried out in airtight vessels by bacteria that do not require oxygen to decompose organic compounds. The growth rates of these micro-organisms are appreciably lower, and the process control of anaerobic treatment more delicate, than with aerobic micro-organisms. Borja and colleagues (1998) studied the anaerobic digestion of wastewater from the washing of olives derived from a two-phase extraction process on the laboratory scale. A fluidized bed reactor with the biomass immobilized on sepiolite at 35 °C was used to treat the wastewater. COD removal efficiencies in the range of 50–90 percent were achieved in the reactor when evaluated at organic loading rates of 0.46 and 2.25 g COD/l per day.

The yield coefficient of methane production was 0.28 l methane STP/g COD removed. Borja *et al.* (1998) also described anaerobic digestion of wastewater derived from the washing of virgin olive oil. The study was carried out in a completely mixed reactor with the biomass immobilized on sepiolite operating at 35 °C. The bioreactor removed 92 percent of the initial COD (3.5 g/l).

Anaerobic treatment is considered to be most suitable for OMW detoxification. Important reasons for this might be the feasibility of treating wastewater with a high organic load, the low energy requirements, the production of biogas, the production of significantly less waste sludge than with aerobic processes, and the ability to restart the plant easily after several months of shut-down (Rozzi and Malpei, 1996; Niaounakis and Halvadakis, 2004).

10.2.3 Combined treatment

An improved technique for treating OMW has been practiced (Mario *et al.*, 2004), based on the combined actions of catalytic oxidation and microbial biotechnologies. In this technique, OMW (COD 10,000–100,000 mg O_2/l) was oxidized up to 80–90 percent by stoichiometric amounts of dilute hydrogen peroxide (35%) and in the presence of water-soluble iron catalysts, at concentrations of up to 1% w/w (Table 10.6). In the combined action, the mineralization activity of a selected microbial consortium was used to degrade residual volatile and non-

Table 10.6: COD removal, chemical and microbiological parameters of olive-mill wastewater after chemical, biological and combined treatments (Mario et al., 2004).

Added H_2O_2 (%)[c]	Chemical treatment[a] COD residual (equiv/l), (COD removal, %)	Biological treatment[b] COD residual (equiv/l), (COD removal, %)	Parameters after both treatments CFU/ml (log)	Total phenol (%)	GI (%)
0	Untreated sample 11.40(0)	Untreated sample 11.40(0)	1.7	1.24	26.0
0	Untreated sample 11.40(0)	5.80(49)	5.48	0.76	45.5
15	4.90(59)	4.30(62)	4.62	0.44	71.5
30	4.35(62)	3.25(71)	5.39	0.32	78.5
60	2.20(81)	1.20(90)	5.24	0.27	81.0
100	1.70(85)	1.15(90)	4.39	0.21	76.5

[a] *Reaction condition: OMW 500 ml (COD 11.4 equiv/l); Fe(III) added as $Fe_2(SO_4)_3$ 100 mM; H_2O_2, 35% added by portion. Parameters monitored after 1 h of reaction.*

[b] *Conditions, aeration (50 ml/min), inoculum of crude OMW (30 ml); parameters monitored after 15 days of treatment.*

[c] *Equiv of H_2O_2 per equiv of initial COD.*

volatile organic compounds into CO_2 and biomass. The biological treatment not only allows the attainment of further breakdown of the COD (up to 90 percent), but also has the potential to eliminate phytotoxicity.

10.3 Uses of olive-mill waste

The waste from olive mills can itself be used in a variety of secondary products. For instance, solid waste can be used as a source of animal feed. Olive cake and other solid residues contain oil, carbohydrates and protein in high amounts, and can be used accordingly. Olive leaves are a well-known source of antioxidant compounds, and are marketed as herbal teas with diuretic, antihypertensive and antioxidant effects. The stones can be removed during processing and used as an energy source due to their high calorific content. In addition, stones have been proposed for use as a soil-less substrate in hydroponics (Melgar *et al.*, 2001).

Adequate treatment of olive-mill effluent can lead to the production of by-products that can increase the economic feasibility of the treatment. OMW may be regarded as a source of inorganic and organic compounds, and recovery of the residue substances can produce by-products for use in agriculture, biotechnology, pharmaceutics and food industries. Some of the uses of olive-mill waste are described below.

10.3.1 Production of bio-fertilizers

OMW should not be applied directly to soil or crops because of its phytotoxic properties. However, it can be converted into a useful fertilizer and soil conditioner, due to its high content of organic matter and plant nutrients, and then used. Cegarra and colleagues (1996) used treated OMW in cultivating horticultural and other crops, and found that yields obtained with OMW fertilization were similar to (and sometimes higher than) those obtained with commercially available fertilizers. Chatjipavlidis *et al.* (1996) studied the exploitation of OMW as a substrate for the cultivation of dinitrogen-fixing micro-organisms *(Azotobacter vinelandi)* capable of transforming the material into an organic liquid of high fertilizing and soil-conditioning value.

Transformation of OMW into bio-fertilizer was conducted in a designated commercial pilot plant in Messinia, Greece. Two-phase OMW processing is described in Figure 10.1. During the first stage, hydrogen peroxide is added under alkaline conditions. In the second stage, the effluent from the first stage is bioconverted, by a N_2-fixing microbial consortium under well-aerated conditions, into the end product. The bio-fertilizer is a thick, yellow, non-phytotoxic liquid; its characteristics are listed in Table 10.7.

The main advantage of the bio-fertilizer is the formation of stable soil aggregates from the exo-polysaccharides produced in the process, which contribute to the improvement of soil tilt and structure (Flouri *et al.*, 1990). It contains almost all the major plant nutrients originating from the olive fruits. It is enriched with organic forms of nitrogen through the mechanism of

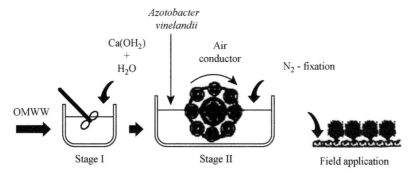

Figure 10.1: Schematic diagram of two-stage bioremediation plant for oil-mill waste (Chatjipavlidis et al., 1996).

Table 10.7: Characteristics of fresh raw material and the end product of olive-mill wastewater (Chatjipavlidis et al., 1996)

Characteristic	Units	Raw material	Bio-fertilizer
Total organic carbon	mg/l	40, 250	37, 600
Total nitrogen	mg/l	1, 360	1, 642
Total solids	mg/l	8.9	9.6
pH		5.4	7.9
Conductivity	μs/cm	10, 000	18, 000
PO_4^{-3}	mg/l	423	550
K^+	mg/l	6, 100	6, 350
Germination index (25% dilution)		0	104

N_2-fixation at the expense of the carbon source.

10.3.2 Production of compost

Composting is one of the main technologies for recycling OMW to allow the recovery of nutrients to the soil. The composting process involves controlled aerobic biological degradation of organic substrates (wastes, residues, etc.). The microorganisms used for this purpose utilize the organic substrate for growth and multiplication. By this process, stabilized, humic substances and minerals are produced that can be used to increase soil fertility and plant production. OMW is added to the solid substrate, which becomes enriched by organic matter, and the OMW is later evaporated or consumed. Manios *et al.* (2004) investigated the commercial production of compost from olive-mill waste. Solid waste from the olive oil industry and olive orchard operations, such as extracted olive-press cake and olive tree leaves and branches, have been used as carriers in a pilot plant.

10.3.3 Production of feed additives

OMW can be converted into different feed supplements. Anthocyanins can be extracted and purified from OMW, while the filter cake can be used as a liquid feed supplement, or dried and later used as animal feed (Codounis *et al.*, 1983). In this process, the eluent from olive extraction plants is first passed through an ultrafiltration unit, which separates the proteins and high molecular weight carbohydrates from simple sugars, anthocyanins, vitamins, amino acids and minerals. The extraction and purification of anthocyanins contained in the permeate can be achieved by chromatographic separation on ion-exchange resins. Recovered anthocyanins can be used as natural food-coloring agents.

10.3.4 Production of antioxidant food supplements

The most important OMW-derived antioxidants are polyphenols, flavonoids, anthocyanins, tannins, oleanolic acid and maslinic acid (Evagelia *et al.*, 2006).

Olive-derived polyphenols present in OMW have been shown to have antioxidant, antibiotic, antimicrobial and antifungal activity. They can be used as nutrition supplements or skin cosmetics. Due to their antimicrobial properties, they can also be used as antimicrobial agents in detergents and in rinsing and cleaning agents.

Hydroxytyrosol can be used as a food preservative, and in pharmacology and cosmetology because of its anti-aging and anti-inflammatory action. Oleanolic acid regulates cholesterol levels in the blood and helps to balance body weight. Maslinic acid has been widely investigated over recent years, and appears to possess anti-inflammatory and antihistaminic activity (Evagelia *et al.*, 2006). Marketable antioxidants found in the pulp of olives include tyrosol, hydroxytyrosol and oluropein. Briante and colleagues (2004) have developed a bioreactor for the production of highly purified antioxidants, which could be converted into pharmacologically active compounds.

10.3.5 Production of phytohormones

Yurekli and colleagues (1999) studied the microbial production of plant growth hormones such as abscissic acid (ABA), giberellic acid (GA_3), auxin (indole-3-acetic acid, IAA) and cytokines (zeatin) grown on OMW. White-rot fungi were used to degrade the pollution load (more then 47 percent COD removal) while increasing the amount of phytohormones. The quantities of various growth hormones (IAA, GA3, ABA and zeatin) obtained from the OWM are presented in Table 10.8.

10.3.6 Production of biopolymers

The production of biopolymers from OMW can decrease the economic constraints on olive-mill wastewater treatment. Production of biopolymers is extremely expensive (US$15–30 per

Table 10.8: Plant growth hormone production from *Trametes versicolor* and *Fantail trogil* grown on olive-mill wastewater (Yurekli *et al.*, 1999)

Particular	Plant growth hormones (μg/ml)			
	IAA	GA₃	ABA	Zeatin
OMW without inoculation	58.48	17.38	4.5	1.28
Inoculated by *T. versicolor*	75.20	180.37	5.32	50.92
Inoculated by *F. trogil*	111.76	475.40	16.28	22.10

kilogram) owing to the cost of substrates. The use of OMW as an alternative substrate can therefore reduce the production cost. Polymeric substances have been described in fresh vegetation water and sludge evaporation ponds (Saiz-Jimenez *et al.*, 1986).

Exopolysaccharides possess rheological properties such as high solubility in water and increased viscosity at low concentration. Pullulan is an extracellular polysaccharide produced by the fungus *Aureobasidium pullulans*, used in the food and pharmaceutical industries (Ramos-Cormenzana *et al.*, 1995). The latter study showed that OMW can be used as an organic substrate to produce pullulan by fermentation.

Xanthan is produced by *Xanthomonas campestris*, which that can also utilize dilute concentrations of OMW as an organic substrate (Lopez *et al.*, 2001). A high content of OMW in the cell inhibits biomass production.

References

Balice, V., Carrieri, C., Cera, O., & Di Fazio, A. (1986). Natural biodegradation in olive mill effluents stored in open basins, *Proceedings of the International Symposium on Olive Byproducts Valorization*, 4–7 March, Seville.

Borja, R., Banks, C. J., Alba, J., & Escobar, J. P. (1995a). The temperature dependence of the kinetic parameters derived for the aerobic treatment of wastewater from the washing of olives prior to the oil production process. *J Envir Sci Health A*, 30(8), 1693–1705.

Borja, R., Alba, J., & Banks, C. J. (1995b). Activated sludge treatment of wash waters derived from the purification of virgin olive oil in a new manufacturing process. *J Chem Technol Biotechnol*, 64, 25–30.

Borja, R., Alba, J., Martin, A., & Mancha, A. (1998). Influencia de la velocidad de carga organica sobre el proceso de digestion anaerobia de aguas de lavado de aceitunas de almazara en reactores de lecho fluidizado. *Grasas y Aceites*, 49, 42–49.

Briante, R., Patumi, M., Febbraio, F., & Nucci, R. (2004). Production of highly purified hydroxytyrosol from Olea Europaea leaf extract biotransformed by hyperthermophilic b-glycosidase. *J Biotechnol*, 111, 67–77.

Carola, C. (1975). By-products. In: M. Martinez (Ed.), *Olive Oil Technology* (pp. 77–87). FAO, Rome.

Cegarra, C., Paredes, A., Roig, M., Bernal, P., & García, D. (1996). Use of olive mill wastewater compost for crop production. *Intl Biodet Biodeg*, 38, 193–203.

Chatjipavlidis, I., Antonakou, M., Demou, D., Flouri, F., & Balis, C. (1996). Bio-fertilization of olive oil mills liquid wastes. The pilot plant in Messinia, Greece. *Intl Biodet Biodeg*, 38, 183–187.

Codounis, M., Katsaboxakis, K., & Papanicolaou, D. (1983). Progress in the extraction and purification of anthocyanin pigment from the effluents of olive-oil extracting plants. *Lebensm Wissensch Technol*, 7, 567–572.

DellaGreca, M., Monaco, P., Pinto, G., Pollio, A., Previtera, L., & Temussi, F. (2001). Phytotoxicity of low-molecular-weight phenols from olive mill wastewaters. *Bull Envir Contam Toxicol*, 67, 352–359.

Evagelia, T., Harris, N. L., & Konstantinos, B. P. (2006). Olive mill wastewater treatment. In: V. Oreopoulou (Ed.), *Utilization of By-Products and Treatment of Waste in the Food Industry* (pp. 133–154). Springer, New York, NY.

Fiorentino, A., Gentili, A., Isidori, M., Lavorgna, M., Parrella, A., & Temussi, F. (2004). Olive oil mill wastewater treatment using a chemical and biological approach. *J Agric Food Chem*, 51, 5151.

Flouri, F., Chatjipavlidis, I., Balis, C., Servis, D., & Tjerakis, C. (1990). *Effect of Olive Oil Wastes in Soil Fertility*. Tratamiento de Alpechines, Cordoba.

Giorgio, L., Andreazza, C., & Rotunno, G. (1981). Esperienze sul funzionamento di un impianto di depurazione per acque di scarico civili e di oleifici. *Ingegneria Sanitaria*, 5, 296–303.

Lopez, M. J., Moreno, J., & Ramos-Cormenzana, A. (2001). *Xanthomonas campestris* strain selection for xanthan production from olive mill wastewaters. *Water Res*, 35, 1828–1830.

Manios, M., Kalogeraki, M., Terzakis, M., Mikros, A., & Manios, V. (2004). Cocomposting olive residuals and green waste on Crete. *Biocycl Intl*, 45(2), 67–70.

Mario, B., Lolita, L., Nicola, D., Lucia, T., Claudia, B., & Giancarlo, R. (2004). Improved combined chemical and biological treatments of olive oil mill wastewaters. *J Agric Food Chem*, 52, 1228–1233.

Melgar, R., Gómez, M., Benítez, E., Polo, A., Molina, E., & Nogales, R. (2001). Posibilidades de utilización de los huesos triturados de aceituna como sustrato de cultivos hidropónicos. Resultados preliminares. I Encuentro Internacional de Gestión de Residuos Orgánicos en el Ámbito Rural Mediterráneo, Pamplona, Spain.

Mendia, L., & Procino, L. (1964). Studio sul trattamento delle acque di rifiuto dei ftantoi oleari. Proceedings of the ANDZS Conference, Bologna, Italy.

Niaounakis, M., & Halvadakis, C.P. (2004). *Olive-Mill Waste Management – Literature Review and Patent Survey*. Athens Typothito-George Dardanos.

Oliveros, E., Legrini, O., Hohl, M., Muller, T., & Braun AM, T. (1997). Large scale development of a light-enhanced Fenton reaction by optimal experimental design. *Water Sci Technol*, 35, 223–230.

Rafael, B., Francisco, R., & Barbara, R (2006). Treatment technologies of liquid and solid wastes from two-phase olive oil mills. *Grasas y Aceites*, 57(1), 32–46.

Ramos-Cormenzana, A., Monteoli-Vasanchez, A., & Lopez, M. J. (1995). Bioremediation of alpechin. *Intl Biodet Biodeg*, 35, 249–268.

Roig, A., Cayuela, M. L., & Sánchez-Monedero, M. A. (2006). An overview on olive mill wastes and their valorisation methods. *Waste Manag*, 26, 960–969.

Rozzi, A., & Malpei, F. (1996). Treatment and disposal of olive mill effluents. *Intl Biodet Biodeg*, 1, 135–144.

Saiz-Jimenez, C., Gomez-Alarcon, G., & de Leeuw, J.W. (1986). Chemical properties of the polymer isolated in fresh vegetation water and sludge evaporation ponds. In: *International Symposium on Olive by Products Valorization, Madrid*. Rome: FAO, pp. 41–60.

Vasso, O., Constantina, T. (2006). Utilization of plant by-products for the recovery of proteins, dietary fibers, antioxidants, and colorants. In: *Utilization of By-Products and Treatment of Waste in the Food Industry*. New York NY: Springer, pp. 209–232.

Walling, C. (1975). Fenton's reagent revisited. *Acc Chem Res*, 8, 125–131.

Wheatley, A. (1990). *Anaerobic Digestion: A Waste Treatment Technology*. Elsevier Applied Science, Barking.

Yurekli, F., Yesilada, O., Yurekli, M., & Topcuoglu, S. F. (1999). Plant growth hormone production from olive oil mill and alcohol factory wastewaters by white-rot fungi. *W J Microbiol Biotechnol*, 15, 503–505.

Further reading

Albuquerque, J. A., Gonzálvez, J., Garcia, D., & Cegarra, J. (2004). Agrochemical characterisation of "alperujo", a solid by-product of the two phase centrifugation method for olive oil extraction. *Biores Technol*, 92, 195–200.

Baeta-Hall, L., Saagua, M. C., Bartolomeu, M. L., Anselmo, A. M., & Rosa, M. F. (2005). Biodegradation of olive oil husks in composting aerated piles. *Biores Technol*, 96, 69–78.

Cayuela, ML. (2004). Producción Industrial de compost ecológico a partir de residuos de almazara. PhD thesis, University of Murcia, Spain.

Cegarra, J., Amor, J. B., Gonzálvez, J., Bernal, M. P., & Roig, A. (2000). Characteristics of a new solid olive-mill by-product ("alperujo") and its suitability for composting. In: P. R. Warman, B. R. Taylor (Eds.), *Proceedings of the International Composting Symposium ICS 99, 1* (pp. 124–140). CBA Press Inc., London.

Madejon, E., Galli, E., & Tomati, U. (1998). Composting of wastes produced by low water consuming olive mill technology. *Agrochimica*, 42, 135–146.

Ordonez, R., Gonzalez, P., Giraldez, J. V., & Garcia-Ortiz, A. (1999). Efecto de la enmienda con alperujo sobre los principales nutrientes de un suelo agricola. In: R. Munoz-Carpena, A. Ritter, C. Tascon (Eds.), *Estudios de la Zona no Saturada*. ICIA, Tenerife.

Rubenchik, L. (1963). *Azotobacter and its Use in Agriculture*. Jerusalem: Academy of Sciences of Ukraine, USSR, Microbiological Institute.

Saviozzi, A., Levi-Minzi, R., Cardelli, R., Biasci, A., & Riffaldi R, A. (2001). Suitability of moist olive pomace as soil amendment. *Water Air Soil Poll*, 128, 13–22.

Vlyssides, A. G., Loizides, M., & Karlis, P. K. (2004). Integrated strategic approach for reusing olive oil extraction by-products. *J Clean Prod*, 12(6), 603–611.

Olive-oil quality biotechnologies

Olive oil is the lipophilic product of olive fruit. When olive fruits are harvested at maturity and processed, olive oil, which has a delicate and unique flavor, is obtained. Oil is one of the main constituents extracted from the olive fruits; others include water, sugars, proteins, anthocyanins and oleuropein, etc. In olive fruits, oil is dispersed within the fruit cells, forming drops that grow in size and may eventually fill the entire cell. The drops of olive oil vary from 39 to 63 µm in diameter. Bigger drops can deform the cell by pressing on the cell membrane. As the amount of oil increases, the color of the olive fruit changes from green to yellow, then reddish-purple and finally deep purple (Kiritsakis, 1998).

As with olives, many fruits, seeds, vegetables and plants contain edible oils, and in most cases the oils are similar in many ways. However, there are a few minor differences that have a significant effect on the characteristics of the oil, and which directly affect the oil quality. For example, in seed oils such as canola there is very little that the grower can do to alter the quality other than choosing the best cultivar and growing it using the best agronomic practices. In seed oils, environmental conditions are the main influence on oil content and chemical composition, because seed oil quality is very consistent in the intact seed. However, regarding olives, handling by the grower and the processing techniques will largely control the quality of the oil. Furthermore, olive oil quality depends a great deal on market preferences, which are based upon the consumers' perceptions of aroma, taste and color – which change over time and in olives from different locations. Olive growers therefore need to have a basic understanding of the consumers' preferences and how these can be catered for consistently.

This chapter presents information regarding the basic composition of olive oil, oil standards, the quality and purity of olive oil and how these are measured, and current research on olive oil quality analysis, with an emphasis on oil cultivated in desert conditions.

11.1 Composition of olive oil

Olive oil is mostly composed of triglycerides or triacylglycerols (TAGs) and some small quantities of free fatty acids (FFA), glycerol, phosphatides, pigments, flavor compounds, sterols, unidentified resinous substances and other constituents (Kiritsakis and Markakis, 1987). The overall constituents of olive oil can be divided in to two categories: the saponifiable fractions, which include TAGs, FFA, phospatides, and the unsaponifiable fractions, which include hydrocarbons, fatty alcohols, sterols. In virgin olive oil the unsaponifiable fractions account for 0.5–1.5 percent of the oil, while in olive-pomace oil the level increases to 2.5 percent (IOOC, 2003).

11.1.1 Triacylglycerols

Triacylglycerols are the chemical form in which most fat exists in food, as well as in the human body. They are also present in blood plasma and, in association with cholesterol, form the plasma lipids. Triglycerols are the major energy reserve for plants and animals. The TAGs of olive oil are comprised of three fatty acids attached to a glycerol backbone, as in other common plant oils. Technically, it is a type of glycerolipid (Figure 11.1). For good-quality oil, three fatty acids should be bound and remain as a TAG.

The glycerol unit can have any three of several fatty acids attached to form a TAG. The carbon chain may be of different lengths, and the fatty acids may be saturated, monounsaturated or polyunsaturated. The major triacylglycerols of the olive oils are oleic-oleic-oleic, palmitic-oleic-oleic, oleic-oleic-linoleic, and palmitic-oleic-linoleic (Fedeli, 1997). Biosynthesis of the triacylglycerols in olives generally follows 1,3 random distribution, which means that fatty acids on the triacylglycerols are randomly distributed in the 1 and 3 positions on the molecule; in the 2 position there is always an unsaturated fatty acid.

11.1.2 Fatty acids

Chemically a fatty acid is a carboxylic acid, often with a carboxyl group at one end with a long unbranched aliphatic tail (chain), which is either saturated or unsaturated. As

Glycerol

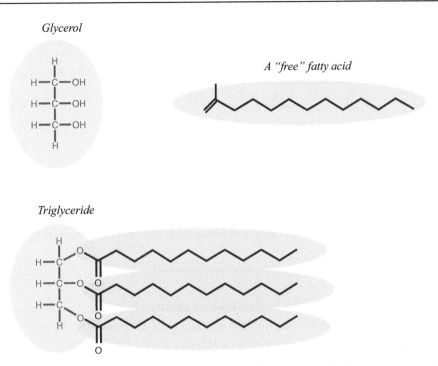

A "free" fatty acid

Triglyceride

Figure 11.1: Structure of a molecule of glycerol, a free fatty acid, and a triglyceride (triacylglycerol) compound of olive oil.
Source: www.oliveoilsource.com.

mentioned above, most of the fatty acids of olive oil are present as triacylglycerols, but a TAG unit may lose one fatty acid to become a diacylglycerol (DAG), or two fatty acids to become a monoacylglycerol (MAG). The main fatty acids presenting as glycerides in olive oil are oleic (C18:1), linoleic (C18:2), palmitic (C16:0) and stearic (C18:0). Oleic acid ($C_{17}H_{33}COOH$, or $CH_3(CH_2)_7CH=CH(CH_2)_7COOH$, also known as oleate; Figure 11.2) is a monounsaturated fatty acid and is dominant in olive oils, comprising 55–85 percent of the total. The IUPAC name of this is *cis*-9-octadecenoic acid. Linoleic acid ($C_{17}H_{31}COOH$, or $CH_3(CH_2)_4CH=CHCH_2CH=CH(CH_2)_7COOH$) is a polyunsaturated fatty acid with two double bonds, and comprises about 9 percent of the total. The IUPAC name of linoleic acid is *cis, cis*-9,12 octadecadienoic acid. Linolenic acid ($C_{17}H_{29}COOH$), which is also a polyunsaturate with triple bonds, makes up 0–1.5 percent of the total. The IPUAC name for this acid is *all-cis*-9,12, 15-octadecatrienoic acid. Palmitoleic (C16:1), myristic (C14:0), arachidic (C20:0), behenic (C22:0), lignoceric (C24:0), heptadecanoic (C17:0), heptadecenoic (C17:1) and eicosenoic (C20:1) acids are the other minor fatty acids found in olive oil.

Fatty acids on which the first double bond exits at the third carbon–carbon bond and sixth carbon-bond from the terminal methyl end (omega, ω) of the carbon are called omega-3 (*n-3*)

(a)

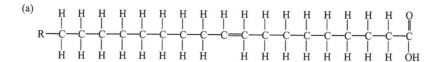

(b)

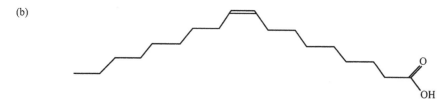

Figure 11.2: Structure of a typical fatty acid (oleic acid, 18:1 n-9): (a) details; (b) scheme.

and omega-6 (*n-6*) fatty acid, respectively. Olive oil contains both omega-3 and omega-6 fatty acids. Linolenic acid (C18:3) and the linoleic (C18:2) fatty acids are respectively the common omega-3 and omega-6 fatty acids in olive oils (Quiles *et al.*, 2003). Omega-3 fatty acids are important in preventing cardiovascular disease, and are particularly high in oily fish such as salmon and flax-seed oil. There is currently some debate about how much omega-3 versus omega-6 should be included in the human diet. According to the *Merck Manual*, an authoritative medical text, essential fatty acids should make up 1–2 percent of the dietary calories for adults, with a suggested ratio of 10:1 for omega-6:omega-3 fatty acids. Olive oil contains about 10 percent linoleic acid (an omega-6 oil) and about 1 percent linolenic acid (an omega-3 oil); therefore, the ratio is 10:1. If we use only olive oil for our dietary fat, and fats represent 30 percent of the calories in our diet, then we would be getting 3 percent of our calories in the form of essential fatty acids in a 10:1 ratio. Other more recent studies suggest that closer to a 5:1 ratio would be more beneficial (*Olive Oil Encyclopedia* (http://www.oliveoilsource.com).

11.1.3 Minor constituents

TAGs in olive oil constitute 95–98 percent of its composition; the remainder of the oil consists of other minor constituents, like mono- and di-glycerols, phosphatides, waxes and esters of sterols, and several other minor non-glyceride (unsaponifiable) constituents. Though these minor constituents are only a small portion of the olive oil, in comparison to TAGs, they have very significant roles. Non-glyceride fatty acid esters, hydrocarbons, sterols, triterpene alcohols, phenols, chlorophyll, flavor compounds and polar phenolic compounds such as hydroxytyrosol are the non-glyceride fractions of olive oils (Table 11.1) (Kiritsakis, 1998). Among the minor constituents, phenols and other related compounds such as tocopherols and sterols are present in only small amounts in the olive oil, but play a pivotal role.

Table 11.1: **Non-glyceride constituents of virgin and refined olive oil**

Non-glyceride constituents (ppm)	Olive oils	
	Virgin	Refined
Terpene alcohols	3500	2500
Sterols	2500	1500
Hydrocarbons	2000	120
Squalene	1500	150
Phenols and related substances	350	80
β-carotene	300	120
Fatty alcohol	200	100
Tocopherols	150	100
Esters	100	30
Aldehydes and ketones	40	10

Source: Fedeli (1988).

11.2 Regulation of olive oils and olive oil quality

The International Olive Council (IOC) (the International Olive Oil Council, IOOC, until 2006) is an intergovernmental organization based in Madrid, Spain, with 23 member states that subscribe to a United Nations charter to develop standards for quality and purity criteria for olive oil. The main aim of the IOC is to promote olive oil around the world by tracking production, defining quality standards, and monitoring authenticity. The main focus of the IOC is regulation of the legal aspects of the olive oil industry, and prevention of unfair competition. The IOC classifies olive oil according to its method of production, its chemistry and its flavor. The standards developed by the IOC are now recognized by most countries in the world.

More than 85 percent of the world's olives are grown in IOC member nations. The IOC officially governs 95 percent of international production and holds great influence over the rest. IOC standards are recognized by the vast majority of the world's oil producers and marketers. The United States is not a member of the IOC, and the US Department of Agriculture (USDA) does not legally recognize its classifications (such as extra-virgin olive oil). The USDA uses a different system, which it defined in 1948 before the IOC existed. However, the California Olive Oil Council (COOC), a private trade group, is petitioning the USDA to adopt IOC rules. Olive oils have been classified into different grades, such as industrial and retail. According to this, olive oil can be classified as virgin oil, refined oil and pomace oil:

- *Virgin oil* is produced by the use of physical means and no chemical treatment. The term *virgin oil*, when referring to production, has a different meaning from virgin oil on a retail label.

- *Refined oil* is oil that has been chemically treated to neutralize both strong tastes (characterized as defects) and the acid content (free fatty acids). Refined oil is commonly regarded as being of lower quality than virgin oil; if oil is labeled extra-virgin or virgin at the point of retail, it cannot contain any refined oil.

- *Pomace oil* is oil that has been extracted from the pomace using chemical solvents, generally hexane, and heat.

11.2.1 IOC standards

The International Standards of the IOC under resolution COI/T.15/NC no 3-25 (revised June 2003) lists nine olive-oil grades. According to the IOC, olive oils must not be adulterated with any other type of oil, must pass sensory analysis by a certified panel of tasters, and must meet the analytical criteria of the standard. The IOC defines olive oil as oil obtained solely from the fruit of olive trees (*Olea europaea*). Furthermore, virgin oils are obtained solely by mechanical means that do not lead to alterations in the oil content. All oils must pass tests for: free fatty acids, peroxide, UV absorbency, water & volatiles, insoluble impurities, flash point, metal traces, and halogenated solvents.

The nine olive grades for retailers are described below. It is mandatory for member nations of the IOC to comply with these, and most other countries also recognize them. These grades are split into two primary categories: olive oil and olive pomace oil.

Olive oil categories

Olive oils are obtained solely from the fruit of the olive tree (*Olea europae* L.), and exclude oils obtained using solvents or re-esterification, and those mixed with oils of other kinds (seed or nut oils). They are divided into six categories.

1. *Extra-virgin olive oil (EVOO)*. This is the highest quality rating for an olive oil. This oil, as evaluated numerically by the mean of a certified taste panel, contains zero (0) defects and greater than zero positive attributes. In other words, more than half of the tasters have indicated that it is not defective and has some fruitiness. Extra-virgin olive oil must also have a free acidity percentage of less than 0.8 percent and a peroxide value less than 20 meq/kg, and conform to all the standards listed in its category. Extra-virgin olive oil should have clear flavor characteristics that reflect the fruit from which it was made. In relation to the complex matrix of variety, fruit maturity, growing region and extraction technique, extra-virgin olive oils can be very different from one another.

2. *Virgin olive oil (VOO)*. This is oil that has achieved a mean sensory analysis rating from tasters with defects in the range of 0 to less than 2.5, has free acidity of less than 2 percent,

and conforms to all the other standards in its category. These are oils with analytical and sensory indices that reflect a slightly lower quality than extra-virgin olive oil.

3. *Ordinary virgin olive oil (OVOO)*. This is oil that has achieved a lower organoleptic rating (defects from the mean of tasters 2.5 to less than 6.0), has a free acidity of less than 3.3 percent, and conforms within its category to all other standards. This is inferior oil with notable defects that cannot be bottled under European Union (EU) laws, so it is sent for refining. The EU has eliminated this category, and other regulating agencies are likely to follow. It will simply be absorbed into the lampante category.

4. *Virgin lamp oil (or lampante virgin olive oil, LVOO)*. This is oil that has severe defects (greater than 6.0) or free acidity of greater than 3.3 percent, and conforms to the other standards within its category. It is not fit for human consumption, and must be refined. These oils come from bad fruit, or from improper handling and processing.

5. *Refined olive oil (ROO)*. This is oil obtained from virgin oils by refining methods that do not alter the initial glyceride structure. It has a free acidity of less than 0.3, and must conform to the other standards within its category. Refined olive oil must not originate from the solvent extraction of pomace. The refining process usually consists of treating poor virgin oil/lampante with sodium hydroxide to neutralize the free acidity, followed by washing, drying, odor removal, color removal, and filtration. In the process, the oil can be heated to as high as 220 °C under a vacuum to remove all the volatile components. Refined olive oil is usually odorless, tasteless and colorless. It is not considered fit for human consumption in many countries, including EU members.

6. *Olive oil (OO)*. This is oil that is a blend of refined and unrefined virgin oils. It must have a free acidity of not more than 1 percent and conform to the other standards within its category. This grade of oil actually represents the bulk of the oil sold on the world market to the consumer. Blends are made in set proportions to create specific styles and prices. Oils in the US labeled as "Extra Light" would most likely be a blend dominated by refined olive oil. Other blends with more color and flavor would contain more virgin or extra-virgin olive oil.

Olive-pomace oil categories

Olive-pomace oil is obtained by treating olive pomace with solvents. It does not include oils obtained in the re-esterification processes, or any mixture with oils of other kinds (seed or nut oils). There are three categories.

1. *Crude olive-pomace oil (COPO)*. This is the solvent-extracted crude oil product as it emerges from the pomace extractor following distillation to separate and recover most of the solvent. EU law also defines any oil containing 300–350 mg/kg of waxes and aliphatic

alcohols above 350 mg/kg as crude pomace oil. It is not fit for human consumption, but is intended for refining.

2. *Refined olive-pomace oil (ROPO)*. This is oil obtained from crude pomace oil by refining methods that do not alter the initial glyceride structure. It has a free acidity of not more than 0.3 percent, and its other characteristics must conform to the standard in its category. Refining is by the same methods used for refined olive oil, except that the source of the raw product comes from pomace by means of solvent extraction. It is not considered fit for human consumption in many countries, and under EU laws.

3. *Olive-pomace oil (OPO)*. This is a blend of refined olive-pomace oil and virgin olive oil that is fit for human consumption. It has a free acidity of not more than 1 percent and must conform to the other standards within its category. In no case can this blend be called "olive oil."

The IOC has also extended the details of the standards to other parameters, such as sterol composition, wax content, UV absorption, and equivalent carbon number (ECN). Providing that the olive fruit is sound, at production most olive oil is extra virgin. When the fruit quality is low, the oil is refined. The classification of olive oil is usually performed just after production. However, stability to oxidation is an important requirement excluded from regulation, and such oxidation can lead to a subsequent loss of extra-virgin quality status (Monteleone *et al.*, 1998). Some parameters that are not included in the IOC and EC standards (EC, 1991; IOOC, 2003), such as phenolic content, are known to have a significant effect on the stability and sensory characteristics of olive oil. In addition to all the chemical characteristics, organoleptic testing (taste tasting) is also a major factor in categorizing oil as being extra-virgin. Indeed, there have been proposals to include phenols in the olive oil standard (Psomiadou *et al.*, 2003).

Organoleptic (sensory) qualities

The IOC has also proposed some organoleptic (sensory) taste qualities for extra-virgin olive oils. The following are the negative and positive attributes, based on the IOC's standard for virgin olive oils.

Positive attributes

- *Fruity*: set of olfactory sensations characteristic of the oil, which depends on the variety and comes from sound, fresh olives, either ripe or unripe. It is perceived directly or through the back of the nose.

- *Bitter*: characteristic taste of oil obtained from green olives or olives changing color.

- *Pungent*: biting, tactile sensation characteristic of oils produced at the start of the crop year, primarily from olives that are still unripe.

Main negative attributes

- *Fusty*: characteristic flavor of oil obtained from olives stored in piles, which have undergone an advanced stage of anaerobic fermentation.

- *Musty-humid*: characteristic flavor of oils obtained from fruit in which large numbers of fungi and yeasts have developed as a result of olives being stored in humid conditions for several days.

- *Muddy-sediment*: characteristic flavor of oil that has been left in contact with the sediment that settles in underground tanks and vats.

- *Winey-vinegary*: characteristic flavor of certain oils reminiscent of wine or vinegar. This flavor is mainly due to the process of fermentation in the olives leading to the formation of acetic acid, ethyl acetate and ethanol.

- *Metallic*: flavor that is reminiscent of metals. It is characteristic of oil that has been in prolonged contact with metallic surfaces during crushing, mixing, pressing or storage.

- *Rancid*: flavor of oils that have undergone a process of oxidation.

Further negative attributes

- *Heated or burnt*: characteristic flavor of oils caused by excessive and/or prolonged heating during processing, particularly when the paste is thermally mixed, if this is done under unsuitable thermal conditions.

- *Hay-wood*: characteristic flavor of certain oils produced from olives that have dried out.

- *Rough*: a thick, pasty mouthfeel sensation produced by certain oils.

- *Greasy*: flavor of oil reminiscent of diesel oil, grease or mineral oil.

- *Vegetable water*: flavor acquired by the oil as a result of prolonged contact with vegetable water.

- *Briny*: flavor of oil extracted from olives that have been preserved in brine.

- *Esparto*: characteristic flavor of oil obtained from olives pressed in new esparto mats. The flavor may differ depending on whether the mats are made from green esparto or dried esparto.

- *Earthy*: flavor of oil obtained from olives that have been collected with earth or mud on them and not washed.

- *Grubby*: flavor of oil obtained from olives that have been heavily attacked by the grubs of the olive fly (*Bactrocera oleae*).

- *Cucumber*: flavor produced when oil is hermetically packed for too long, particularly in tin containers; attributed to the formation of 2-6 nonadienal.

11.3 Oil quality and purity parameters

There are several ways of defining the quality of olive oils, and perhaps there is no single universal definition that adequately satisfies all situations. In general terms, quality is defined as "the combination of attributes or characteristics of a product that have significance in determining the degree of acceptability of that product by the user." The purity of an aliment refers to the fact that "it has not been subjected to technologies different from those traditionally used, nor has any substance extraneous to its nature been added" (Gould, 1992).

Olive oil quality can be defined from different perspectives – for example, commercial, nutritional or organoleptic. The nutritional value of olive oil arises from high levels of oleic acid and other minor but important components, such as phenolic compounds, whereas the aroma is strongly influenced by volatile compounds (Angerosa, 2002). Nutritional value and pleasant flavor have contributed to an increase in consumption of olive oil, which has fostered cultivation of olives outside the traditional olive-oil producing region of the Mediterranean and in newer areas where cultivar adaptability, different climatic conditions and different agronomic practices can alter the oil quality to some extent (Patumi *et al.*, 2002).

The IOC and the EEC have defined the quality of olive oil, based on parameters that include free fatty acid (FFA) content, peroxide value (PV), UV absorbency and sensory score. In particular, the quantity of FFA is an important factor for classifying olive oil into commercial grades (Boskou, 1996). The general classification of olive oils into different commercial grades is based mainly on FFA and sensory characteristics (taste and aroma). The commercial grades distinguish oils obtained from the olive fruit solely by mechanical or physical means (virgin) from those that contain refined oils (Kalua *et al.*, 2007).

11.3.1 Basic quality parameters

Basic quality parameters are related to the primary quality of the olive oil. The quality standards of the IOC and other concerned agencies are based on these parameters, which can be divided in to two types: hidden and sensorial. Free fatty acids, UV absorbency, peroxide value, moisture and volatile matters, halogenated solvent content, insoluble impurities and trace metals are the hidden quality parameters, whereas organoleptic assessment is the sensorial quality parameter.

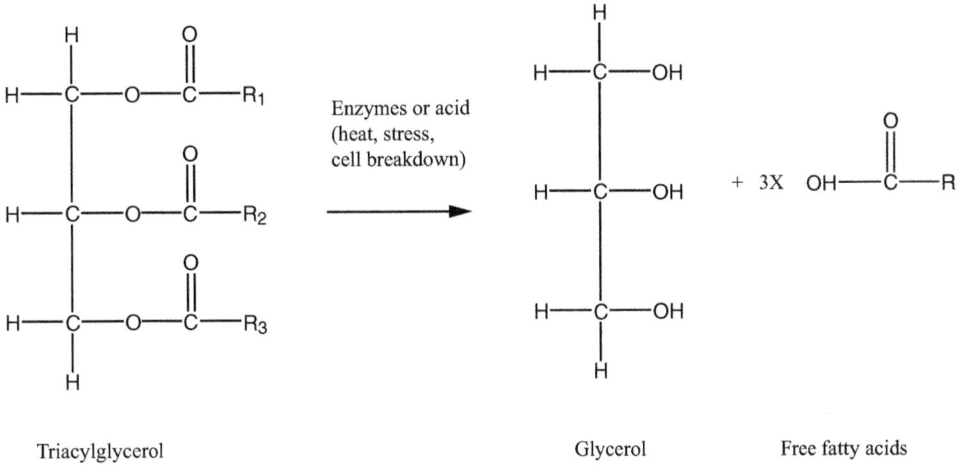

Figure 11.3: Schematic diagrams showing the breakdown of triacylglycerol to form free fatty acids. Source: Lawson (1995).

Table 11.2: General classification of olive oil based on FFA

Olive oil classification	FFA limit (as oleic acid) (%)
Extra-virgin olive oil	0.8 (max.)
Virgin olive oil	2.0 (max.)
Ordinary virgin olive oil	3.3 (max.)
Lampante virgin olive oil	3.3 (max.)
Refined olive oil	0.3 (max.)
Olive oil	1.0 (max.)
Crude olive pomace oil	No limit
Refined olive pomace oil	0.3 (max.)
Olive pomace oil	1.0 (max.)

Source: IOOC (2003).

Free fatty acids

Free fatty acids (FFAs) in the oil are fatty acids that have been lost from the TAG or break away from oil molecules (Figure 11.3). Since biologically synthesized fats are neutral, the presence of FFAs in the olive oil indicates that degradation has occurred through poor handling during processing, and is considered a sign of deterioration of oil quality. FFAs can also influence the organoleptic value of the oil. Additionally, FFAs are water-soluble, and may be lost during processing. Therefore, it is best to test every batch of oil for FFAs. The IOC has specified different limits for FFAs for different categories of olive oils (Table 11.2). Factors that increase the FFA content in olive oil include bruising of the fruit, insect damage, and incorrect post-harvest storage. This type of mechanical damage causes cell destruction, exposing the oil to lipase from within the cell, or to other types of hydrolytic activity, thus leading to broken TAGs and increased FFA concentration. Studies have also shown that the level of FFAs in the olive oil also depends upon the cultivar, as some cultivars ripen earlier than others and thus might be more susceptible to damage (Koutsaftakis *et al.*, 2000). Although most fresh olive oils have a FFA content well below the 0.8 percent needed for the extra-virgin grade, high FFA oils can be blended with low FFA oils to reduce the FFA to within the desired level. When the FFA level is high, there are likely to be other problems.

Measurement of FFAs

The measurement of FFAs is also known as the measurement of oil acidity. The most common method for the determination of the FFA level uses acid/base titrations with phenolphathlein as an indicator. The oil is dissolved in an organic solvent and titrated with an alkaline solution such as sodium hydroxide. The results are given as % FFA as oleic acid, as by IOC regulation. Briefly, 5 g of oil is weighed in an Erlenmeyer flask and 10 ml of 95% ethanol and 0.5 ml of phenolphthalein are added; titration is then carried out drop by drop with 0.5% NaOH solution until the color changes. Acidity is calculated as a percentage by mass as follows:

$$\text{Acidity}(\%\text{oleic acid}) = \text{V.c}\,\text{M}/10\,\text{m}$$

where V = volume in ml of standard volumetric NaOH solution; c = concentration (in moles per liter) of the standard volumetric NaOH; M = molar mass (in g per mole of oleic acid); and m = mass (in g) of the test portion.

UV absorbance

The UV absorbance test is generally performed to identify the age or refinement of oils. It is also a more precise indicator of oxidation, especially in oils that have been heated in the refining process. It is therefore considered to be an extremely rapid method of showing the presence of refined oil in a sample, and is also used to highlight aging of the oil. The test measures changes in the structure of fatty acids – something that occurs during aging or heating of oil. It measures the quantity of certain oxidized compounds that resonate at wavelengths of 232 and 270 nm in the ultraviolet spectrum in a spectrophotometer (Lawson, 1995). Absorbency at 232 nm is caused by hydroperoxides (at the primary stage of oxidation) and conjugated dienes (at the intermediate stage of oxidation); absorbency at 270 nm is caused by carbonylic compounds (at the secondary stage of oxidation) and conjugated trienes (technologica treatment). The Delta index (Δ) K, which is a criterion of discrimination between oil qualities, detects the treatment of oil with color-removing substances, and the presence of refined or pomace oil, by measuring the difference between the absorbances at 270 nm and 266–274 nm. The IOC standards for UV at 270 nm and ΔK are presented in Table 11.3.

Table 11.3: Classification of olive oils according to UV absorbency

Oil category	UV absorption at 232 nm[*]	UV absorption at 270 nm	Delta (Δ) K
Extra virgin olive oil	≤ 2.50[**]	≤ 0.22	≤ 0.01
Virgin olive oil	≤ 2.60[**]	≤ 0.25	≤ 0.01
Ordinary virgin olive oil		≤ 0.30[***]	≤ 0.01
Refined olive oil		≤ 1.10	≤ 0.16
Olive oil		≤ 0.90	≤ 0.15
Refined olive-pomace oil		≤ 2.00	≤ 0.20
Olive-pomace oil		≤ 1.70	≤ 0.18

Source: IOOC (2003).

[*]This determination is solely for application by commercial partners on an optional basis.

[**]Commercial partners in the country of retail sale may require compliance with these limits when the oil is made available to the end consumer.

[***]After passes of the sample through activated alumina, absorbency at 270 nm shall be equal to or less than 0.11.

Measurement of UV absorbance

The UV absorbance of oil is assessed using a spectrometer, which measures the quantity of oxidized compounds that resonate at wavelengths of 232 and 270 nm in the ultraviolet spectrum (UV). For UV absorption testing, the olive oil is diluted (0.5%, wt/vol) in iso-octane and the UV spectra K_{232} and K_{270} are calculated based on the absorption at 232 and at 270 nm, respectively. For extra-virgin olive oil, the value at 232 nm should be less than 2.5; at 270 nm it should be less than 0.22. The ΔK is calculated by the following equation:

$$\Delta K = \text{Abs } 270 - (\text{Abs } 266 + \text{Abs } 274)/2.$$

Peroxide value

Oils generally become oxidized, or auto-oxidation occurs, when they are exposed to oxygen in the air. This is considered to be undesirable because it affects the sensory qualities of the oil, as rancid odors are produced as a consequence of oxidation. The rate of oxidation depends on the degree of saturation of the fatty acids, as well as the quantity of antioxidants present. In oil, oxidation generally occurs in the presence of oxygen that is also accessible, in the presence of unsaturated bonds in the fatty acid structure, and in the presence of catalysts. Oxidation takes place by a free radical mechanism, and this involves the process of initiation, propagation and termination (Figure 11.4).

Looking at Figure 11.4, during the initiation phase (1) an unsaturated fatty acid reacts with oxygen, involving a catalyst such as light, heat, metal ions (copper or iron), peroxides, or the enzyme lipoxygenase. This leads to the formation of a free radical. These free radicals are very reactive because they have an unpaired electron, and they react with oxygen immediately

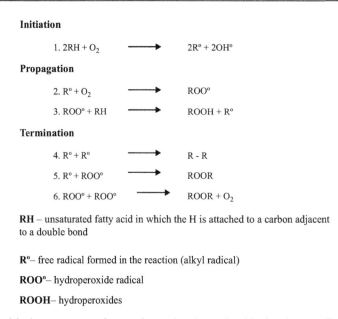

Initiation

1. $2RH + O_2 \longrightarrow 2R° + 2OH°$

Propagation

2. $R° + O_2 \longrightarrow ROO°$

3. $ROO° + RH \longrightarrow ROOH + R°$

Termination

4. $R° + R° \longrightarrow R - R$

5. $R° + ROO° \longrightarrow ROOR$

6. $ROO° + ROO° \longrightarrow ROOR + O_2$

RH – unsaturated fatty acid in which the H is attached to a carbon adjacent to a double bond

R°– free radical formed in the reaction (alkyl radical)

ROO°– hydroperoxide radical

ROOH– hydroperoxides

Figure 11.4: Initiation, propagation and termination of oxidation in a radical mechanism. Source: Ayton (2006).

upon exposure. A hydroperoxide radical is formed from the reaction of oxygen with these free radicals (2). This leads to propagation. During propagation, hydroperoxide radicals react with other fatty acids to create hydroperoxides (ROOH) (3) (Koprivnjak and Conte, 1998). Once hydroperoxides have been formed, initiation reactions proceed easily.

As hydroperoxides decompose, a wide variety of aldehydes, ketones and hydrocarbons are formed. These are the compounds responsible for rancid odors and flavors (Lawson, 1995). The termination process occurs when two radicals are combined so that their electrons "pair up" to form a chemical bond (4–6), a measure of hydroperoxides in oxidized oil. The hyperperoxides are measured quantitatively on the basis of their ability to liberate iodine from acidic solutions of potassium iodide.

The peroxide value (PV) indicates the level of oxidation of oil and fat that has occurred. In other words, PV is the quantity of those substances in the sample that oxidize potassium iodide under the operating conditions. The PV is due to hydroperoxides (primary stage of oxidation). The oxidation may be either enzymatic or chemical. Therefore, PV is another important test that should be performed on every batch of oil. The IOC has standards for PV that specify less than 20 meq of active oxygen/kg oil for extra-virgin olive oil. Along with exposure to temperature, air, light, and the presence of fatty acids, contact with water and metal surfaces such as copper, and the ion sediment content of the oil also affect the rate of oxidation. Damage to fruit and delays between harvest and processing are also risk factors. For this reason, oil should be stored

$$ROOH + 2H^+ + 2I^- \longrightarrow I_2 + ROH + H_2O$$

$$I_2 + 2S_2O_2^{2-} \longrightarrow S_4O_6^{2-} + 2I^-$$

where ROOH is the lipid hydroperoxide

Figure 11.5: The reaction that takes place when measuring peroxide values in oil (Robards *et al.*, 1988).

in a cool inner vessel, such as stainless steel or glass, and away from light. Oxygen should be removed by sparking with nitrogen prior to sealing the vessel. Once the process of oxidation, which leads to rancidity, begins, it progresses very quickly. Generally, a PV level higher than 10 meq/kg is considered to indicate a less stable oil with a shorter shelf-life.

Measurement of PV

One method used to determine lipid oxidation is to measure the peroxide value, which is indicative of the amount of hydroperoxide in the oil. The standard method for measuring the peroxide value is to titrate a mixture of the oil, chloroform, acetic acid and saturated potassium iodide solution with sodium thiosulfate, as described by the IOOC according to ISO 3960 (IOOC, 1995) and expressed as the milliequivalent (meq) of active oxygen per kilogram. The reaction that takes place during measurement is shown in Figure 11.5. This reaction is convenient in that although there is a seemingly complicated sequence of reactions, it serves as its own indicator – the iodide forms a blue color with a starch indicator, and the end-point can easily be detected when this blue color disappears upon titration with sodium thiosulfate.

In brief, 5 g of oil is placed in an Erlenmeyer flask and 50 ml of a mixed solution of acetic acid and isooctane (60:40 v/v) is added to it; 0.5 ml saturated potassium iodide solution is then added and the solution is mixed. Titration is performed with sodium thiosulfate solution 0.01 mol/l. When the solution turns light yellow, about 0.5 ml of starch solution (5 g/l) is added. The end-point is determined when the color of the solution changes from brown to transparent. The peroxide value is calculated by the following equation:

$$\text{Peroxide value (meq/kg oil)} = 1000\,(V - V_0) \times c/m$$

where V = volume of sodium thiosulfate solution used (in ml); V_0 = volume of sodium sulfate used for the blank determination (in ml); c = concentration of the sodium thiosulfate in moles per liter; and m = mass of the test portion (in g).

Heat stability

Though olive oil is mostly used for salad dressings, in the Mediterranean countries it is also commonly used for cooking and frying. During this process, owing to the high temperature and the absorption of oxygen and water, TAGs in vegetable oils suffer a series of reactions –

hydrolysis, oxidation, isomerization and polymerization – and form a wide range of compounds, such as hydroperoxides, alcohols, hydrocarbons, aldehydes, free fatty acids, esters, lactones, ketones, furans and other products (Van de Voort, 1994; Frankel, 1998). During this process, changes in other minor components (e.g., sterols or steryl esters) as well as the elimination of water or organic acids occur, which gives the oil an "off" flavor (Gomez-Alonso *et al.*, 2003). However, studies have shown that, compared to soybean and rapeseed oil, olive oil is found to be less affected after the frying process (Kiritsakis, 1998). Lower amounts of new chemical species are found in olive oil after heating compared to other oils, and this is believed to be related to the high oleic acid (monounsaturated) and low linoleic and linolenic (polyunsaturated) contents of olive oil. The amount of linolenic acid in the oil is considered to be the factor that increases the off flavor of oil during frying. A report presented by Solinas *et al.* (1987) shows that olive oil is more stable during frying and a lower amount of acrolein is formed compared to soybean, peanut, sunflower and cotton seed oils. A later study also suggests that the phenolic compounds naturally contained in virgin olive oil improve its resistance to oxidative deterioration (Frankel, 1998). The rapid decrease in the concentration of hydroxytryrosol (3,3-DHPEA) and its secoiridoid derivatives (3,4-DHPEA-EDA and 3,4-DHPEA-EA), the main phenolic compounds of virgin olive oil during frying (Gomez-Alonso, 2003), also supports this evidence.

Moisture and volatiles

Olive oil can contain water and volatile compounds as a result of the extraction method (water from vegetable tissue, etc.). Besides being foreign matter, water impairs the quality of oils. The quantity of moisture and volatile matter can be evaluated by the loss in weight in the product on heating at $103\,°C \pm 2$ in a drying stove for 30 minutes at a time, and measuring the weight differences until no further difference is detected. It is expressed as a percentage (%) of the total weight. According to international standards, moisture and volatile matters should be limited to less than 0.2 percent in extra-virgin olive oil, virgin olive oil and ordinary virgin olive oils. However, this limit can rise to 0.3 percent in lampante virgin oil, but should be less than 0.1% in refined olive oil and olive oil grades (IOOC, 2003).

Insoluble impurities

Poor manufacturing practices during the production of olive oil can produce a high level of insoluble impurities. Dirt, minerals, resins, oxidized fatty acids, alkaline soaps of palmitic and stearic acid, and proteins are the main insoluble impurities in olive oils. These remain suspended in the oil. The level of insoluble impurities can be determined by dissolving some of the oil in *n*-hexane or light petroleum ether, removing the insoluble matter by filtration, and expressing its proportion as a percentage of the total. There should be less than 0.1 percent insoluble impurities in extra-virgin olive oil, virgin olive oil and ordinary virgin olive oil. In lampante virgin olive oil this figure can rise to 0.2 percent; however, in refined olive oil and graded olive oil this value should be less than 0.05 percent (IOOC, 2003).

Trace metals

Trace metals in olive oil may originate from the soil and fertilizers, or from contamination during processing and storage. In refined olive oil the concentration is lower than in virgin oils, due to the refining process. The presence of transition metals such as copper and iron is also related to the oxidative stability of olive oils because of their catalytic effect on the decomposition of hydroperoxide. According to IOC standards, the iron concentration must not exceed 3 ppm while copper, arsenic and lead should not exceed 0.1 ppm in olive oils. Atomic absorption spectrometry is the best method of determining the trace metals in olive oils (IOOC, 2003).

Flash point

The flash point of oil is the temperature at which the oil will burst into flames, and is slightly different from the smoke point (the temperature at which the oil spontaneously begins to burn). The flash point is related to the free acidity content of the oil. Refined olive oil, pomace oil and seed oils have a lower flash point than virgin olive oil. The IOC states that virgin olive oils should have a flash point of around 210–220 °C, while most seed oils begin to burn at 190–200 °C (IOC, 2006).

Halogenated solvents

Halogenated solvents (e.g., chloroform, Freon, perchloroethylene, trichloroethylene and tetra-chloroethylene) may be present as residues in solvent-extracted olive oils, or result from environmental contamination. It has been discovered that, in olive mills, the use of drinkable water for extraction by pressing may cause the formation of halogenated compounds (especially chlorides and bromides), which are extremely soluble in oils and thus become increasingly concentrated. Moreover, there is contamination of the water table by used industrial water, in addition to pollution of the atmosphere caused by halogenated solvents. To assess the level of halogenated solvents, the headspace volatile gases are measured using gas chromatography, and expressed in mg/kg oil. The IOC limits the maximum content of each halogenated solvent to 0.1 mg/kg, and the maximum content of the sum of all halogenated solvents to 0.2 mg/kg (IOC, 2006).

Organoleptic qualities

Organoleptic or sensory qualities are the most important parameters in ensuring the acceptance of olive oil for consumption. Aroma, flavor, pungency and bitterness are the primary sensory components of olive oils. Volatile compounds (aldehydes, ketones, esters, saturated and unsaturated alcohols and others) are responsible for the aroma of the oil, while the taste is influenced by all the components. The assessment of sensory qualities is based on the positive and negative descriptors of olive oil sensory components. The positive descriptors are fruity, green, spicy, citrus, fragrant, tropical soft, over-ripe, bitter and pungent, and are determined by the quality of the fruit. Negative descriptors are caused by human error, and include fusty, musty, muddy, winey, metallic, rancid, burnt (Mailer and Beckingham, 2006). According to IOC standards, extra-virgin olive oil should not have any sensory defects and should have some fruitiness. Sensory assessment should begin with stipulating the taste condition of the olive oil, followed by assessment of the aroma by sniffing, and then detection of flavor, pungency and bitterness. Organoleptic tests should be carried out by 8–12 people who have been trained to IOC standards.

Most of these sensory compounds are polar or water-soluble, and are unfortunately lost to a greater or lesser degree during extraction – especially in the older extraction method of a three-phase decanter, where a huge amount of water is added during the process. In refined oil, almost all of these sensory compounds would be lost, including bad flavors.

Organoleptic test for virgin olive oil

A virgin olive oil taste test is carried out by professionals in order to classify and separate extra-virgin olive oil from virgin olive oil prior to sale. The two are easily confused. Two different oils with acidity lower than 0.8 percent could, at first glance, both be considered to be extra-virgin. In fact, a virgin olive oil and an extra-virgin olive oil are both natural products and real olive juice. The difference is in the level of olive oil acidity, and in the score gained in sensorial analysis. For extra-virgin olive oil, the median of defects is zero and the median of the fruity attribute is greater than zero. For virgin olive oil, the median of defects is greater than zero but lower than or equal to 2.5, and the median of the fruity attribute is greater than zero.

The ideal temperature for a taste test is 28 °C. In between tasting each different sample of olive oil, and in order to eliminate the taste of the previous one, tasters eat a small piece of apple and drink a sip of water.

The actual tasting is directly from a glass, which is rotated to wet the sides fully. The tasters lift the glass to their nose and sniff rapidly in order to receive the aroma from the oil. They should notice the fruity attributes and the aromas of apple, grass and tomato. The tasters take a sip of oil, which should touch all areas of the mouth so that all the various tastes and sensations can be noted. At this stage, the bitter attributes should be detected (on the rear part of the tongue), pepper notes (in the throat) and sweet notes (on the front part of the tongue). Extra-virgin olive oil has irreproachable taste and aroma – in other words, it has zero defects. If any defects *are* found, even if almost imperceptible (winey, fusty, rancid, musty), the oil cannot be passed as "extra virgin." It is extremely difficult for a non-trained consumer to notice these subtle negative attributes, which is why professional taste testers and origin denominations play an important part in ensuring that the olive oil bottle we buy really does contain extra-virgin oil.

The descriptive analysis should use a six-point intensity ordinal rating scale from 0 (no perception) to 5 (extreme) to quantify the intensity of different sensory attributes (fruity, bitter, pungent, and others like sweet apple, other ripe fruit, green, etc.). Overall grading uses a nine-point scale ranging from 1 (lowest quality) to 9 (optimal quality). Depending on the average score of the panel, the oil is classified as extra virgin (≥ 6.5), virgin (< 6.5 and ≥ 5.5), ordinary virgin oil (< 5.5 and ≥ 3.5) or virgin lampante olive oil.

11.3.2 Other quality parameters

Other quality parameters are concerned more with the purity of the oil than the quality itself, and are important in detecting the adulteration of virgin olive oils with other vegetable oils, olive-pomace oils or refined oils.

Fatty acid profile

The fatty acid profile (FAP) of the oil is a measure of the proportions of individual fatty acids in the oil, and is therefore an important factor in oil quality. The ratio of the different fatty acids in the oil influences the stability of the oil, as well as determining its nutritional value. Some fatty acids are considered to be better than others; in the case of olive oil, oleic acid is more desirable than the others from the nutritional point of view. Oils that have high levels of monounsaturated oleic acid are considered to be of the highest nutritive value (in fact, oleic acid is named after the olive, "*olea*"). Ideally, the oil should also have a low level of palmitic acid, which is the major saturated fat. The IOC has produced a list of the permitted levels for each of the fatty acids in order for the oil to be designated extra-virgin (Table 11.4). Since the range is quite large, almost all olive oils will match these guidelines; however, there are exceptions to this rule, especially for linolenic acid. According to the IOC standard, linolenic acid should be present at a level of less than 1 percent. Although linolenic acid is considered to be nutritionally beneficial, because of its polyunsaturated nature and three double bonds, it is the most reactive fatty acid and is particularly unstable and susceptible to oxidation (rancidity). Higher levels of linolenic acid are likely to contribute to reduced storage stability of the oil.

The fatty acid profile of the oil is mostly influenced by the cultivar and the environment. Although the IOC allows a wide range of fatty acids in extra-virgin olive oil, most growers prefer cultivars that have higher levels of the more desirable fatty acids.

The fatty acid profile is generally determined by assessing fatty acid methyl esters by gas chromatography, as suggested by IOC regulations. In brief, 0.1 g oil is weighed and diluted with 2 ml heptane and 0.2 ml of 2N methanolic KOH. The combined solution is shaken vigorously for 30 s and left to stratify until the upper solution becomes clear. The upper solution is collected, and evaporated to dryness under a nitrogen gas flow. The methyl ester is re-suspended in 1 ml of heptane and injected into the GC for determination. For the best results, the following conditions for GC are preferred:

- a column temperature of $120–190 \pm 5\,°C$

Table 11.4: Permissible fatty acid ranges, according to the IOC and EC, as determined by GC for different categories of olive oil

Fatty acid	Allowable range (%)		
	Extra-virgin olive oil, virgin olive oil	Olive oil and refined olive oil	Olive-pomace oil, refined olive-pomace oil
Myristic (C14:0)	≤ 0.05	≤ 0.05	≤ 0.05
Palmitic (C16:0)	7.5–20.0	7.5–20.0	7.5–25.0
Palmitoleic (C16:1)	0.3–3.5	0.3–3.5	0.3–3.5
Heptadecanoic (C17:0)	≤ 0.3	≤ 0.3	≤ 0.3
Heptadecenoic (C17:1)	≤ 0.3	≤ 0.3	≤ 0.3
Stearic (C18:0)	0.5–5.0	0.5–5.0	0.5–5.0
Oleic (C18:1)	55–83.0	55–83.0	55–83.0
Linoleic (C18:2)	3.5–21.0	3.5–21.0	3.5–21.0
Linolenic (C18:3)	≤ 1.0*	≤ 1.0	≤ 1.0
Arachidic (C20:0)	< 0.6	< 0.6	< 0.6
Gadoleic or eicosenoic (C20:1)	< 0.4	< 0.4	< 0.4
Behenic (C22:0)	≤ 0.2	≤ 0.2	≤ 0.3
Lignoceric (C24:0)	≤ 0.2	≤ 0.2	≤ 0.2
Trans fatty acids			
C:181 T	0.0–0.05	0.0–0.20	0.0–0.40
C:182 T + C18:3 T	0.0–0.05	0.0–0.30	0.0–0.35

Source: CODEX Stan (2003).

- an injection temperature of 250 °C

- a detector temperature of 300 °C

- linear velocities of the carrier gases of 30 cm/s for N_2; 20–30 cm/s for H_2, and 300 cm/s for air

- an injection volume of 4 μl of re-suspended methylester solution.

Studies have shown that the fatty acid profile of olive oil is related to the cultivar (Kiritsakis, 1998; Wiessbein *et al.*, 2008), and may also vary according to the age of the tree. Studies have also shown that fruit maturity affects the FAP. The concentration of palmitic acid decreases during ripening, probably because of the dilution effect, and the fact that although the absolute palmitic acid content may remain constant, the level of oleic acid increases due to active triacylglycerol biosynthesis (Gutierrez *et al.*, 1999). Linoleic acid levels might increase during ripening due to the formation of oleic acid and the activation of enzyme oleate desaturase, which transforms

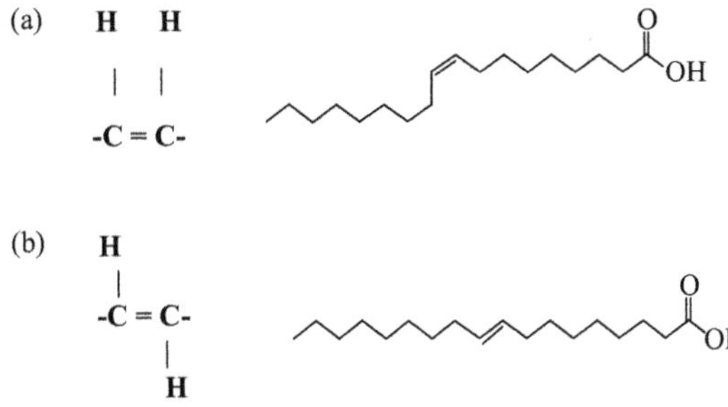

(a)
H   H
| |
-C = C-

(b)
H
|
-C = C-
|
H

Figure 11.6: *Cis* (a) and *trans* (b) configuration of double bond.

oleic acid into linoleic acid. However, other fatty acids, such as palmitoleic, stearic and linolenic, remain relatively stable throughout the normal harvesting period (Ayton, 2006).

Trans-fatty acid isomers

Although olive oil contains more than 80 percent unsaturated fats – mostly monounsaturated (oleic) although there are small amounts of polyunsaturated fats (linoleic and linolenic) – the majority of the unsaturated fats in olive oils have the *cis* configuration, where the hydrogen atoms are on the same side of the double bonds of the carbon chain (Figure 11.6a). However, during some processes, such as bleaching and deodorization, there is the chance of conversion to *trans* fatty acids. *Cis* fatty acids have regular metabolic activity, but *trans* fatty acids (fatty acids containing the *trans* isomer in their bond; Figure 11.6b) appear to be metabolized differently than their *cis* isomers, with adverse health effects. *Trans* fats are of concern because their consumption leads to raised low density lipoprotein (LDL – "bad" cholesterol) and reduced high density lipoprotein (HDL – "good" cholesterol).

Olive oil, owing to its special fatty acid composition (Table 11.5), indeed has a beneficial impact on controlling cholesterol levels, and the role of olive oil in the prevention of cardiovascular disease is unique. Epidemiological studies have suggested that heart disease is in direct proportion to the concentration of cholesterol in the blood. Apart from the beneficial impact of *cis* unsaturated fats, the presence of a higher percentage of unsaturated fatty acids in olive oil could make it vulnerable to rapid rancidity (oxidization) compared with other oils that contain *trans* fats (naturally or by hydrogenation) because of the presence of the double bond. However, the presence of high amounts of antioxidative substances such as phenols, tocopherols and other natural antioxidants prevents olive oil from lipid oxidation within the body, eliminating the formation of free radicals that may cause cell destruction. It is important, though, that strict precautions should be taken, especially in bottling, storage and transportation of olive oil, to ensure it remains in good condition. *Trans* isomers of fatty acids in olive oil can also be detected by gas chromatography. Heat and hydrogenation twist the shape (*trans*) so that it does not fit correctly with enzymes. The new labeling laws in the US require that, since 2006, products be

Table 11.5: Permissible phytosterol content for extra-virgin olive oil, according to IOC

Phytosterol fingerprint	Content (%)
Cholesterol	≤ 0.5
Brassicasterol	≤ 0.1
Campesterol	≤ 4.0
Stigmasterol	< campesterol in edible oils
Δ-7-stigmasterol	≤ 0.5
β-sitosterol	
Δ-5-avenasterol	
Δ-5-23-stigmastadienol	≤ 93
Clerosterol	
Sitostanol	
Δ-5-24-stigmastadienol	

Source: IOOC (2003).

labeled with their *trans* fatty acid content (Vossen, 2007). The permitted amounts of *trans* fatty acids in olive oil are presented in Table 11.5.

Olive oil also experiences conversion of *cis* isomers to *trans* isomers during heating, as with other vegetable oils; however, the conversion to *trans* isomers is less in olive oils compared with other vegetable oils due to the presence of a high level of phenolic compounds (Gamel, 1995).

Total triacylgycerols and 2-position fatty acid composition

Studies have suggested that the triacylglycerol structure, and not only the fatty acid profile, is of special importance regarding physiological effects in oils (Mu and Porsgaard, 2005). Many studies concerning lipid metabolism and the effect within the human body emphasize the importance of structure–activity relationships. Analysis of triacylglycerols has always been of great interest in the food industry. In the case of olive oil, the structure of TAGs may have a particular significance in the detection of adulteration, and may provide valuable information related to the origin of the oil. The middle carbon of the TAG molecule (the 2 or beta position) in natural virgin olive oil always contains the unsaturated fatty acids, such as oleic or linoleic acids. Re-esterified oils that are processed artificially do not conform to this same fatty acid distribution. Furthermore, adulteration of virgin olive oil can be detected by the determination of the equivalent carbon number (ECN) of the oil. According to the IOC, standard edible virgin olive oils should have a maximum 0.2 ΔECN 42 value (i.e., between the real and theoretical ECN 42 triacylglycerol contents). The theoretical triacylglycerol values are calculated from the fatty acid composition and the 1, 3-random distribution theory (Cortesi *et al.*, 1990). Lampante virgin olive oil, refined olive oil and olive oil can have a value of up to 0.3, whereas refined olive pomace, olive-pomace and crude olive-pomace oils generally have 0.5, 0.5 and 0.6 ΔECN 42 values, respectively (IOOC, 2003). The ECN value can be calculated as follows:

$$ECN = CN - 2n$$

where CN is the carbon atom of the fatty acids in the triacylglycerol molecules, and n is the number of double bonds.

As greater concern has developed regarding the triacylglycerol content in oils and its significance for health and oil quality, determination of triacylglycerols in olive oil has become more common. Despite this interest, there has been little progress in the compositional analysis of TAGs. In the past, chromatographic techniques (mostly GC) were used; however, GC requires derivitization of fatty acids in order to determine the volatility of the triacylglycerols. Furthermore, this method is time-consuming, and lacks information regarding the fatty acid composition of a certain TAG in a particular mixture of TAGs (Schiller *et al.*, 2002). HPLC and TLC both require considerable experience, and do not always provide clear information because chromatographic differences depend only to a limited extent on the fatty acid composition (Touchstone, 1995). Other methods, such as electron impact MS, could be more powerful; however, derivatization is still necessary. Modern soft ionization mass spectrometry with matrix assisted laser desorption time of flight ionization (MALDI-TOF) provides good opportunities for analyzing the TAGs in oils (Asbury *et al.*, 1999). Preparation of MALDI analysis of a sample is fast and easy. The extent of fragmentation is normally low, and therefore detection of molecular ions is better. The most important advantage of MALDI, however, is that both the liquid and the matrix are readily soluble in organic solvents, which provides highly homogenous matrix/analyte cocrystals that help with reproduction. MALDI-TOF/MS has been found to be the best technique for determining the TAGs in olive oil.

MALDI-TOF analysis for triglycerol fingerprinting

TAG fingerprinting in olive oil is possible by MALDI-TOF/MS, as described by Kaufman and Wiesman (2007). In brief: an oil sample is dissolved in hexane to 0.1–1 mg/ml. A MALDI matrix solution is prepared by dissolving 2.5-dihydroxybenzoic acid in 90% methanol to about 20 mg/ml. A small volume of sample and matrix is mixed together in the ratio of 1 : 2, and 1 μl is then applied directly to a stainless-steel MALDI target. Samples are analyzed on a MALDI-TOF mass spectrometer using 337 nm radiation from a nitrogen laser. An accelerating voltage of 20 kV is used. The spectra are recorded in the reflectron mode within a mass range of *m/z* 450–2400, as described earlier by Lay *et al.* (2006).

Calculations are then made to determine the theoretical molecular mass of the different TAGs. The combinations of the major fatty acids are then calculated accordingly. All the possible combinations for TAGs composed of these fatty acids are added together, and their molecular weight is calculated. The molecular weight of a sodium ion (Na^+) is added to the TAG mass, since this is the predominant ion in the matrix used for this experiment. It is assumed that during the ionization process this ion is added to the TAGs – in most cases, one ion per TAG molecule.

The calculation described above is displayed in the following equation:

$$\text{Mw (TAG)} = \text{Mw}\{\text{Glycerol} - 3(\text{OH})\} + \text{Mw}\{(\text{FA}_1 + \text{FA}_2 + \text{FA}_3) - 3(\text{H})\} + \text{Mw}[\text{Na}^+]$$

Total unsaponifiable matter

Unsaponifiable matter consists of the total components of the oil that do not turn to soap in the process of saponification (addition of lye). Non-glyceride constituents of the oils, mainly sterols, aliphatic alcohols, pigments and hydrocarbons, are the main constituents of unsaponifiable matter. Natural olive oil must contain less than 1.5 percent (15 g/kg oil) unsaponifiable matter; olive-pomace oil may contain up to 30 g/kg oil total unsaponifiable matter (IOOC, 2003).

Squalene

Squalene ($C_{30}H_{50}$) is an important hydrocarbon in olive oils. Squalene is a biochemical precursor of the phytosterols, and constitutes up to 40 percent by weight of non-saponifiable matter in olive oil. Olive oil is the highest squalene producer among vegetable oils (Hamann *et al.*, 1988), and squalene makes up to 90 percent of the hydrocarbon in olive oil, at levels ranging from 200 to 750 mg/kg oil or even higher (800–1200 mg/kg oil). The squalene content in olive oil depends on the cultivar and the extraction technology, and is drastically reduced during the refining process. Squalene is also known for its contribution to the oxidative stability of olive oil (Psomiadou and Tsimidou, 1999). Studies have suggested that the high squalene content of olive oil as compared to other human foods is a major factor in the cancer risk-reducing effect of olive oil (Newmark, 1997), because of its chemopreventive effects (Rao *et al.*, 1998).

Phytosterols

Phytosterols are compounds originating in the plant kingdom, and are structurally and functionally similar to the animal-derived cholesterols. Phytosterols are complex compounds that carry out biochemical functions within cellular membranes. Their composition is typical of the botanic species from which the oil originates. The term *phytosterols* is usually used to refer to both sterols (having a double bond at position 5 of the sterol ring structure) and stanols (a single bond at position 5). Phytosterols present in olive oil as minor components of the non-glycerin fraction, but are considered to be important quality parameters because of their connection with human health – especially since their recognition as cancer-preventive biologically active substances (Canabate-Diaz *et al.*, 2007). Phytosterols also apparently help to reduce the total plasma and low density lipoprotein (LDL) cholesterol, and as a result are being considered as ingredients of functional foods (Ostlund, 2002). Phytosterols make up the greatest proportion of the non-saponifiable fraction of olive oil, and their composition varies depending on the type of olive oil. The content and composition of phytosterols can also vary according to the agronomic and climatic conditions, fruit or seed quality, oil extraction and refining procedures and storage conditions. Testing for phytosterol content is particularly important when olive oil is for export.

Permitted phytosterol contents in extra-virgin olive oil, according to the IOC, are presented in Table 11.5.

In crude olive oil, the predominant phytosterols are sitosterol (90 percent) and stigmasterol. Compositional analysis of the phytosterol fraction of olive oil can be used to assess the degree of purity of the oil and the absence of other plant oils. Phytosterol determination also permits characterization of the type of olive oil in question – extra virgin, virgin, refined, etc. (EEC, 2003).

Determination of phytosterols

Determination of phytosterols in oils is quite a long process. A series of analyses has to be performed to obtain the sterol composition of the olive oil. The phytosterols in olive oil can be determined using GC-MS, as described previously by Damirchi *et al.* (2005) and later, with slight adjustment, by Kaufman and Wiesman (2007).

Preparation of the unsaponifiables Briefly, 5 g of oil is placed in a 250 ml flask and 500 µl of 0.2% α-cholestanol solution is added. The combined solution is evaporated to dryness with nitrogen, and 5 g of this dry filtered material is returned to the same flask. 2N ethanolic potassium hydroxide solution (50 ml) is added, and saponification is carried out by boiling and stirring the sample. The sample is heated for 20 minutes before adding 50 ml of distilled water and allowing the sample to cool to approximately 30 °C. The contents are then transferred to a separating funnel using distilled water rinses (about 50 ml in total), approximately 80 ml of ethyl ether is added, and the sample is shaken vigorously for 30 s before being allowed to settle. The lower aqueous phase is separated, and collected in a second separating funnel. Two more extractions are performed from the water-alcohol phase, using 65 ml of ethyl ether each time.

The ether extracts are pooled into a single separating funnel and washed with distilled water (50 ml each time) until the wash gives a neutral reaction. The wash water is then removed, and the sample is dried with anhydrous sodium sulfate before being filtered via anhydrous sodium sulfate into a previously weighed 250-ml flask. The funnel and filter are washed with small aliquots of ethyl ether.

The ether is evaporated to a few milliliters and then dried with nitrogen; drying is completed in an oven at 100 °C for approximately 15 minutes and the sample is then weighed after cooling in a desiccator.

Separation of the sterol fraction An approximately 5 percent solution of the unsaponifiables in chloroform is prepared, and a 0.25 mm silica gel TLC plate (Merck, Darmstadt, Germany) is streaked with 0.3 ml of this solution. At the same time, 2–3 µl of the sterol reference solution is streaked to aid identification of the sterol band after developing.

The plate is placed in the developing chamber and allowed to elute until the solvent (toluene-acetone, 95 : 5 v/v) reaches 1 cm from the upper edge of the plate; the plate is then left inside the hood to evaporate the solvent. The plate is sprayed with 2,7-dichlorofluorescein solution, and the sterol band is identified under ultraviolet light. This band is scraped from the silica gel, and the final comminuted material is placed in the filter funnel, 10 ml of hot chloroform is added and filtered under vacuum, and the filtrate is collected in a conical flask attached to the filter funnel.

The collected residue is washed three times with ethyl ether (10 ml each time) and the filtrate is collected in the same flask attached to the funnel; this filtrate is evaporated to a volume of 4–5 ml and the residual solution is transferred into a 10-ml test tube. The solution is then dried by heating in a gentle flow of nitrogen, dissolved again with a few drops of acetone, and evaporated to dryness again. The sample is placed in an oven at 105 °C for approximately 10 minutes, and then allowed to cool in a desiccator before being weighed. The residue contained in the tube consists of the sterol fraction.

Preparation of the trimethylsilyl ethers   A silylation reagent consisting of a 9 : 3 : 1 (v/v/v) mixture of pyridine/hexamethyl disilazane/trimethyl chlorosilane in the ratio of 50 μl for every milligram of sterol, is added to the test tube containing the sterol fraction, and the tube is shaken until the sterols are completely dissolved. The sample is left for at least 15 minutes at ambient temperature and then centrifuged for a few minutes. The obtained clear solution is then ready for gas chromatography.

Gas chromatography operating conditions should be as follows: column temperature 260 °C; injector temperature 280 °C; detector temperature 290 °C; linear velocity of the helium carrier gas 35 cm/s; split ratio 25 : 1; sample injection 1 μL.

After gas chromoatography, peak identification is performed. Individual peaks are identified on the basis of retention times and by comparison with a mixture of sterol TMSE analyzed under the same conditions.

The sterols are eluted in the following order: cholesterol, brassicasterol, 24-methylene cholesterol, campesterol, campestanol, stigmasterol, Δ-7-campesterol, Δ-5, 23-stigmastadienol, clerosterol, β-sitosterol, sitostanol, Δ-5-avenasterol, Δ-5, 24 stigmastadienol, Δ-7-stigmastenol, Δ-7-avenasterol. The concentration of each individual sterol, expressed in mg/kg of oil, is calculated as follows:

$$\text{Sterol concentration (mg/kg)} = A_x . m_s 1000/A_s m$$

where A_x is the peak area for sterol x; A_s is the peak area of the α-cholestanol peak; m_s is the mass of α-cholestanol added, in milligrams; and m is the mass of the samples used for determination, in grams.

Individual sterol concentration is recorded as mg/kg of oil, and summed as "total sterols."

Color pigments

The color of olive oil is dependent on the pigments in the fruit from which it was extracted – green olives give green oil because of the high chlorophyll content, and ripe olives give yellow oil because of the carotenoid (yellow red) pigments. In general, the color of virgin olive oil generally ranges from greenish-yellow to gold. The color of the oil is influenced by the exact combination and proportions of pigments. A simple equation would be:

$$\text{Color} = \text{Chlorophyll (green)} + \text{Carotenoids (yellow red)} + \text{other pigments.}$$

Color is not an official standard, but it certainly matters to the consumer.

The measurement of chlorophyll concentration is not required, according to the IOC standard; however, the level of chlorophyll is considered as one of the most important factors in olive oil. Chlorophyll plays a vital role in determining the olive oil color, and the color plays a key role in acceptability among consumers. In fact, many consumers preferred a deep green color in olive oil, as in virgin oils (Del Giovine and Fabietti, 2005). The chlorophyll level also affects the oxidative stability of the olive oil, because it is implicated in antioxidation in the dark and the photo-oxidation mechanism in the light. Color is an important attribute to consumers, who associate the green hues from the chlorophyll in the oil with freshness of product (Ryan *et al.*, 1998). Chlorophyll and carotenoid contents can be evaluated at 670 and 470 nm, respectively, from the absorption spectrum of the oil sample. For this, a 7.5 g sample is dissolved in 25 ml of cyclohexane and the result is expressed as mg/kg oil. As with polyphenols, climate also has an important role in the chlorophyll concentration in olive oil, as does the ripeness of the fruit. Some researchers have found that the concentration of chlorophyll is high – at up to 80 mg/kg of oil – early in the ripening period, and very low – about 2 mg/kg oil – when fruit is very ripe (Salvador *et al.*, 2001).

Polyphenols

Polyphenols (PP), or phenolic compounds, are perhaps the most important of the minor components in olive oil, owing to their powerful antioxidant effect on the oil and the resulting contribution to shelf-life stability. *Polyphenol* is a general term used to describe natural substances that contain a benzene ring with one or more hydroxyl groups containing functional derivatives that include esters, methyl esters and glycosides (Tsimidou, 1998). According to Harborne and Dey (1989), phenolic compounds are grouped into:

- phenols, phenolic acids, phenylacetic acids;
- cinnamic acids, coumarins, isocoumarins and chromones;

- lignans;

- a group of ten flavonoids;

- lignins;

- tannins;

- benzophenones, xanthones and stilbenes;

- quinines; and

- betacyanins.

Most phenolic compounds are found in nature in a conjugated form, mainly with a sugar molecule. The relationship between oxidative stability and the concentration of polyphenols has also been well established (Aparicio and Luna, 2002). The redox properties of polyphenols allow them to act as hydrogen donors and singlet oxygen quenchers, hence their role as antioxidants (Jesus Tovar *et al.*, 2001). Polyphenols are also responsible for the bitterness perceived in olive oil.

In general, the level of polyphenols in the olive oil can vary from 0 to 1000 ppm or more. Usually, the range is 60–400 ppm. Oil is categorized as being low, medium or high in polyphenols when their level is around 50–200, 200–400 and more than 400 ppm, respectively. To date, there are no international standards for PPs in olive oil. High PP levels can lead to increased bitterness, so oil with a medium PP level (200–400 ppm) can be easier to manage than that with a high PP content, which leads to a strongly flavored, robust oil. Some olive varieties have relatively low PP levels but relatively high tocopherol levels (tocopherol is also an antioxidant), which is considered equally good. Monitoring the PP level can provide a guide to flavor intensity, particularly in the early harvest period.

Presence of polyphenols is a special characteristic of olive oil, because most other nut and seed oils contain no polyphenols. About 20 polyphenols have been identified in virgin olive oils; however, the predominant phenolic compounds in such oils are tyrosol and hydroxytyrosol, followed by traces of substituted cinnamic acids such as caffeic, oleuropein, and traces of flavoids. Luteolin, apigenin, cumaric acid and vanillic acids, are the other common phenolic compounds in olive oils. Of the phenolic compounds in olive oil, hydroxytyrosol and tyrosol are considered to be the main ones; these contribute to a bitter taste and astringency, and are resistant to oxidation, and are the ones now being cited in the press as being desirable health components of olive oil. The antioxidant properties of hydroxytyrosol have been well studied in different models (Manna, 1999). The natural antioxidants of flavenoid polyphenols in olive oil are also shown to have a host of beneficial effects, ranging from healing sunburn to reducing cholesterol levels, blood pressure, and the risk of coronary disease. The polyphenol content in olive oil is determined by the olive variety, the time of picking, the oil processing method, whether the oil is refined, and the length of time for which the oil has been stored.

There are some specific types of olives, such as the Tuscan varieties, that have higher polyphenol values. These oils are valuable in that when blended with low polyphenol oils they will extend the shelf-life by preventing rancidity.

Studies have shown that most olives picked earlier in the year will have more polyphenols. Olives picked later in the winter have fewer polyphenols and a more mellow taste. Polyphenol concentrations increase with fruit growth until the olives begin to turn purple, when they begin to decrease. Years ago farmers valued the more mellow taste and tried to wait to pick their olives, but this led to the risk of freezing or loss of their crop to the elements. Nowadays, the strong earlier-harvest taste has become popular.

The processing method is the factor that most affects the quantity of polyphenols in olive oil. Compared with virgin olive oils, refined oils contain only very small amounts of polyphenols. As oil sits in storage tanks or in the bottle, the polyphenols will slowly be oxidized and used up. Therefore, if an oil with more polyphenols is desired, it is important to buy one that displays a date guaranteeing that it is fresh, and one that has been stored properly. A study has also revealed that olive oil from highly irrigated trees contains less polyphenols than that from moderately irrigated trees (Beregguer *et al.*, 2006).

Determination of polyphenols

The level of polyphenols in oil is generally determined as the total polyphenols by the colorimetric method using Folin-Ciocalteu reagent, as described by Gutfinger (1981). In brief, 1 g of oil is dissolved in 5 ml of hexane, 2 ml of aqueous methanol (60 : 40 v/v) is added and the sample is mixed vigorously for 2 minutes. The methanol phase is then pipetted off and put in a beaker. The process is repeated twice. The combined methanolic solution is evaporated to dryness using a vacuum evaporator at 40 °C, and the residue is re-suspended in 1 ml methanol. An aliquot (0.05 ml) of concentrated phenol solution is transferred to a 5 ml volumetric flask and 2.5 ml of distilled water is added, followed by 0.125 ml of 2N Folin-Ciocalteu reagent and 0.5 ml of 35% Na_2HCO_3 solution. The flask is then filled with distilled water up to the mark. The absorbance is measured after 1 hour under a UV spectrophotometer at 725 nm. Although this method is simple and easy, it does not provide qualitative information regarding single phenolics which are of particular importance. To obtain qualitative and quantitative information, an LC-MS method can be employed.

Determination of polyphenol composition by LC-MS Briefly, 5 μl of the concentrated phenol solution as prepared in total polyphenol determination (but without the addition of the Folin-Ciocalteu reagent) is injected into a reverse-phase column (5 μm, 4 × 250 mm) in an HPLC system with mobile phase (A) methanol and (B) 0.1% aqueous acetic acid at a 1 ml/min flow rate with an increasing mobile phase (A) of 5–35% (0–30 min); 35–65% (30–35 min); 65–100% (35–50 min) and 100% (50–55 min) at 280 nm. After separation, the detector can be switched to MS (Bruker MS Esquire 300 Plus). Before analyzing, sample solutions of the standards are injected into the LC-MS system and recorded in the system's library. Each LC peak of the sample is qualitatively and quantitatively determined in comparison with the library records. For further analysis, fragmentation methods can also be utilized.

Tocopherols

Tocopherols are collectively known as vitamin E, and represent an important class of phenolic antioxidants that occur naturally in vegetable oils and function to maintain oil quality by terminating free radicals (Pocklington and Dieffenbache, 1988; Yoshida *et al.*, 2003). α-, β-, γ- and δ- are the four homologue forms of tocopherols that generally exist in oil as minor ingredients. In particular, the α-isomer of tocopherol is found in olive oil in only the free (non-esterified) form. Most natural olive oil contains 12–150 ppm tocopherols, of which 90 percent is α-tocopherol (Kiritsakis, 1998); however, commercial olive oil contains 100–300 mg of tocopherol per kilogram of oil (Boskou, 1996). The tocopherol content is significantly reduced towards the end of the olive harvest period, and also varies according to the cultivar. Because of its valuable activity against oxidation, a certain level of α-tocopherol is added to refined olive oils to improve their stability (Blekas *et al.*, 1995).

The concentration of tocopherols in olive oil is determined using the procedure suggested by the IOC and IUPAC using HPLC. Briefly, 1 g of oil is weighed out into a 20 ml glass vial and 10 ml of HPLC-grade hexane is added. The vial is vortexed until all the oil is dissolved, and a fraction of this solution is filtered through 0.45 μm Millipore disks (Teknokarma, Spain) and transferred to 1.5 ml vials before injection into the HPLC under a detection wavelength of 295 nm. The mobile phase consists of an isocratic solution of 75 percent hexane and 25 percent ethyl acetate with a flow rate of 1 ml/min. For the standard, stock solutions should be prepared by dissolving α-, β-, and γ-tocopherol standards in methanol. The absorbance of each of the standard solutions is measured at 292 (α), 296 (β) and 298 (γ) nm, respectively, and the concentration of tocopherols (μg/ml solution) is calculated by dividing the absorbance values by prescribed factors of 0.0076 (α), 0.0089 (β) and 0.0091 (γ), respectively. The concentration of tocopherols in the samples is determined in relation to the peak areas of the standards, and the results are adjusted for the average specific gravity of olive oil.

Erythrodial and uvaol

There is a growing tendency for mixing olive-pomace oil and refined olive oils into virgin olive oils. The compounds erythrodiol and uvaol are the two main triterpenic alcohols of olive oils present in small quantities in pressed oils, but at much higher levels in extracted oils (Blanch *et al.*, 1998). Most of these compounds are found in the skin of the fruit. The major compound of this group is erythrodial, also known as home-olestranol, which is the glycerol derived from oleanolic acid by reducing the carboxyl group of alcohol. These compounds can be detected by GC. The limit of these compounds in edible olive oil is ≤ 4.5 ppm (IOC, 2006).

Aliphatic alcohols

Aliphatic alcohols are part of the unsaponifiable matter, and although their level doesn't usually exceed 350 ppm in pressed olive oils, it is higher in pomace oils (IOC, 2006). However, climatic conditions and high temperatures may cause a high alcanol content in olive oil, and because of this the criteria for measuring aliphatic alcohols have been removed from international standards. Some olive-growing areas produce higher levels of alcanol content. Since 2002, the measurement of aliphatic alcohols has only been applicable to the category of lampante virgin olive oil (IOOC, 2003). Solvent-extracted pomace oils contain a higher concentration of aliphatic alcohols, because their levels are higher in the fruit skin.

Waxes

Waxes are esters of fatty alcohols and fatty acids. The refining process can eliminate aliphatic alcohols, but it is more difficult to remove waxes. Olive-pomace oil contains more waxes than virgin olive oil because pomace contains a greater portion of fruit skin, which is where most of the waxes originate; because of this difference, the wax content is used to identify adulteration with pomace oil.

The presence of wax in olive oil can be determined by gas chromatography. The main waxes found are the esters C-36, C-38, C-42, C-44 and C-46 (Kiritsakis, 1998). There is an international standard for virgin olive oils, refined olive oils, olive oils, refined olive-pomace oils and olive-pomace oils, fixed at ≤ 250, ≤ 350, ≤ 350, > 350, and > 350 mg/kg, respectively (IOC, 2006).

Induction time

Induction time has no official standard, but it is a useful measurement for comparing the relative stability of different oils, and is therefore considered to be a good tool for evaluating the resistance of olive oil to oxidation. To do this, the sample is heated and exposed to oxygen to initiate oxidation, and the formation of hydroperoxide is measured, either by titration or electronically. A Metrohm 679 Rancimat – a machine that accelerates the oxidation of oil – is commonly used to determine the induction time of the oil. A fixed temperature of $130\,°C$ and airflow of 20 l/hour are used. Volatile components that develop as a result of oxidation are measured in this process. When the oil begins to oxidize, the change in conductivity of the water used to trap these volatile components is recorded. The results are reported as induction time in hours (Kiritsakis *et al.*, 2002).

11.4 Effects of the desert environment on olive oil quality

Intensive studies have been carried out in recent years, mainly in Israel, regarding the quality of desert-grown olive oils. The first of these studies was carried out by Wiesman and colleagues on the Barnea variety cultivated under intensive irrigation in Ramat Negev, in the central Negev Desert (Weissman *et al.*, 2008). Details of the climatic and geographical data are given in Chapter 3. Since irrigation water is the main limiting factor in the desert or drylands, and the most common water available for these areas for olives and other crops is saline water, the study was carried out to compare irrigation using water with different salt contents: fresh water (1.2 dS/m EC), saline water pumped from the local underground aquifer (7.5 dS/m EC), and moderately saline water (mixed saline and fresh water, 4.2 dS/m EC). The moderate and high saline irrigation treatments were also adjusted with leaching, as described in Chapter 5. The outcome of this study showed that there is hardly any alteration in the quality of olive oil grown in the desert environment compared to more traditional areas (Table 11.6), and no adverse effects on quality were found with any of the irrigation treatments. Most of the parameters tested were found to be better with both moderately saline water irrigation and higher saline water irrigation than with fresh water irrigation under desert conditions. Free fatty acids (FFAs), the percentage of which is considered to be one of the main quality parameters, were found to be slightly lower in both saline water treatments. The peroxide value (PV), polyphenols (PPs), tocopherols (TPs) and oleic acid (18:1) were found to be higher following both moderately saline (4.2 dS/m EC) and higher saline (7.2 dS/m EC) water irrigation treatments. Although most of the quality parameters of oil were observed in the product grown with higher salinity irrigation compared to that grown with moderately saline irrigation, vegetative growth was found to be better with moderately saline irrigation. Moderately saline irrigation inhibited the growth significantly only in the first year after planting, and from the second year onward retardation of vegetative growth compared with freshwater treatment was not significant; however, growth retardation in the trees grown with higher salinity irrigation was greater (results not shown). This result clearly indicates that oil from olives produced in the desert environment does not suffer in quality.

In this study, acidity, peroxide value, fatty acid composition, polyphenol composition, phenol composition, tocopherol content, total and compositional content of sterol, and organoleptic tests were carried out on a series of many promising varieties of different origin, namely Barnea, Souri and Maalot (Israeli); Frantoio and Leccino (Italian); Arbeqina, Picual and Picudo (Spanish); Kalamata and Koroneiki (Greek); Picholine (French); and Picholin de Morocco (Moroccan). Based on the overall results obtained regarding the quality, the yield and the availability of the water source, we came to the conclusion that moderately saline irrigation is optimal in the desert environment; therefore, we focused on this type of irrigation. This study was again carried out in a comparative manner with moderately saline (4.2 dS/m EC level) irrigation and fresh water (1.2 dS/m EC level) irrigation; however, the data presented from here onwards only refer

Table 11.6: **Effect of desert environment on olive oil qualities (variety Barnea), 2000–2001**

Salinity level (dS/m)	FFA (%)	PV (meq/kg)	TP (ppm)	PS (µg/kg)	FA profile						
					16:0	18:0	18:1	18:2	18:3	20:0	
1.2	0.82	7.6	41.3	1509	15.9	2.4	55.8	20.9	0.8	0.7	
4.2	0.71	9.3	163.7	2056	16.3	2.5	59.1	18.0	0.6	0.1	
7.5	0.76	9.8	111.2	2450	18.8	2.9	60.4	17.2	0.6	0.6	

FFA, free fatty acid; PV, peroxide value; PP, total polyphenols; TP, tocopherols; FA, fatty acid; SA, saturated fatty acids; UN, unsaturated fatty acids.
Source: Wiessbein et al. (2008).

Table 11.7: Effect of desert environment on the fatty acid profile of olive oil of different varieties

Variety	Fatty acid profile (%)						Saturated	Unsaturated
	16:0	18:0	18:1	18:2	18:3	Others		
Barnea	14.52	2.79	65.30	16.30	0.30	0.61	17.31	81.60
Souri	16.49	2.32	63.00	17.47	0.21	0.54	18.81	80.47
Maalot	15.78	1.85	60.00	19.00	0.98	1.92	17.63	79.00
Frantoio	19.32	1.89	59.22	16.94	0.80	1.00	21.21	76.16
Leccino	16.20	1.01	65.62	14.06	0.35	0.98	17.21	79.68
Arbeqina	19.17	2.70	54.10	19.61	0.87	3.70	21.87	73.71
Picual	15.35	3.18	70.55	9.92	0.30	0.97	18.53	79.47
Picudo	15.47	1.30	58.49	19.78	0.50	1.85	16.78	80.27
Kalamata	12.73	1.27	68.91	15.35	0.15	0.65	14.00	84.25
Koroneiki	14.82	2.52	71.53	10.12	0.25	0.86	17.34	81.65
Picholine	14.82	1.48	71.30	9.04	0.75	1.85	16.30	80.34
Picholin de Morocco	13.87	0.95	65.92	17.44	0.45	0.75	14.82	83.36
Significance	***	***	***	***	***	***		

*** Significant at 0.01% level as determined by Tukey-Kramer.

to moderately saline irrigation. The studies were carried out between 2001 and 2005 on trees grown at the Ramat Negev experimental research station. Samples of oil were obtained using a commercial olive oil extraction mill (dual-phase decanter system) stationed at the research station, and analysis of each sample was performed within 15 days of oil extraction. The results are presented below.

11.4.1 Fatty acid profile

The data from the study showing the effect of the desert environment on the fatty acid profile of olive oil from different selected varieties are presented in Table 11.7. The results are the average data for 2001 through 2005.

Although the data show a high variation in the fatty acid profile among the varieties tested, all the fatty acids, including linoleic (18:2) and linolenic (18:3), are within the permitted range for virgin olive oil as suggested by the IOC. In line with previous studies, oleic acid (18:1) was found to be predominant among the fatty acids, ranging from 54 to 71 percent of the total. Oleic acid is the most desirable in olive oil, from the nutritional point of view (IOC standard 53–85 percent); linolenic acid, with three double bonds, is the most chemically reactive, and is therefore undesirable from the viewpoint of stability (IOC standard < 1.0 percent); and palmitic acid, a saturated fatty acid, is also not desirable (IOC standard 7.5–20.0 percent). All these were

Table 11.8: Desert olive oil quality from different cultivars (values are the average of five years 2001–2005)

Variety	Acidity (% oleic acid)	Peroxide value (meq/kg)
Barnea	0.48	5.76
Souri	0.63	5.45
Maalot	0.78	6.99
Frantoio	0.70	7.16
Leccino	0.26	5.57
Arbeqina	0.77	5.25
Picual	0.32	6.72
Picudo	0.53	4.13
Kalamata	0.36	5.51
Koroneiki	0.64	8.0
Picholine	0.22	4.54
Picholin de Morocco	0.41	5.72
Significance	***	NS

*** *Significant at 0.01% level, NS not significant, as determined by Tukey-Kramer.*

found to be within permitted limits in desert-grown oil. These data clearly indicate that, from the FFA profile point of view, desert-produced olive oil has no quality defects. Furthermore, the significant variation among the varieties provides an opportunity for selection of the superior varieties for the desert environment.

11.4.2 Acidity and peroxide values

The data presented in Table 11.8 show the average acidity and peroxide values of the olive oil from different cultivars grown in the Negev Desert over a five-year period (2001 to 2005).

The results show that there is high variation in the acidity of the olive oil of different varieties grown in the Negev Desert under moderately saline irrigation. However, no significant difference was observed in the peroxide values.

The data clearly indicate that the FFA levels of all the tested varieties were lower than 0.8 percent, with some varieties (Picholine and Leccino) having levels of less than 0.3 percent. This demonstrates that, with regard to the acidity level (the most basic quality parameter for olive oil), there is no difficulty in meeting the IOC standard for desert-grown olive oil.

According to IOC standards, the peroxide value should be less than 20 meq/kg oil, and the data show that all the values were far lower than this, at less than 10 meq/kg. This means that the

Table 11.9: Tocopherol composition of desert olive oil from different cultivars (values are the average of five years 2001–2005)

Variety	Tocopherol (μg/g)				
	α-	β-	γ-	δ-	Total
Barnea	368	2.2	62	0.7	432.9
Souri	273	4.2	81	2.0	360.2
Maalot	82	2.1	24	1.0	109.1
Frantoio	211	1.2	27	0.0	239.2
Leccino	311	2.4	67	3.0	383.4
Arbeqina	168	3.4	24	1.2	196.6
Picual	284	4.1	46	2.0	336.1
Picudo	145	1.8	34	1.2	182.0
Kalamata	306	3.8	44	1.0	354.8
Koroneiki	285	2.8	43	2.5	333.3
Picholine	333	2.3	45	1.1	381.4
Picholin de Morocco	108	1.4	18	0.5	127.9
Significance	***	***	***	***	***
Average of all varieties	239.5	2.6	42.9	1.4	286.4
% of total	83.6	0.9	15.0	0.5	

*** *Significant at 0.01% level, as determined by Tukey-Kramer.*

desert-grown olive oil will have a good shelf-life. High levels at the bottling stage are not a good indication of a long shelf-life.

11.4.3 Tocopherols and polyphenols

Studies have shown that α-, β-, γ- and δ-tocopherol are the four isomers of tocopherol, known collectively as vitamin E, that are present in olive oils. They occur in the range 12–150 ppm in most virgin olive oils, with the α-isomer accounting for about 90 percent of the total tocopherols (Kiritsakis, 1998) and some studies reporting a rate as high as 95 percent (Psomiadou *et al.*, 2003; Cunha *et al.*, 2006) in oil from traditional olive-growing areas such as Spain, Greece and Portugal. Five-year average data relating to the various tocopherol levels of olive oils of different cultivars grown in the Negev Desert are presented in Table 11.9. The data show that the desert-grown olives contain a far higher level of total tocopherols, compared to previous literature results regarding the traditional olive-growing areas, ranging from 109–433 ppm (μg/g) of olive oil.

The data clearly also indicate that the percentage of α-tocopherol in desert-grown olive oils is lower (83 percent) than in oils from traditional growing areas, while the percentage of the γ-isomer is higher (15 percent of the total). Furthermore, there is significant variation regarding the different isomers of the tocopherols, as well as the total, among the tested varieties, with Barnea showing the highest level of total tocopherols (432 μg/g) and Picholin de Morocco the lowest (109 μg/g).

The breakdown of the polyphenols, as determined by LC-MS, using the internal standards of hydroxytyrozol, tyrozol, luteolin, apigenin, cinnamic acid, cumaric acid, vanillic acid and caffeic acid, is presented in Table 11.10. The two-year average data (2004–2005) show a strong variation regarding the constituents of the different phenol compounds among the varieties tested. In line with the literature, hydroxytyrozol was found to be dominant except in Souri (where tyrozol was dominant), and tyrozol was next, except in Barnea and Arbequina (where luteolin came second). In most varieties, caffeic acid was not detected. Compared to the literature data, in all cases the total amount of tocopherols was found to be higher in desert-grown olives – a very encouraging result. Of the varieties tested, Picudo, Kalamata and Souri oils showed a comparatively low level of polyphenols.

Koroneiki, Leccino, Barnea, Arbeqina and Picual were found to have more than 400 ppm of total polyphenols in the oils. The higher polyphenol levels in desert-grown olive oil may be connected with the irrigation and moisture status of the tree. Earlier studies have shown that olive fruits with a moisture deficit produce oil with increased total polyphenol content (Tovar *et al.*, 2002). In the present study, each plant had received sufficient irrigation water (6560 m^3/y per ha), with an additional 2000 m^3/y per ha for leaching. However, owing to the effect of salinity, some kind of minor moisture deficiency might have occurred in the internal part of the olive tree that could be the cause of the general increment in total polyphenols in the olive oil produced under these conditions. Moreover, it is generally known that phenols are bitter in taste, contributing pungency and bitterness to the oil; however, even though high polyphenol contents were found in the desert oils, organoleptically they did not demonstrate any specific bitterness or pungency that adversely affected quality (for details of organoleptic tastes, see below).

11.4.4 Phytosterols

The phytosterol composition of the olive oil grown in the Ramat Negev plantation (Arbequina, Picual, Leccino and Koroneiki varieties) in the 2005 season is presented in Table 11.11. All four varieties grown under desert conditions contained more than 2000 ppm of total phytosterols, which is slightly higher than in oil from traditional areas (El-Agaimy *et al.*, 1994). Among the varieties tested, Koroneiki had the highest amount of total phytosterols. In accordance with the literature, the main phytosterol found in oil from olives grown in desert conditions with moderately saline water irrigation (4.2 dS/m EC) was the β-sitosterol (90–97 percent of the total sterols). The other main phystosterols found were Δ-5-avenasterol, clerosterol and campesterol. The slightly higher total phytosterol content in desert olive oil might again be connected to the saline irrigation water.

Table 11.10:   Phenol composition of desert olive oil from different varieties cultivated with moderately saline irrigation (EC 4.2 dS/m) (values are the average of two years, 2004 and 2005)

Variety	Phenols (ppm)								
	Hydroxytyrozol	Tyrozol	Luteolin	Apigenin	Cinnamic acid	Cumaric acid	Vanillic acid	Caffeic acid	Total
Barnea	102.1	59.0	253.0	54.6	20.3	nd	nd	nd	489.0
Souri	78.6	110.2	nd	nd	nd	10.3	6.8	nd	205.9
Leccino	342.9	131.3	29.9	20.6	20.0	0.4	31.4	0.4	576.9
Arbeqina	210.0	32.9	131.4	29.8	20.6	5.7	46.2	nd	476.6
Picual	201.1	151.7	34.4	7.8	19.6	21.8	23.4	0.6	460.4
Picudo	44.2	53.9	15.2	7.5	nd	15.5	2.2	1.4	139.9
Kalamata	45.4	75.0	29.6	nd	nd	3.4	21.6	nd	175.0
Koroneiki	286.2	161.7	32.1	9.5	20.9	nd	30.7	nd	641.1
Significance	S	S	S	S	S	S	S	S	

S, significant at 0.05% level as determined by Tukey-Kramer; nd, not detected.

Table 11.11: Distribution of sterol in desert olive oil from different varieties cultivated with moderately saline irrigation (EC 4.2 dS/m)

Phytosterols (ppm)	Olive varieties			
	Arbeqina	Picual	Leccino	Koroneiki
Cholesterol	1.66	1.37	0.99	0.70
Brassicasterol	0.07	0.155	0.56	0.32
24-methylene-cholesterol	1.35	1.04	3.01	2.89
Campesterol	17.89	23.77	26.99	12.27
Campestanol	0.70	3.05	0.61	3.17
Stigmasterol	8.58	5.96	6.34	6.79
Δ-7-campesterol	0.40	0.76	1.00	0.96
Δ-5,23-sigmastadienol	0.72	1.48	0.17	1.55
Clerosterol	28.65	12.35	9.02	8.87
β-sitosterol	1965.90	2814.32	2057.15	3031.90
Sitostanol	4.29	1.66	3.51	5.13
Δ-5-avenasterol	128.60	53.02	110.10	22.69
Δ-5,25-sigmastadienol	5.12	4.18	4.99	5.56
Δ-7-stigmastenol	2.20	2.53	2.00	2.15
Δ-7-avenasterol	3.05	2.57	3.08	3.29
Total	2169.12	2928.22	2229.53	3108.27

11.4.5 TAG composition

The TAG fingerprints of the olive oils from some of the prominent olive varieties grown in desert conditions under moderate saline irrigation, measured using MALDI-TOF/MS, are presented in Figure 11.7. As described in the literature, the major TAGs found were oleic-oleic-oleic (OOO), oleic-oleic-palmitic (OOP) and oleic-oleic-linoleic (OOL). Other TAGs found were linoleic-linoleic-oleic (LLO), palmitic-oleic-linoleic (POL), linoleic-linoleic-palmitic (LLP), palmitic-palmitic-oleic (PPO) and palmitic-palmitic-oleic (PPO). In addition to these qualitative results, MALDI-TOF/MS also provided quantitative data for each of the TAGs using a simpler method than the prevailing techniques (Kaufman and Wiesman, 2007); these are presented in Table 11.12. The results show that, as reported in other studies regarding oil samples from traditional olive-growing areas, the main TAGs found in all the tested varieties were OOO (30 percent, range 21–41 percent), followed by OOP (24 percent, range 21–27 percent) and OOL (14 percent, range 11–18 percent). The lowest TAG observed was PPL (3 percent, range 2–6 percent). This study clearly indicates that the distribution of TAGs in desert-grown olive oil cultivated with moderately saline irrigation (4.2 dS/m EC) is similar to that from other prominent olive-growing regions.

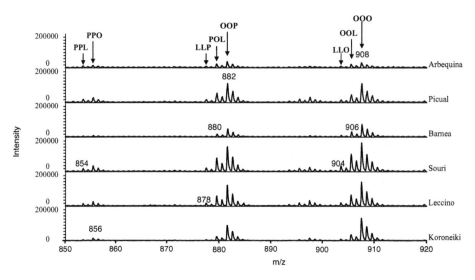

Figure 11.7: ALDI-TOF/MS fingerprints of desert olive oil from some of the varieties grown under moderately saline (4.2 dS/m EC) irrigation.

Table 11.12: MALDI-TOF/MS TAG composition of desert olive oils from different varieties

Varieties	TAG profile (%)							
	PPL	PPO	LLP	POL	OOP	LLO	OOL	OOO
Arbequina	3.78	6.69	7.13	17.22	20.79	7.96	15.83	20.60
Picual	5.62	5.69	5.13	13.90	25.84	4.75	11.36	27.17
Barnea	3.01	4.32	–	9.94	24.42	5.21	15.52	37.59
Souri	3.72	5.27	4.48	14.58	22.48	7.03	18.06	24.38
Leccino	3.83	4.94	4.68	13.46	24.57	5.20	14.84	28.48
Koroneiki	1.94	4.81	1.88	8.41	27.42	2.95	11.25	41.34
Average	3.36	5.29	4.66	12.96	24.25	5.52	14.48	29.93

P, palmitic acid; O, oleic acid; L, linoleic acid.

11.4.6 Organoleptic qualities

The quality grade of an olive oil is determined not only on the basis of a number of chemical and physical parameters, but also on an additional sensory evaluation of it. The sensory attributes of the olive oils produced from several selected European olive varieties cultivated in the Negev Desert with saline irrigation are presented in Table 11.13, defining all the oil as of extra-virgin quality. A summary of the evaluation of the local Negev expert panel suggests that none of the oils showed any defects. Regarding positive descriptors, Leccino and Picual oils were evaluated as being slightly higher quality than Koreneiki and Arbequina oils.

Table 11.13: Sensory attributes of desert olive oil cultivated with saline irrigation, as tested by a panel of local experts in 2004

Parameter	Olive variety			
	Arbeqina	Picual	Leccino	Koroneiki
Overall grading[a]	6.5	7.5	8.0	7.0
Flavor description[b]				
Positive description				
Fruity	2.5	4.5	5.0	4.0
Bitter	3.5	3.0	3.5	3.0
Pungent	3.0	3.5	4.5	3.0
Others	2.5	4.0	4.5	3.5
Negative description (defects)				
Fusty	–	–	–	–
Musty	–	–	–	–
Metallic	–	–	–	–
Rancid	–	–	–	–

[a] *1, lowest quality; 9, optimal quality.*
[b] *1, lowest quality; 5, optimal quality.*

Comparable sensorial testing of Souri and Phicoline olive oils extracted from trees irrigated with saline water and fresh water in the Negev Desert experimental station (Table 11.14) showed that the use of saline water may affect some aspects of the aroma and taste of the oil. Indeed, it is possible to produce extra-virgin olive oil under Negev Desert conditions by applying both fresh and saline water. The results obtained further suggest that there are some fine and delicate differences between the olive oils cultivated with the two different sources of water. Saline water may contribute to the fruity and bitter characteristics of the oils, and this effect may be related to a small increase in the volatile compounds in the saline-cultivated olive oils. Furthermore, the panel experts evaluating the oils described a small increase in "other components" in the saline olive oils in both the tested varieties, and indeed the overall grading of the oils was also slightly increased for oils cultivated under saline conditions in the Negev Desert compared to oils cultivated with fresh water irrigation.

Basically, the results of organoleptic olive oil testing carried out with the oils produced in the Negev environment are in agreement with literature reports, suggesting that genetic material is the main factor contributing to the quality of the olive oil (a detailed description is presented in Chapter 7). However, the contribution of environmental conditions, with an emphasis on the controlled stress effect achieved by intensive cultivation in the desert (described in detail

Table 11.14: Comparison of sensory attributes of Negev Desert cultivated olive oils irrigated with saline water and fresh water, as tested by a panel of local experts in 2004

Parameter	Olive variety			
	Souri/fresh	Souri/saline	Picoline/fresh	Picoline/saline
Overall grading[a]	7.5	8.0	7.0	7.5
Flavor description[b]				
Positive description				
Fruity	3.5	4.0	2.5	3.5
Bitter	4.0	4.5	3.0	3.5
Pungent	3.5	3.5	3.5	4.0
Others	3.5	4.0	2.0	2.5
Negative description				
(defects)				
Fusty	–	–	–	–
Musty	–	–	–	–
Metallic	–	–	–	–
Rancid	–	–	–	–

[a] *1, lowest quality; 9, optimal quality.*
[b] *1, lowest quality; 5, optimal quality.*

in Chapter 6), should also be considered. The data collected during the last decade regarding Negev Desert conditions suggest that a high-quality oil with improved flavor can be produced, as reported by Kiritsakis (1998).

To further clarify the sensorial effects of desert conditions on olive oil, advanced technologies based on an "artificial nose" and using a head space system must be applied in order to provide analytical data regarding the volatile compounds released from olive varieties cultivated with saline water in Negev Desert conditions. Such a study is already in progress, and may soon yield enough data to allow differentiation of these oils, and subsequent branding, in the international market.

11.5 Summary

Olive oil has been the subject of many studies, particularly in the past few decades, because of both increasing health concerns and the usefulness of the oil. The studies have agreed that olive oil should be qualified as a healthy food, and that the quality of the olive is not the only issue. Many factors affect the quality of olive oil. These factors can be grouped into those that act during oil formation in the fruit, during fruit collection, and during the processing and storage of oil. Genetic (varietal), climatic and environmental factors might affect all of the above. However, genetics and environmental conditions may have a direct effect during the stage of oil formation in the developing olive fruit. Olive oil biosynthesis is a product of plant metabolism that is directly affected by all these factors.

The assessed quality of olive oil is mainly dependent upon market preferences, but aroma, taste, color and composition are the basic parameters consumers always look for. Producing a good-quality product or the consumers' preferred product is always the prime challenge for producers these days, and this is more important in olive oil than for many other products. The majority of plant oil parameters are common to most of the commercial edible oils, and since the price of olive oil is almost five-fold higher than that of most other edible oils, like soy or canola, there is a tendency to adulteration. For this reason, measurement of the quality parameters of olive oils is very crucial.

Experience has shown that olive trees can be grown successfully in most arid regions where irrigation is available, with some precautions. The results summarized in this chapter clearly indicate that, from the quality aspect as well, dryland areas are suitable for olive oil cultivation. Some quality parameters, such as polyphenols, tocopherols and phytosterols, are found to be present at even higher levels in desert-grown olive oils. Organoleptic taste tests by an expert panel have suggested some advantages of desert-produced oils, and this general evaluation should be further developed by using advanced analytical means. Bearing in mind its high quality, certain desert olive-growers have already launched their product in the national and international markets, and this trend is likely to increase in the near future.

References

Angerosa, F. (2002). Influence of volatile compounds on virgin olive oil quality evaluated by analytical approaches and sensor panels. *Eur J Lipid Sci Technol*, 104, 639–660.

Aparicio, R., & Luna, G. (2002). Characterisation of monovarietal virgin olive oils. *Eur J Lipid Sci Technol*, 104, 614–627.

Asbury, G. R., Al-Saad, K., Siem, W. F., Hannan, R. M., & Hill, H. H. (1999). Analysis of triacylglycerols and whole oils by matrix-assisted laser desorption/ionization time of flight mass spectrometry. *J Am Soc Mass Spectrom*, 10, 983–991.

Ayton, J.G., (2006). The effect of harvest timing and irrigation on the quality of olive oil. MSc Thesis, University of Western Sydney, Australia.

Beregguer, M. J., Vossen, P. M., Grattan, S. R., Cannel, J. H., & Pilito, V. S. (2006). Tree irrigation levels for optimum chemical and sensory properties of olive oil. *Hort Sci*, 41, 427–432.

Blanch, G. P., Villen, J., & Herraiz, M. (1998). Rapid analysis of free erythrodial and uvaol in olive oils by coupled reversed phase liquid chromatography-gas chromatography. *J Agric Food Chem*, 46, 1027–1030.

Blekas, G., Tsimidou, M., & Boskou, D. (1995). Contribution of a-tocopherol to olive oil stability. *Food Chem.*, 52, 289–294.

Boskou, D. (1996). *Olive Oil Chemistry and Technology*. AOCS Press, Urbana, IL.

Canabate-Diaz, B., Carretero, A. S., Fernandez-Gutierrez, A., Vega, A. B., Frenich, A. G., Marinez-Vidal, J. L., & Martos, J. D. (2007). Separation and determination of sterols in olive oil by HPLC-MS. *Food Chem*, 102, 593–598.

Cortesi, N. R., Rovellini, R., & Fedeli, E. (1990). Triglycerides of olive oil. *Riv Ital Delle Sost Grasses*, 67, 167.

Cunha, S. C., Amaral, J. S., Fernandes, J. O., Beatriz, M., & Olivera, P. P. (2006). Quantification of tocopherols and tocotrienols in Portuguese olive oils using HPLC with three different detections. *J Agric Food Chem*, 54, 3351–3356.

Damirchi, S. A., Savage, G. P., & Dutta, P. C. (2005). Sterol fractions in hazelnut and virgin olive oils and 4, 4′-dimethylsterols as possible markers for detection of adulteration of virgin olive oil. *J Am Oil Chem Soc*, 82, 717–725.

Del Giovine, L., & Fabietti, F. (2005). Copper chlorophyll in olive oils: identification and determination by LIF capillary electrophoresis. *Food Control*, 16, 267–272.

EC (1991). Commission regulation (EEC) no. 2568/91 of 11 July 1991 on the characteristics of olive oil and olive-residue oil and on the relevant methods of analysis. *Official J Eur Union* L348. (0001-0083).

EEC European Union Commission (2003). Regulation EEC/1989/2003, *Official J Eur Union*, L295.

El-Agaimy, M. A., Neff, W. E., El-Sayed, M., & Awatif, I. I. (1994). Effect of saline irrigation water on olive oil composition. *J Am Oil Chem Soc*, 71, 1287.

Fedeli, E. (1997). Lipids of olives. *Progr Chem Fats Lipids*, 15, 57–74.

Frankel, N. (1998). *Lipid Oxidation*. The Oily Press Ltd, Dundee.

Gamel THA (1995). Effect of phenolic extracts from rosemary plant and vegetable water on the stability of blend olive oil and sunflower oil. MSc Thesis, Mediterranean Agronomical Institute of Chania, Crete, Greece.

Gomez-Alonso, S., Fregapane, G., Salvador, M. D., & Gordon, M. H. (2003). Changes in phenolic composition and antioxidant activity on virgin olive oil during frying. *J Agric Food Chem*, 51, 667–672.

Gould, W. A. (1992). *Total Quality Management for the Food Industries*. CTI Publications Inc, Baltimore, MA.

Gutfinger, T. (1981). Polyphenols in olive virgin oils. *J Am Oil Chem Soc*, 58, 966–968.

Gutierrez, F., Jimenez, B., Ruiz, A., & Albi, M. (1999). Effect of olive ripeness on the oxidative stability of virgin olive oil extracted from the varieties Picual and Hojiblanca and on the different components involved. *J Agric Food Chem*, 47, 121–127.

Hamann, Y. A., Chaouch, A., & Lesgards, G. (1988). Analysis of unsaponifiable constituents of olive oil by HPLC using electrochemical detection. *Ann Falsif Exp Chim Toxicol*, 81, 11–13.

Harborne, J. B., & Dey, P. M. (1989). *Methods in Plant Biochemistry*. Academic Press, London.

International Olive Oil Council (IOOC) 2008. http://www.internationaloliveoil.org/. Principe de Vergara 154, 28002, Madrid, Spain.

IOC (2006). Trade standard applying to olive oil and olive-pomace oil. In COI/ T.15/NC no. 3/Rev. 2.

IOOC (2003). Trade standard applying to olive oil and olive-pomace oil. In COI/ T.15/NC no. 3/Rev. 1.

Jesus-Tovar, M., Jose, M. M., & Paz, R. M. (2001). Changes in the phenolic composition of virgin olive oil from young trees (*Olea europaea* L.cv. Arbequina) grown under linear irrigation strategies. *J Agric Food Chem*, 49, 5502–5508.

Kalua, C. M., Allen, M. S., Bedgood, Jr. D. R., Bishop, A. G., Prenzler, P. D., & Robards, K. (2007). Olive oil volatile compounds, flavor development and quality: a critical review. *Food Chem*, 100, 273–286.

Kaufman, M., & Wiesman, Z. (2007). Pomegranate oil analysis with emphasis on MALDI-TOF/MS triacylaglycerid finger printing. *J Agric Food Chem*, 55, 10405–10413.

Kiritsakis, A. K. (1998). *Olive Oil from the Tree to the Table* (2nd edn). Food and Nutrition Press Inc, Trumbull, CT.

Kiritsakis, A. K., & Markakis, P. (1987). Olive oil, a review. *Adv Food Res*, 31, 453–482.

Kiritsakis, A., Kanavouras, A., & Kiritsakis, K. (2002). Chemical analysis, quality control and packaging issue of olive oil. *Eur J Lipid Sci Technol*, 104, 628–638.

Koprivnjak, O., & Conte, L. (1998). Specific components of virgin olive oil as active participants in oxidative processes. *Food Technol Biotechnol*, 36, 223–234.

Koutsaftakis, A., Kotsifaki, F., Stefanoudaki, E., & Cert, A. (2000). A three year study on the variations of several chemical characteristics and other minor components of virgin olive oils extracted from olives harvested at different ripening stages. *Olivae*, 80, 22–27.

Lawson, H. (1995). *Food Oils and Fats*. Chapman-Hall, Melbourne.

Lay, J. O., Liyanage, R., Durham, B., & Brooks, J. (2006). Rapid characterization of edible oils by direct matrix assisted laser desorption/ionization time-of-flight mass spectrometry analysis using triacylglycerols. *Rapid Commun Mass Spectrom*, 20, 952–958.

Mailer R, Beckingham C (2006). Testing olive oil quality: chemical and sensory methods. Primefacts 231. NSW DPI-Australia. (www.dpi.nsw.gov.au).

Manna, C., Galletti, P., Cucciola, V., Montedoro, G., & Zappia, V. (1999). Olive oil hydroxytyrosol protects human erythrocytes against oxidative damage. *J Nutr Biochem*, 10, 159–165.

Monteleone, E., Caporale, G., Carlucci, A., & Pagliarini, E. (1998). Optimisation of extra virgin olive oil quality. *J Sci Food Agric*, 77, 31–37.

Mu, H., & Porsgaard, T. (2005). The metabolism of structured triacylglycerols. *Progr Lipid Res*, 44, 430–448.

Newmark, H. L. (1997). Squalene, olive oil and cancer risk – a review and hypothesis. *Cancer Epidemiol Biomark Prev*, 16, 1101–1103.

Olive Oil Encyclopedia (2007) http://www.oliveoilsource.com/ (accessed 2 December 2007).

Ostlund, Jr. R. E. (2002). Phytosterols in human nutrition. *Annu Rev Nutr*, 22, 533–549.

Patumi, M., Dandria, R., Marsilio, V., Fontanazza, G., Morelli, G., & Lanza, B. (2002). Olive and olive oil quality after intensive monocone olive growing (*olea europaea* 1., cv. kalamata) in different irrigation regimes. *Food Chem*, 77, 27–34.

Pocklington, W. D., & Dieffenbacher, A. (1988). Determination of tocopherols and tocotrienols in vegetable oils and fats by high performance liquid chromatography. *Pure Appl Chem*, 60, 877–892.

Psomiadou, E., & Tsimidou, M. (1999). On the role of sequalene in olive oil stability. *J Agric Food Chem*, 47, 4025–4032.

Psomiadou, E., Konstantinos, X., Blekas, K. G., Tsimidou, M. Z., & Boskou, D. (2003). Proposed parameters for monitoring quality of virgin olive oil (Koroneiki cv). *Eur J Lipid Sci Technol*, 105, 403–409.

Quiles, J. L., Huertas, J. R., Ochoa, J. L., Batlino, M., Mataix, J., & Mans, M. (2003). Dietary fat (virgin olive oil or sunflower) and physical training interactions on blood lipid in the rat. *Nutrition*, 19, 363–368.

Rao, C. N., Newmark, H. L., & Reddi, B. S. (1998). Chemopreventive effect of squalene on colon cancer. *Carcinogenesis*, 19, 287–290.

Robards, K., Kerr, A., & Patsalides, E. (1988). Rancidity and its measurement in edible oils and snack food. *Analyst*, 113, 213.

Ryan, D., Robards, K., & Lavee, S. (1998). Assessment of quality in olive oil. *Olivae*, 72, 23–41.

Salvador, M., Aranda, F., & Fregapane, G. (2001). Influence of fruit ripening on Cornicabra virgin olive oil quality. A study of four successive crop seasons. *Food Chem*, 79, 45–53.

Schiller, J., Sub, R., Petkovic, M., & Arnold, K. (2002). Triacylglycerol analysis of vegetable oils by matrix-assisted laser desorption and ionization time of flight (MALDI-TOF) mass spectrometry and $^{P}31$ NMR spectroscopy. *J Food Lipids*, 9, 185–200.

Solinas, M., Angeros, F., & Cucurachi, A. (1987). Relation between the autoxidation products of fats and oils and increase in rancidity by organoleptic evaluation II. Quantitative determination. *Riv Ital Sost Grasse*, 64, 137.

Touchstone, J. C. (1995). Thin-layer chromatographic procedures for lipid separation. *J Chromatogr*, 671, 169–195.

Tovar, M., Romero, M., Alegre, S., Girona, J., & Moltiva, M. (2002). Composition and organoleptic characteristics of oil from Arbequina Olive (*Olea europaea* L.) trees under deficit irrigation. *J Sci Food Agric*, 82, 1755–1763.

Tsimidou, M. (1998). Polyphenols and quality of virgin olive oil in retrospect. *Ital J Food Sci*, 10, 99–115.

Van de Voort, F. R., Ismail, A. A., Sedman, J., & Emo, G. (1994). Monitoring the oxidation of edible oils by Fourier transform infrared spectroscopy. *J Am Oil Chem Soc*, 71, 273.

Vossen, P. (2007). *International Olive Council (IOC) and California Trade Standards for Olive Oil*. University of California, Cooperative Extension. (available at http://ucce.ucdavis.edu).

Wiessbein, S., Wiesman, Z., Ephrath, J., & Zilberbush, M. (2008). Vegetative and reproductive response of olive varieties to moderate saline water irrigation. *Hort Sci*, 43, 320–327.

Yoshida, H., Hirakawa, Y., Murakami, C., Mizushina, Y., & Yamada, T. (2003). Variation in the content of tocopherols and distribution of fatty acids within soya bean seeds (*Glycine max* L.). *J Food Comp Anal*, 16, 429–440.

Further reading

CODEX Stan-33 (2003). Codex Alimentarius Commission WHO/FAO Standard for olive oils and olive pomace oils, Rev 2.

Fedeli, E. (1988). The behaviour of olive oil during cooking and frying. In: G. Varela, A. E. Bender, I. A. Morton (Eds.), *Frying of Food: Principles, Changes, New Approaches* (pp. 52–81). Ellis Horwood, Chichester.

CHAPTER 12

Non-conventional olive oil industries: products and biotechnologies

Olive oil is not a recent discovery; it has been used for centuries as an edible oil and for various other purposes. The Ancient Greeks and Romans were the first to use olive oil for non-edible purposes (non-conventional). They used the oil to care for their skin, spreading it all over their bodies before and after bathing, initially only as a cleanser but later as a moisturizer, scented with herbs and flowers. The Egyptians manufactured perfumes and ointments for wounds and curative pomades, and olive oil – usually from colorless, tasteless, unscented olives harvested well before they were ripe – was often utilized as a base. Such ointments were used to treat bleeding wounds, insect bites, headaches, tired eyes, or for disinfection of any part of the body. Olive oil was also used in cases of poisoning, stomach problems and even pregnancy.

Modern investigations have proved that olive oil contains fatty acids, triglycerides, tocopherols, squalene, carotenoids, sterols, polyphenols, chlorophylls, and volatile and flavor compounds. All these components contribute to its beneficial character, making it a highly desirable component in many various products besides its conventional edible purposes. Olive oil is considered in general to be only a very weak irritant, and over the years, as methods and skills improved along with knowledge, people developed different products using all the parts of the olive tree. The potential non-traditional uses of olive oil and/or its related materials can be broadly divided into two groups: health-related uses (cosmetics, pharmaceuticals, food supplements, etc.) and others (biodiesel, fertilizers and methane, etc.).

12.1 Cosmetics
Emulsifiers

Most cosmetic products include both water-soluble and fat-soluble materials, which are mixed together to form a uniform phase otherwise known as an emulsion. The formation of an emulsion often requires the use of an emulsifier. Different synthetic emulsifiers based on alcohols, fatty acids or ethoxylated esters can be employed, though some have been found to be harmful to human skin and their viscosity is sometimes too high. Amari and Schubert (2005) developed a method for the preparation of a natural emulsifier consisting of the fatty acids of olive oil. The advantage of using this functional ingredient derived from olive oil is the high similarity of fatty acids contained in olive oil and the acids in the skin, which leads to high compliance with the human skin, good absorption, and a light feeling on the skin. The emulsifier can be used in emulsified, hyperfluid, fluid and consistent oil/water systems, and in emulsified and non-emulsified products for use on the skin, hair and mucous membranes.

Massage and bath oils

Traditionally, in the Mediterranean, people used to apply olive oil directly on the skin to cleanse, nourish and moisturize it. Nowadays, markets are flooded with bath and massage oil products designed to be added to a running bath, or applied after showering as a deep moisturizing treatment. These products consist of different blends of oils and aromatic constituents. Some are enhanced with concentrated pure essential oils. Olive oil acts as a carrier oil to lubricate the mixture and blend, dilute and diffuse the essential oils through the compound. A bath oil recipe based on olive oil was suggested by Belle (1980). This contains a cleansing element, which cleans the body and at the same time restores essential vitamins and proteins to the body via the skin. It comprises a mixture of hydrochloric acid, ascorbic acid, protein powder, olive oil, vitamin D and vitamin A.

Moisturizers

Moisturizers perform as barriers between the skin and environmental conditions. Even though these products cannot permanently affect or change the skin, they are effective at temporarily keeping depleted skin from feeling dry and uncomfortable. Lack of moisture in the skin causes skin to become dry and more prone to wrinkling. The skin has natural moisturizing factors (NMFs) which, along with lipids, prevent evaporation and provide lubrication to the skin's surface. NMFs and lipids comprise an expansive group of ingredients that include ceramide, hyaluronic acid, cholesterol, fatty acids, triglycerides, phospholipids, glycosphingolipids, amino acids, linoleic acid, glycosaminoglycans, glycerin, mucopolysaccharide and sodium PCA (pyrrolidone carboxylic acid). All of these are present in the intercellular structure of the epidermis, both between the skin cells and within the lipid content on the surface of skin. The cosmetic formulation is characterized by a lipid composition that is close to that of human sebum, and it appears to help stabilize and maintain this complex intercellular–skin matrix. Moisturizers typically work either by employing a humectant and/or an occlusive agent against moisture loss, or by working at the cellular level, rebuilding or preventing damage to the skin's natural barrier. Emollients such as oil and oleaginous materials, once applied to the skin, act as a barrier.

Olive-oil-based moisturizers contain some or all of the following ingredients: olive oil; hydro-carbon oils and waxes; silicone oils; other vegetable, animal or marine fats or oils; glyceride derivatives; fatty acids, fatty acid esters, alcohols or alcohol ethers; lecithin, lanolin and derivatives; polyhydric alcohols or esters; wax esters; sterols; phospholipids and the like; and, generally, emulsifiers. Some are all natural, while others may contain fragrances, perfumes, extracts, preservatives, pest-repellents, sunscreens, etc. These same general ingredients can be formulated into a cream, lotion or gel by utilization of different proportions of the ingredients and/or by inclusion of thickening agents such as gums or other forms of hydrophilic colloids. Some moisturizers may be in the form of a water-in-oil emulsion (where the oil phase is the continuous phase and the aqueous phase is dispersed within it) or, preferably, an oil-in-water emulsion (where the aqueous phase is the continuous phase and the oil phase is dispersed within it). This way, the treatment hydrates the skin and then traps moisture due to the occlusive agent. The oil-in-water emulsion method of preparation, as disclosed by Geria (1991), consists of admixing the two phases by slowly adding the heated oil phase to the aqueous phase with high shear mixing while maintaining an elevated temperature, until a uniform mixture is formed. A physically stable emulsion will not separate into two layers on standing. Glover and Gurman (2005) suggested a cosmetic composition comprising an olive-oil-based compound, a quaternary ammonium salt which is believed to penetrate and hydrate the skin in a molecular level, sodium pyrrolidone carboxylic acid which increases skin softness and elasticity, and an emulsifier. Other inventions in this field (for example, Veney 1981, 1983; Tritsarolis, 1993; Farooqi *et al.*, 2002; Predovan, 2004) use different compositions and formulations, made by simply heating and mixing ingredients together (Table 12.1).

Soaps

Olive oil is used in the soap industry to manufacture pure or blended vegetable oil soap.

Pure olive oil soap produces a hard soap that dries quickly, is mild and non-drying to the skin, has a rich creamy lather, and lasts longer than most other vegetable, animal and mineral oil

Table 12.1: Recipe for a home-made olive-oil-based moisturizer

Ingredients	Quantity
Olive oil	4 tablespoons
Lanolin	2 teaspoons
Glycerine	2 teaspoons
Beeswax	1 teaspoon
Emulsifying wax	1 teaspoon
Borax	1.25 teaspoons
Water	3 tablespoons

From: http://www.indianceleb.com/infopedia/beauty/moisturizing-methods-and-its-benefits.

soaps. Olive oil's chemical-additive-free nature makes it suitable for those who have allergic reactions to substances used in mass-produced soaps. These soap bars are unique and luxurious, and the cost of the ingredients and production keeps the product at the higher end of the market. Cheaper olive oils and blends can be used; however, the trade-off for reduced price is a loss of quality.

Olive oil natural soaps are sodium or potassium salts of fatty acids, originally made by combining oil with a caustic agent such as lye (sodium hydroxide) and using water as a catalyst. Hydrolysis of the fats and oils occurs, yielding glycerol and crude soap in a process called saponification. In the industrial manufacture of soap, the olive oil is heated with sodium hydroxide. Once the saponification reaction is complete, sodium chloride is added to precipitate the soap. The water layer is drawn off the top of the mixture, and the glycerol is recovered using vacuum distillation. The crude soap obtained from the saponification reaction contains sodium chloride, sodium hydroxide and glycerol. These impurities are removed by boiling the crude soap curds in water and re-precipitating the soap with salt. After the purification process has been repeated several times, the soap may be mixed with colors, fragrances and moisturizers. The method of production may vary, as may some of the ingredients, but the main components and the saponification reaction are constant.

Body butter

Olive body butter feels very much like shea butter, and behaves very similarly in cosmetic applications. It is ideal for massage, or to restore the skin's softness and natural moisture balance. Olive butter adds moisturizing properties when used in combination with other products, such as in skin- or hair-care products and handcrafted soaps. It has a mild aroma with little coloration, and contains all the properties associated with olive oil. This body butter is very simple, consisting of a blend of olive oil, beeswax and perfume oil (for scented butter).

Suntan products

Budiyanto and colleagues (2000) found that when olive oil is topically applied after UVB exposure, it can effectively reduce UVB-induced skin tumors – possibly owing to its antioxidant effects. These antioxidant effects, as previously stated, are due to polyphenol compounds present in olive oil. Skin damage is related to the destructive activity of free oxygen related radicals produced by skin cells. Papadakos (2001) proposed a sun oil with UV protection; its basic ingredient is olive oil with the addition of vitamins E and A. Soubhie (2006) disclosed a composition suitable for skin treatment of burns, scalds and sunburns due to its analgesic and antiseptic activity and promotion of skin healing. This topical composition consists essentially of olive oil, beeswax, lemon juice and boric acid, wherein the total amounts by weight of olive oil and beeswax are such that the mixture is formulated into a cream or a thick liquid. The ingredients and composition of this product are shown in Table 12.2. There is no particular restriction on the manner of mixing these ingredients.

Table 12.2: Ingredients and composition of the olive-oil-based product for treating burns developed by Soubhie (2006)

Ingredients	Quantity (g)
Olive oil	200
Beeswax	60
Lemon juice	40
Boric acid	6

Shampoo

The chemical mechanisms that underlie hair cleansing are similar to those of traditional soap, with the exception of the surfactant. Soap surfactants have high affinity to sebum, while shampoos remove less of the natural oil. In general, soap formulas can be used on the hair and shampoo formulas will work on the body, but this won't yield the best results.

In addition, olive oil has been found to alleviate dandruff, thus appearing in anti-dandruff shampoos. This condition is associated with a dramatic decrease in free lipid levels, with significant decreases in ceramides, fatty acids and cholesterol; thus the epidermal water barrier is impaired in the scalp of dandruff sufferers, and the perturbed barrier leaves sufferers more prone to the adverse effects of microbial and fungal toxins and environmental pollutants, thus perpetuating the impaired barrier (Harding *et al.*, 2002).

Some of the commercial cosmetic products made from olives and their oil are illustrated in Figure 12.1.

12.2 Pharmaceuticals

Olive oil has been used in enemas, liniments, ointments, plasters and soaps. It has also been utilized in oral capsules and solutions, and as a vehicle for oily injections; in topically applied lipogels of methyl nicotinate; to soften ear wax; and in a lipid emulsion for use in caring for pre-term infants, along with soybean oil.

Infused oils

Infused oils are widely used in aromatherapy, medical, cosmetics and food industries. Infusion is a simple and effective method of extracting oil-soluble components from herbal material. The benefit of using infused oils as opposed to plain carrier oils is that the infused oil contains the therapeutic properties of both the carrier oil and the herbs that were infused into it. Whilst savory herb oil infusions are consumed as foods, most medicinal homeopathic and cosmetic herb oil infusions are used topically. Olive oil is often used as the menstruum of choice in oil

Figure 12.1: A selection of commercial cosmetic products made from olives and their oils.
Source: http://seattlepi.nwsource.com/lifestyle/277059_olive11.html.

infusions, because of its greater stability at room temperature and the beneficial activity many attribute to its minor components (Aburjai and Natsheh, 2003). The medicinal efficacy of an oil infusion is a function of the quality and type of the menstruum, the quality and concentration of the herbal material, and the process used in extraction.

The preparation of infused oil is not complicated; it requires very mild heating, or a few weeks' storage period in a dark, warm place; disinfected dry or partially dry herbs; olive oil (preferably extra-virgin); and a clean container. After the infused oil has been prepared, the herbs must be removed to prevent the oil from becoming rancid.

Nutraceuticals

Nutraceuticals represent products that are isolated or purified from foods, although they are generally sold in medicinal forms that are not usually associated with the original form of the food. They have some demonstrable physiological benefits or properties that contribute to the prevention of chronic disease, and are used in dosages that exceed those that could be obtained from normal foods (Barnes and Prasain, 2005). A nutraceutical may, then, be a single natural nutrient in powder or tablet form, not necessarily a complete food but equally not a drug (Hardy, 2000).

Table 12.3: Composition of the anti-viral formulation (Voorhees and Nachman, 2002)

Ingredient	Quantity
Olive leaf extract	80–88%wt
Neem leaf extract	5–10%wt
Homeopathic blend	6–8%wt
Pharmaceutically acceptable excipients for proper tableting of the composition	10–20%wt

Interest in the potential health benefits of olive extracts dates back to the mid-nineteenth century, when reports were made of the ability of an extract (made from boiling the leaves) to reduce fever, and to prevent or cure malaria symptoms even more effectively than quinine (Hanbury, 1854; Capretti and Bonaconza, 1949).

Strictly speaking, a dietary supplement is the result of relatively simple procedures (i.e., extraction with water or tincture of alcohol). New patents appear daily for natural food antioxidants, and for preparation of functional foods, pharmaceutical solutions or cosmetics, based on olive extracts. The patents are quite varied in their formulation (Guinda, 2006).

As technologies have advanced, methods for the extraction and isolation of antioxidant compounds to serve as nutraceuticals have developed, either as a dedicated process or through the reuse extraction by-products to yield additional income. During the mechanical extraction process, the major proportion of the phenolic compounds is found in the aqueous phase, while only a minor percentage (< 1 percent) is located in the olive oil (Vierhuis *et al.*, 2001). Vlyssides *et al.* (2004) dealt with this in their research, showing that during the production of 1 kg of olive oil 7.5 kg of wastewater is produced, which contains 0.08 kg phenolic compounds.

Voorhees and Nachman (2002) have suggested a pharmacological composition that is effective against a broad spectrum of viruses and bacteria, due to its components, and alleviates the symptoms of cold and influenza. When combined together in certain ratios, the herbs in the formula presented in Table 12.3 have viricidal components that prevent or abate incipient colds and 'flu. The major ingredient in this formulation is an olive-leaf extract that is extracted as described previously.

Oral hygiene

According to Alvarez Hernandez (2006), the use of olive oil in the preparation of a product intended for oral hygiene, to eliminate or reduce bacterial plaque or bacteria present in the buccal cavity, contributes numerous advantages, such as:

- providing non-abrasive, effective cleaning of the cavity and teeth, removing the lipophilic microorganisms;

Table 12.4: Ingredients for a toothpaste preparation (Alvarez Hernandez, 2006)

Component	Percentage by weight
Olive oil	1–70%
Abrasive	10–20%
Moisturizer	20–50%
Surfactant	1–2%
Thickener	0.5–2%
Sweetener	Sufficient quantity
Preservative	Sufficient quantity
Water	Sufficient quantity

- reducing the absolute quantity of bacterial plaque, with a significant improvement in periodontal health;

- reducing the occurrence of halitosis by lowering the amount of volatile sulfurated compounds (VSCs), which are compounds that are produced by microorganisms and cause bad breath.

This can be used in other products, such as toothpastes, mouthwashes, oral sprays or inhalers, and chewing gums. Each can be obtained by employing conventional techniques. The components of a toothpaste containing olive oil are listed in Table 12.4.

Another product containing olive oil and directed at oral hygiene is a soft gelatin capsule designated to eliminate bad breath and clean the oral cavity, as described by Yang (2005). The capsule is taken orally and chewed. Once chewed and dissolved, the natural oil and flavor refresh the oral cavity. This product comprises core and shell ingredients. The core ingredients are:

- 40–90%wt of base oil, including olive oil, safflower oil and medium chain triglycerides

- 1–40%wt of natural or artificial flavors for refreshing the oral cavity

- 0.1–10%wt of at least one artificial sweetener.

The shell ingredients consist of gelatin as the base material, glycerin, D-sorbitol, citric acid, surfactant and edible pigments. The capsule is manufactured seamlessly by dropping the core and shell solution in a double-layer stream.

Ointments and balms

Ointments are different from creams or balms, although they appear very similar. Creams are better absorbed through the skin, whereas ointments create a layer that protects the skin while simultaneously delivering healing properties. Both may be used for skin softening, skin protection, as a water repellent, and for the application of medications into the system through endermic absorption. Ointments containing olive oil and applied topically are used to treat skin damage, such as contact dermatitis, atopic dermatitis, xerosis, eczema (including severe hand and foot eczema), rosacea, seborrhea, psoriasis, thermal and radiation burns, other types of skin inflammation, and aging (Perricone, 2001).

Ointments consist of a mix of different oils and waxes as the principal ingredients. Beeswax is added to the oil to give the ointment its consistency. The preparation method consists of heating the oil and wax until they melt together, and setting up the correct texture by simply cooling the mixture. Olive oil is considered to be an excellent base for ointments, since it is very stable when subjected to heat. If the ointment has medical purposes, different infused oils may be used as the base oil or as an additive. Essential oils can be added to give the ointment a pleasant fragrance. This is done by the time the ointment is close to setting. Geria (1991) suggested a preparation procedure for topical medicaments including anesthetics, analgesics, anti-inflammatories, antibiotics, hydroxyl acids and antifungals; compounds for the treatment of sunburn, dermatitis, seborrheic dermatitis, dandruff and psoriasis; and sunscreens. The composition and method of this preparation are directed to adding moisture to dry skin and applying a thin, long-lasting occlusive film which sustains the presence of a medication over an extended period of time without the need for an oily coating or protective bandage (Table 12.5).

Table 12.5: Home-made recipe for olive-oil-based herbal lip balm and calendula ointment

Ingredient	Quantity	
	Herbal lip balm	Calendula ointment
Olive oil	35 g	30 g
Yellow beeswax (mp 62 °C)	10 g	10 g
Jojoba oil	0.5 ml (25 drops)	
Flax seed oil	2 ml (100 drops)	
Rosemary extract (antioxidant)	2 drops	
St John's wort-infused oil	1 ml (50 drops)	
Lavender oil	2 drops	
Calendula tincture	2 ml (100 drops)	
Friars balsam tincture	2 ml (100 drops)	
Infused calendula flowers (in olive oil)		10 g
Fragrance (essential oil)	3 drops	

Source: http://www.pindariherbfarm.com/educate/makeoint.htm.

12.3 Food supplements

Olives, and olive oil in particular, have been widely used in the food industry. Besides the obvious uses, they are included in spreads, butter, sauces and more. A substantial amount of research has been done regarding the beneficial attributes of olive oil for human beings. The nutrient characteristics of olive oil are due to its high energy value, high content of mono-unsaturated fatty acids (MUFA), a nutritionally favorable saturated fatty acid to polyunsaturated fatty acid ratio (SFA/PUFA), and the presence of minor constituents with antioxidant properties (Finotti *et al.*, 2001). Olive oil may be added to other food products in smaller amounts, for flavoring rather than nutritional purposes.

Oil blends

Kincs and colleagues (2007) have developed a product that deals with the production of improved edible oils by blending different types of oils. This is done to achieve a desired composition of fatty acids and an oxidative stability otherwise achieved by hydrogenation. In this way, it is possible to provide edible oils that have a low content of PUFA without an increase in the content of *trans* fat. To achieve a desired oil blend, no special preparation methods are required besides the actual blending of oils; however, specific amounts of different types of oils are essential to achieve the correct quality. The blended oils may be used in various foods and food preparations, such as baked, fried or frozen foods; margarines; salad dressings and mayonnaise; cheeses; spreads; condiments; and others. A blending recipe for canola and extra-virgin olive oil, which can be blended in varying percentages to form eight vegetable oil blends (blends 6A through 6H) is given in Table 12.6. These two types of oil are very rich in oleic acid (C18:1).

Spreads

Spreads usually consist of emulsions of the water and fat phases, where the fat phase includes both liquid oil and a structuring fat. The fats employed as structuring agents in these formulations are generally more saturated, richer in *trans* fatty acids, and have more rather short or long fatty acid chains. Some of these fats go through a hydrogenation process, by which unsaturated fatty acids become saturated and thus solidify. Another type of spread preparation constitutes of a mixture of an edible oil of a natural origin and a monoglyceride. In this case, neither water nor other hydrogenizing agents are used.

In general, oils with a favorable SFA/PUFA ratio that are free of *trans* fatty acids tend to be liquid at room temperature, and are thus not widely used in making spreads. Olive oil, however, is an excellent candidate for such use, along with other oils of high oleic acid content.

In the case of high olive-oil content spreads, less solidifying agents (or none at all) are required.

An example of an olive-oil spread composition as suggested by Livingston (2000) is shown in Table 12.7; the fatty acid composition of the spread appears in Table 12.8.

Table 12.6: Blends of high oleic canola oil and olive oil (Kincs et al., 2007)

Fatty acid content of starting material	C8:0	C10:0	C12:0	C16:0	C18:0	C18:1	C18:2	C18:3
High oleic canola oil	–	–	–	3.6	1.6	69.0	19.4	3.7
Olive oil (virgin)	–	–	–	9.0	2.7	80.3	6.3	0.7
	6A	6B	6C	6D	6E	6F	6G	6H
Percentage of starting materials in blends:								
High oleic canola oil	60%	55%	50%	45%	40%	35%	30%	25%
Olive oil	40%	45%	50%	55%	60%	65%	70%	75%
Fatty-acid content in blends:								
Caprylic acid C8:0	–	–	–	–	–	–	–	–
Capric acid C10:0	–	–	–	–	–	–	–	–
Lauric acid C12:0	–	–	–	–	–	–	–	–
Palmitic acid C16:0	5.8	6.0	6.3	6.6	6.8	7.1	7.4	7.7
Stearic acid C18:0	2.0	2.1	2.2	2.2	2.3	2.3	2.4	2.4
Oleic acid C18:1	73.5	74.1	74.7	75.2	75.8	76.3	76.9	77.5
Linoleic acid C18:2	14.2	13.5	12.9	12.2	11.5	10.9	10.2	9.6
Linolenic acid C18:3	2.5	2.4	2.2	2.1	1.9	1.8	1.6	1.5
Oleic acid content	73.5	74.1	74.7	75.2	75.8	76.3	76.9	77.5
Polyunsaturates content	16.7	15.9	15.1	14.2	13.4	12.6	11.8	11.0

Table 12.7: Ingredients of olive-oil spread (Livingston, 2000)

Ingredient	Compositon
Oil phase (60 wt% on product)	
Olive oil	40 wt% on fat
Rapeseed oil	20 wt% on fat
Soybean oil	20 wt% on fat
Hydrogenized soybean oil	20 wt% on fat
Aqueous phase (40 wt% on product)	
Spray-dried milk powder	1% on product
Salt	0.9% on product
Sorbic acid	0.12% on product
Lactic acid	0.04% on product
Monoglyceride	0.08% on product
Lecithin	0.08% on product
Beta-carotene	Trace
Flavor	Trace
Vitamin A	Trace
Vitamin D	Trace
Balance water	

Food flavoring agents

Flavoring components from olive oil may be used, as suggested by Ganguli and colleagues (2000), to enrich other oils in a method by which a gas acts as a carrier for flavor transfer. The flavor of virgin olive oil is relatively strong as opposed to other cooking oils, such as sunflower, rapeseed or soya oils. These refined oils have good frying properties but lack any taste of their own. By imparting some of the olive-oil flavor to these oils, a relatively cheap substitute for olive oil may be achieved.

12.4 Other products

12.4.1 Bio-fertilizers and/or compost

Olive oil industries produce large amounts of waste, which can be used for recovering compounds that can be used as valuable substances by developing new processes. This is of great relevance for both the environment and the economy. Reusing total biological products may double the market price, compared to the conventional olive-oil-only product (Vlyssides *et al.*, 2004). The waste consists of either solid and liquid phases, or a two-phase extraction process by-product known as "alperujo." Fernandez-Bolanos *et al.* (2006) have developed a process that allows integral recovery of by-products from alperujo, enhancing the recovery field of valuable compounds. The recycling strategy for the waste is presented schematically in Figure 12.2.

Table 12.8: Fatty-acid composition of olive-oil spread, by gas-liquid chromatography (Livingston, 2000)

Fatty acids	Composition (%)
C12:0	0.14
C14:0	0.35
C16:0	11.39
C16:1	0.56
C18:0	6.36
C18:1	55.52
C18:2	18.76
C18:3	3.52
C20:1	0.19
C22:0	0.45

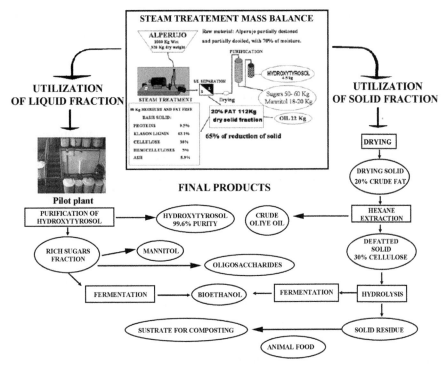

Figure 12.2: Strategy for integral recovery and revalorization of alperujo (Fernandez-Bolanos *et al.*, 2006).

The system consists of a hydrothermal treatment (steam treatment to temperatures in the range of 160–240 °C), where an autohydrolysis process occurs and the solid olive by-products are partially solubilized. This method makes solid–liquid separation easier, and allows the recovery of many added-value compounds (such as hydroxytyrosol, mannitol, oligosaccharides and fermentable sugar) from the water-soluble fraction. At the same time, the final solid residue can be used. This residue is considerably reduced after treatment, and several compounds of value (such as oil, cellulose and proteins) are concentrated. Nowadays, the system is being adapted in order to achieve simpler and less severe operating conditions. This will allow the recovery of other valuable compounds already identified in olive fruits or their by-products, and at the same time employ conditions that are more familiar to olive oil producers.

The bioconversion into useful products of the olive-oil cake (OOC) and olive-mill wastewater (OMWW), as stated by the same research team, has been widely researched around the world, as shown in Table 12.9. OMWW has been tested as a growth medium for the production of different organic products from fungi, yeast or bacteria, while some of these fungal and microbial biomasses simultaneously metabolize phytotoxics. Owing to its nutrient content (carbohydrates and proteins), the OOC maintains good holding and swelling capacity, and can be an excellent substrate for solid-state fermentation (SSF) with filamentous fungi and yeast. Alperujo seems to be a promising substrate for the growth of yeast. The isolation, purification and recovery of the products are of high cost, and therefore bioconversion is economically attractive only if high-value products are produced.

Composting is a low-cost and environmentally friendly method for turning solid oil extraction waste into organic fertilizers with a high content of organic matter. Following this method, solid waste from the oil extraction process is collected, mixed, pressed, and sent to a compost unit for the production of an ecological soil conditioner with excellent nutrient properties, as shown in Table 12.10. The solid waste in this method includes the sludges produced in the centrifuge, the chemical-organic sludges produced by the oxidative process, the sludges produced by the biological process, the olive tree leaves, the olive stones, and residues from the reed beds.

12.4.2 Alternative energy

A recent approach has investigated the use of OOC or OMWW as basic materials for the production of alternative energy. Mill waste is suitable as an alternative energy source because it has a negligible sulfur content and hence a reduced impact on the environment, is of low cost compared to fossil fuels, is easy to store, and has no transportation requirements if used in an olive oil production facility. The current worldwide annual olive oil production of 2,307,500 tons may generate 9,230,000 tons of waste annually, in which the oil content varies between 3 and 10 percent (Hepbasly *et al.*, 2003). These figures indicate the financial potential of mill waste.

Table 12.9: Value added-products obtained by bioconversion of olive oil residues (Fernandez-Bolanos *et al.*, 2006)

Residue	Description process/Biocatalyst	Product	Reference
OMWW	*Clostridium* spp. (Medium with 50% v/v OMWW)	Butanol (2.8–8 g/L)	Warmer *et al.*, 1988
OMWW	*Arthobacter* spp.	Indolacetic acid	Tomati *et al.*, 1990
OMWW	*Pseudomonas aeruginosa* (OMWW as the sole carbon source)	Biosuriaclant: rhamnolipid	Mercade and Manresa (1994)
OMWW	*Proptonibactefium shermanii*, on predigested OMWW with *Aspergillus niger*	Vitamin B_{12}	Munoz, 1998
OMWW	Recombmant strain *Escherichia coli* P-260, by expression of the enzyme 4-HPA hydrolase of *Klebsiella pneumoniae*	Synthesis of pigments, colorants, alkaloids and polymers, which structure base is a quinone	Martin *et al.*, 1998
Olive oil cake (OOC)	SSF: *Rhizomucorpusillus, R. rhizopodiformis*	Lipase (applied in bakery, pharmaceuticals)	Cordova *et al.*, 1998
OOC	SSF: Delignification (with four fungi), saccharification with *Trichoderma* spp., and biomass formation with *Candida ulilis* and *Saccharomyces cerevisiae*	Crude protein enriched from 5.9 to 40.3% Source for animal fodder	Haddadin *et al.*, 1999
OMWW	*Funalia trogit* ATCC200300 *Tmmetes mrsicolot* ATCC200801	Plant growth hormones: Gibberellic acid, absdsic acid and indolacetic acid and cytokinin	Yurekli *et al.*, 1999
OMWW	*Xanlhomonas oampeslty*, in a medium with OMWW (50–60% v/v)	Xanthan gum, for food and non-food applications as thickener or viscosifier	Lopez *et al.*, 2001
OMWW	*Paenibacillus jamilae* CP-7, in aerobic condition in a medium with OMWW (80% v/v)	Exopolysaccharide, antitumor agent with inmunomodulatory properties	Ruiz-Bravo *et al.*, 2001
OMWW	*Azolobacter ohroococcum* (OMWW as the sole carbon source)	Bioplastic: Homopolymers of β-hydroxybutyrate and β-hydroxyvalerate	Pozo *et al.*, 2002

Table 12.9: (*Continued*)

Residue	Description process/Biocatalyst	Product	Reference
OMWW (undiluted)	*Botryosphaaria thodina* mycelium growth	β-glucan $\beta(1 \rightarrow 3)$, $\beta(1\text{-}6)$	Crognale *et al.*, 2003
OMWW	SSF: *Panus tigrinus*, on OMWW-based media	Laccase and Mn-peroxidase with interest by ligninolytic activity	Fenice *et al.*, 2003
OOC	SSF: *Asporcpllus oyzae*	α-amylase, used in bakery, breweries, textile industry, clinical sector	Ramachandran *et al.*, 2004
OMWW	*Lentinula edodes* mycelium growth	Xylan and β-glucan (lentinan), with pharmacological properties as antitumoral agent	Tomati *et al.*, 2004
OOC	SSF: *Ceratocystis moniliforms, Moniliella suaveolens, Thichodetma haizianum*	Flavor active δ- and γ-decalactones	Laufenberg *et al.*, 2004
Alperujo	Growth of six phenotypically distinct groups of yeast, by a dynamic fed-batch microcosm system	Promising fermented product	Giannoutsou *et al.*, 2004
OMWW	Anaerobic fermentation to obtain volatile fatty acids, as substrate for polyhydroxyalkanoates production	Biodegradable polymers	Dionisi *et al.*, 2005
OOC	SSF: *Aspergillus oryzae*	Neutral protease	Sandhya *et al.*, 2008

Table 12.10: Chemical and biological characteristics of an ecological soil conditioner (Vlyssides et al., 2004)

Parameter	Value
Moisture, %	21.5
pH	7.7
Ash, % of dry matter	25.85
Organic matter, % of dry matter	74.15
Total organic carbon, % of dry matter	38.2
Total nitrogen, g/100 g	1.505
Total phosphorous, mg/kg	445
Humic substances, % of dry matter	5.84
Total phenolic compounds, mg/kg	190
CFU/g	3.6×10^8
Total Ca, g/100 g	3.25
Total Cu, mg/kg	26.7
Total Zn, mg/kg	49.7
Total Mn, mg/kg	135
Fe, % of dry matter	0.88
Total Mg, % of dry matter	0.45
Pb, mg/kg	Not detected

Olive oil residue is a renewable source of energy, with a relatively high heating value of around 18 MJ/kg. It can be used effectively in thermochemical conversion technologies, such as combustion, gasification and pyrolysis, which use elevated temperatures to convert the energy content of biofuel materials. During thermochemical conversion, thermal degradation of the various components (cellulose, hemicellulose and lignin) of the organic fuel materials occurs (Arvanitoyannis *et al.*, 2007).

Pyrolysis is the thermochemical process that converts biomass into liquid (bio-oil or bio-crude), charcoal and non-condensable gases, acetic acid, acetone and methanol by heating the biomass to about 450–550 °C in the absence of air, as shown in Figure 12.3. This process yields three types of products – a solid product (char), which usually has a porous structure and a surface area that makes it appropriate for use as active carbon; liquid products that contain chemical compounds that can be used as feedstock for synthesis of fine chemicals, adhesives and fertilizers; and finally gas, which holds a high calorific value and may be used as a fuel.

Combustion is the controlled burning of waste at high temperatures. In this process, waste is reduced to 10 percent of its original volume and 25 percent of its original weight. For this process to succeed, incinerators must operate at high temperatures (800–900 °C). This method

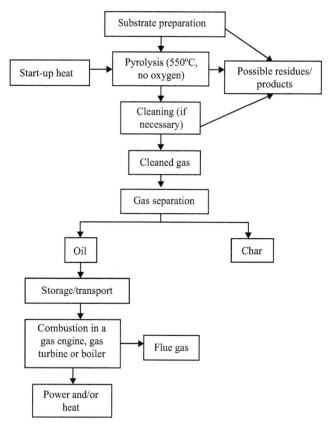

Figure 12.3: The pyrolysis process (Arvanitoyannis *et al.*, 2007).

may be used for the disposal of mill waste material, after pre-drying. The combustion process is shown in Figure 12.4.

The potential use of OOC as a source of energy was investigated by Alkhamis and Kablan (1999). Olive cake with an initial 30 percent moisture content was dried in a general-purpose oven, ground in a mill for particle size reduction, and mixed at different percentages. The results suggested that the olive cake can be used as an excellent source of renewable energy and as a catalyst to oil shale combustion because of the calorific value of the mixture.

Biomass gasification

Biomass gasification is a new physicochemical method, used especially for the de-oiled two-phase olive mill waste. This process transforms solid biomass into synthetic gas, a mixture of CO and H_2, at 900–1200 °C in an oxygen-restricted environment. Synthetic gas is used for obtaining important chemical products such as CH_3OH or NH_3, and for the preparation of synthetic fuel. The gasification process is shown in Figure 12.5.

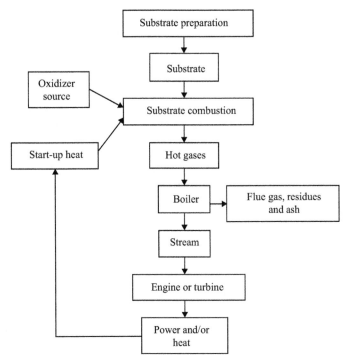

Figure 12.4: The combustion process (Arvanitoyannis *et al.*, 2007).

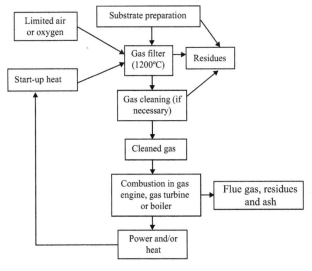

Figure 12.5: The gasification process (Arvanitoyannis *et al.*, 2007).

Caputo and colleagues (2003) suggested a centralized-combined gasification and combustion technology. OMWW and OOC were thoroughly mixed and then fed to a rotary dryer to reduce the moisture level from 69 to 15 percent and utilize the produced hot gases. The dried blend was then fed to a gasifier, and the released gas was fed to a combined gas-stream cycle, gas turbine cycle or internal combustion engine in order to produce a low-heating-value gas (5860 kJ/kg). Another experiment was conducted where the dried mixture of OMWW and OOC was combusted in a boiling atmospheric fluidized bed combustor. The released fumes were fed into a steam generator, where the steam produced could be effectively used for the generation of electric energy. Another combustion method was also proposed for the blend, using a conventional boiler and steam turbine cycle. The mixture was dried in a rotary dryer until the moisture was reduced to 10 percent; it was then fed to a boiler and the resulting steam went through a steam turbine cycle. The results indicated considerable energy recovery when treating both olive husk and OMW from olive mills.

Another biotechnological approach has recently been developed in Israel by a pioneering bio-engineer, Dr. Yuri Wladislavsiky. In this latest technology, the olive waste is heated and dried. After drying, it is introduced into the reactor before entering two processes, pyrolysis and gasification; these involve the biomass being heated to 800 °C, at which temperature its molecules break down. In the process, a combination of high-calorie gases (including methane and carbon monoxide) is produced which, because the gases are lighter than air, flow upwards through a pipe into a standard gas turbine to generate electricity in the usual way. The other by-product is coke, which can be converted into the active type of coke that can be sold for use to power air conditioners or as filters for various substances (Figure 12.6). This technology has already attracted much attention from many companies, not only for economic but also for environmental reasons.

Biodiesel

Biodiesel is an alternative fuel that is applied in unmodified diesel engines, having a minimal effect on the engine's performance. It can be produced from new or used vegetable oils or animal fats, which are non-toxic, biodegradable, renewable resources. Biodiesel and other alternative fuels for diesel engines have become increasingly important because of increased environmental concerns and for socioeconomic reasons. Dorado *et al.* (2003) investigated greenhouse gas emissions from an engine fueled with waste olive-oil methyl ester, and found a notable decrease in CO, CO_2, NO and SO_2 emissions. In this process, residual olive oil is restored through solvent extraction and then transferred to a biodiesel production workstation. Biodiesel is prepared via a chemical reaction called transesterification, by which an organic acid ester is converted into another ester of the same acid using an alcohol and in the presence of a catalyst which is usually a strong base (see Figure 12.7).

Figure 12.6: A bioreactor for producing methane from olive-mill waste, developed by Dr Yuri Wladislavsiky.
Source: http://www.olives101.com/2007/09/25/turning-olive-pits-into-energy.

Methane

Another alternative energy source that can be produced from olive-mill waste is methane. This material can be used as a fuel in the production of electricity. Mill waste is rich in carbohydrates, and is hence an ideal substrate for the production of energy in the form of ethanol, hydrogen and biogas (methane) through aerobic and anaerobic microbial digestion. Fiestas and colleagues (1996) investigated methane production in an anaerobic biomethanization process applied to OMWW, and concluded the process may yield 37 m^3 of methane per m^3 of OMWW – an energy output of 325 kW/h, of which 30 percent can be converted into electrical energy and 63 percent into calorific energy by co-generation processes. This is sufficient to cover the energy needs of both an olive oil production plant and an anaerobic reactor for the integral purification of wastewater.

The system is based on the application of the following processes:

1. *Biotransformation.* In this process, the olive oil that was emulsified in the alpechin is recovered, and undesired phenolic compounds that inhibit biological activity are eliminated.

2. *Biomethanization.* This is the process in which anaerobic bacteria transform organic matter to methane and other by-products. The process takes place in a bioreactor where

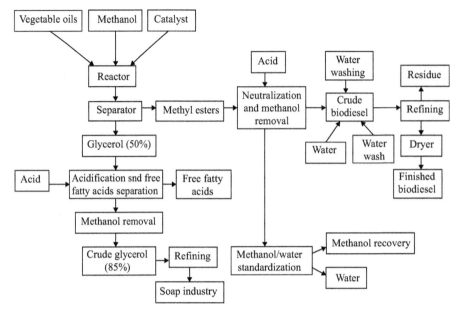

Figure 12.7: Flow diagram for biodiesel production (Arvanitoyannis *et al.*, 2007).

the operating conditions (such as temperature, laminar flow, load density and others) have been carefully chosen, based on extensive research.

3. *Aerobic biological treatment.* This process is applied to reduce the remaining high chemical oxygen demand level in the anaerobic effluent.

References

Aburjai, T., & Natsheh, F. M. (2003). Plants used in cosmetics. *Phytother Res*, 17, 987–1000.

Alkhamis, T. M., & Kablan, M. M. (1999). Olive cake as an energy source and catalyst for oil shale production of energy and its impact on the environment. *Energy Conv Manag*, 40, 1863–1870.

Alvarez Hernandez, M. (2006). Use of olive oil in the preparation of a product for oral hygiene for eliminating or reducing bacterial plaque and/or bacteria in the mouth. US Patent No.7,074,391 (11.7.2006).

Amari, S., & Schubert, C. (2005). Natural emulsifying agent. US Patent No. 2005/0002882 (6.1.2007).

Arvanitoyannis, I. S., Kassaveti, A., & Stefanatos, S. (2007). Current and potential uses of thermally treated olive oil waste. *Intl J Food Sci Technol*, 42, 852–867.

Barnes, S., & Prasain, J. (2005). Current progress in the use of traditional medicines and nutraceuticals. *Curr Opin Plant Biol*, 8, 324–328.

Belle, J. (1980). Bath composition. US Patent No. 4,223,018 (16.9.1980).

Budiyanto, A., Ahmed, N. U., & Wu, A. (2000). Protective effect of topically applied olive oil against photocarcinogenesis following UVB exposure of mice. *Carcinogenesis*, 21, 2085–2090.

Capretti, G., & Bonaconza, E. (1949). Effects of infusions or decoctions of olive leaves (*Olea europaea*) on some physical constants of blood and components of metabolism. *Giorn Clin Med*, 30, 630–642.

Caputo, A. C., Scacchia, F., & Pelagagge, P. M. (2003). Disposal of by-products in olive oil industry: waste-to-energy solutions. *Appl Thermal Eng*, 23, 197–214.

Cordova, J., Nemmaoui, M., Ismaili-Alaoui, M., Morin, A., Roussos, S., Raimbault, M., & Benjilali, B. (1998). Lipase production by solid state fermentation of olive cake and sugar cane bagasse. *J Mol Catal Benzym*, 5, 75–78.

Crognale, S., Federici, F., & Petruccioli, M. (2003). β-Glucan production by *Botryosphaeria rhodina* on undiluted olive-mill wastewaters. *Biotechnol Letts*, 25, 2013–2015.

Dionisi, D., Carucci, G., Papini, M. P., Riccardi, C., Majone, M., & Carrasco, F. (2005). Olive oil mill effluents as a feedstock for production of biodegradable polymers. *Water Res*, 39, 2076–2084.

Dorado, M. P., Ballesteros, B. E., Arnalc, J. M., Gómezc, J., & López, D. F. J. (2003). Exhaust emissions from a diesel engine fueled with transesterified waste olive oil. *Fuel*, 82, 1311–1315.

Farooqi A. H. A., Sharma, S., & Kumar, S. (2002). Herbal skin care formulation and a process for the preparation thereof. US Patent No. 6,368,639 (9.4.2002).

Fenice, M., Sermanni, G. G., Federici, F., & D'Annibale, A. (2003). Submerged and solid-state production of laccase and Mn-peroxidase by *Panus tigrinus* on olive mill wastewater-based media. *J Biotechnol*, 100, 77–85.

Fernandez-Bolanos, J., Rodriguez, G., Rodriguez, R., Guillen, R., & Jimenez, A. (2006). Potential use of olive by-products: Extraction of interesting organic compounds from olive oil waste. *Grasas y Aceites*, 57(1), 95–106.

Fiestas, J. A., de Ursinos, R., & Borja-Padilla, R. (1996). Biomethanization. *Intl Biodet Biodeg*, 38, 145–153.

Finotti, E., Beye, C., Nado, N., Quaglia, G. B., Milin, C., & Giacometti, J. (2001). Physico-chemical characteristics of olives and olive oil from two mono-cultivars during various ripening phases. *Nahrung/Food*, 45(5), 350–352.

Ganguli, K. L., Hofman, C., Van Immerseel, A. R., Van Putte, K. P. (2000). Flavoring food compositions. US Patent No. 6,117,469 (12.9.2000).

Geria, N.M. (1991). Anesthetic/skin moisturizing composition and method of preparing same. US Patent No. 5,002,974 (26.3.1991).

Giannoutsou, E. P., Meintanis, C., & Karagouni, A. D. (2004). Identification of yeast strains isolated from a two-phase decanter system olive oil waste and investigation of their ability for its fermentation. *Biores Technol*, 93, 301–306.

Glover, D.A., & Gurman, N.Y. (2005). Cosmetic compositions with long lasting skin moisturizing properties. US Patent No. 2005/01699879 (4.8.2005).

Guinda, A. (2006). Use of solid residue from the olive industry. *Grasas y Aceites*, 57(1), 107–115.

Haddadin, M. S., Abdulrahim, S. M., Al-Khawaldeh, G. Y., & Robinson, R. K. (1999). Solid state fermentation of waste pomace from olive processing. *J Chem Technol Biotechnol*, 74, 613–618.

Hanbury, D. (1854). On the febrifuge properties of the olive (*Olea Europaea,* L.). *Pharm J Provincial Trans*, 353–354.

Harding, C. R., Moore, A. E., Rogers, J. S., Meldrum, H., Scott, A. E., & McGlone, F. P. (2002). Dandruff: a condition characterized by decreased levels of intercellular lipids in scalp stratum corneum and impaired barrier function. *Arch Dermatol Res*, 294, 221–230.

Hardy, G. (2000). Nutraceuticals and functional foods: introduction and meaning. *Nutrition*, 16, 688–697.

Hepbasli, A., Akdeniz, R. C., Vardar-Sukan, F., & Oktay, Z. (2003). Utilization of olive cake as a potential energy source in Turkey. *Energy Sources. Part A: Recov Util Environ Effects*, 25(5), 405–417.

Kincs, F., Abrassart, C., Nakhasi, D., & Daniels, R. (2007). Edible oils and methods of making edible oils. US Patent No. 2007/0065565 (22.3.2007).

Laufenberg, G., Rosato, P., & Kunz, B. (2004). Adding value to vegetable waste: oil press cakes as substrates for microbial decalactone production. *Eur J Lipid Sci Technol*, 106, 207–217.

Livingston, R. M. (2000). Edible spread based on olive oil as the major fat component. US Patent No. 6,159,524 (12.12.2000).

Lopez, M. J., Moreno, J., & Ramos-Cormenzana, A. (2001). The effect of olive mill wastewater variability on xanthan production. *J Appl Microbiol*, 90, 829–835.

Martin, M., Ferrer E., Sanz, J., & Gibello, A. (1998). Process for the biodegradation of aromatic compounds and synthesis of pigments and colorants, alkaloids and polymers, with the use of the recombinant strain *Escherichia coli* P260. Patent No. WO98/04679 (5.02.1998).

Mercade, M. E., & Manresa, M. A. (1994). The use of agroindustrial by-products for biosurfactant production. *J Am Oil Chem Soc*, 71, 61–64.

Munoz, S. (1998). Procedimiento para la produccion de vitamina B12 a partir de residuos contaminantes de la industria de la aceituna. Patent No. ES2,122,927 (16.12.1998).

Papadakos, J. (2001). Preparation of sun protection oil. US Patent No. 6,280,709 (28.8.2001).

Perricone, N. V. (2001). Treatment of skin disorders with olive oil polyphenols. *PCT Int Appl*, 16, WO0176579.

Pozo, C., Martinez-Toledo, M. V., Rodelas, B., & Gonzalez-Lopez, J. (2002). Effects of culture conditions on the production of polyhydroxyalkanoates by *Azotobacter chrococcum* H23 in media containing a high concentration of alpechin (wastewater from olive oil mills) as primary carbon source. *J Biotechnol*, 97, 125–131.

Predovan, J. (2004). Skin cream. US Patent No. 2004/0101507 (27.5.2004).

Ramachandran, S., Patel, A. K., Nampoothiri, K. M., Chandran, S., Szakacs, G., Soccol, C. R., & Pandey, A. (2004). Alpha amylase from a fungal culture grown on oil cakes and its properties. *Brazil Arch Biol Technol*, 47, 309–317.

Ruiz-Bravo, A., Jimenez-Valera, M., Moreno, E., Guerra, V., & Ramos-Cormenzana, A. (2001). Biological response modifier activity of an exopolysaccharide from *Paenibacillus jamilae* CP-7. *Clin Diag Lab Immunol*, 8, 706–710.

Soubhie, E. (2006). Composition for the treatment of burns, sunburns, abrasions, ulcers and cutaneous irritation. US Patent No. 7,141,252 (28.11.2006).

Tomati, U., Di Lena, G., Galli, E., Grappelli, A., & Buffone, R. (1990). Indolacetic acid production from olive waste water by *Arthrobacter* spp. *Agrochimica*, 34, 228–232.

Tomati, U., Belardinelli, M., Galli, E., Iori, V., Capitani, D., Mannina, L., Viel, S., & Segre, A. (2004). NMR characterization of the polysaccharidic fraction from *Lentinula edodes* grown on olive mill waste waters. *Carbohydr Res*, 339, 1129–1134.

Tritsarolis, D. (1993). Facial skin moisturizing composition. US Patent No. 5,242,952 (7.9.1993).

Veney, R. G. (1981). Cosmetic composition. US Patent No. 4,255,452 (10.03.1981).

Veney, R. G. (1983). Cosmetic composition and method of making the same. US Patent No. 4,395,424 (26.7.1983).

Vierhuis, E., Servili, M., Baldioli, M., Schols, H. A., Voragen, A. G. J., & Montedoro, G. (2001). Effect of enzyme treatment during mechanical extraction of olive oil on phenolic compounds and polysaccharides. *J Agric Food Chem*, 49, 1218–1223.

Vlyssides, A. G., Loizides, M., & Karlis, P. K. (2004). Integrated strategic approach for reusing olive oil extraction by-products. *J Clean Prod*, 12, 603–611.

Voorhees, J., & Nachman, L. (2002). Composition for treating symptoms of influenza. US Patent No. 2002/0110600 (15.8.2002).

Yang, J. H. (2005). Gelatin soft capsule having the properties of removal of oral smell and cleaning of oral cavity. US Patent No. 6,852,309 (8.2.2005).

Yurekli, F., Yesilada, O., Yurekli, M., & Topcuoglu, S. F. (1999). Plant growth hormone production from olive oil mill and alcohol factory wastewaters by white rot fungi. *World J Microb Biot*, 15, 503–505.

Further reading

Brenes, M., Hidalgo, F. J., Garcia, A., Rios, J. J., Garcia, P., Zamora, R., & Garrido, A. (2000). Pinoresinol and 1-acetoxypinoresinol, two new phenolic compounds identified in olive oil. *J Am Oil Chem Soc*, 77, 715–720.

Fleming, H. P., Walter, W. M., & Etchells, J. L. (1973). Antimicrobial properties of oleuropein and products of its hydrolysis. *Appl Microbiol*, 26(5), 777–782.

McArthur, C.M. (1976). Hair dressing cosmetic. US Patent No. 3,932,611 (13.1.1976).

Nachman, L, (1998). Method for producing extract of olive leaves. US Patent No. 5,714,150 (3.2.1998).

Ninfali, P., Aluigi, G., Bacchiocca, M., & Magnani, M. (2001). Antioxidant capacity of extra-virgin olive oil. *J Am Oil Chem Soc*, 78, 243–247.

Sandhya, C., Sumantha, A., Szakacs, G., & Pandey, A. (2005). Comparative evaluation of neutral protease production by *Aspergillus oryzae* in submerged and solid-state fermentation. *Process Biochem*, 40, 2689–4.

Tsimidou, M., Papadopoulos, G., & Boskou, D. (1992). Determination of phenolic compounds in virgin olive oil by reversed-phase HPLC with emphasis on UV-detection. *Food Chem*, 44, 53–60.

Wahner, R. S., Mendez, B. A., & Giulietti, A. M. (1988). Olive black water as raw material for butanol production. *Biol Wastes*, 23, 215–220.

Part 3
Desert Olive Oil Branding

CHAPTER 13

Conventional desert olive oil cultivation: case study

13.1 Re'em: a conventional intensively cultivated desert olive farm

Re'em Farm is a conventional commercial desert olive farm that produces both olive oil and table olives. Re'em Farm is located at Ramat Negev in the high dryland area of the central Negev in Israel. Established in 1995, Israel's first intensively cultivated desert olive farm currently has a total area of 90 ha of olive trees drip irrigated with the wastewater from a fish farm and from local drilling sites. The farm expands each year, using plants from its on-site nursery of 21,000 seedlings (Figure 13.1). The mature trees are mainly Barnea (42 percent) and Picual (40 percent) cultivars, with some Souri (8 percent) and Manzanillo (10 percent). The trees are planted at a distance of 7 m between trees and 4 m between rows. Alluvial soils dominate the adult plot, while the plot for young plants is characterized by aeolian soils. This farm includes a fish farm.

13.2 Fish culture: a supplement at Re'em Farm

Re'em's fish culture is based on geothermal groundwater (from the Ashelim and Mashabei Sade local drilling sites), using a super-intensive method. The water delivered from the drill is about 45 °C and its salinity level is 3.5 dS/m. Bass and Barramundi are the promising fish species presently cultured in the Negev. Owing to the climate and plenty of warm underground water, fish are farmed throughout the year, assuring a constant supply to consumers. In order to improve the economic viability of olive growing, a desert water-use chain has been established.

Figure 13.1: Young olive trees, recently planted at 7 × 4 m spacing, at Re'em Farm, Ramat Negev.

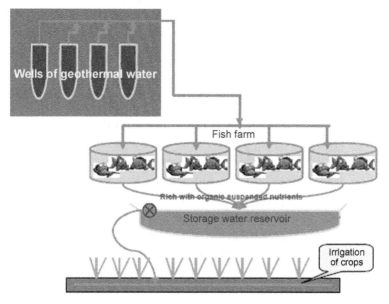

Figure 13.2: A desert water-use chain increases economic feasibility (adapted from Rothbard and Peretz, 2002).

Geothermal groundwater passes through the fish farm on cement raceways, and is later used for drip irrigation of the olive trees (Figure 13.2).

The influence exerted by water temperature on most biochemical and physiological processes makes it the most important physical factor for aquatic organisms (Reynolds and Casterlin,

Table 13.1: Current status of the olive crop at Re'em Farm, Ramat Negev, and future plans

Year	Size (ha)
1995	63
2005	73
2006	83
2007	90
2008	120
2009	150

1979). Hence, control of metabolic activity by temperature conditions is generally considered essential for fish to optimize the use of their ecosystem. In order to maintain high biomass rates during the winter ($10 \, kg/m^3$ average), large volumes of geothermal water ($38 \, ^\circ C$ when entering the raceways) continually pass through the fish-growing compartments, providing heat and flushing out suspended solids and harmful nitrogenous wastes. Such a system is efficient in the summer, when the trees need plenty of fertilizer and a good water supply; however, in the winter the agricultural demand for water decreases, and excessive amounts of water are disadvantageous to olive cultivation. As a consequence, a water recirculation system has been developed that involves the filtration and removal of solid waste, thus maintaining optimal water quality. The water that is not recycled is flushed into the surrounding soil. In recent years storage pools have been constructed that collect excess aquaculture wastewater in the winter, and this is later used for irrigation early in the spring to avoid evaporation losses.

13.3 The olive plantation at Re'em

Of the 90 ha olive plantation at Re'em Farm, 63 ha consists of mature olive trees grown for both oil and table olives. The current status of the farm and future plan are presented in Table 13.1.

The future crop development (to 150 ha by 2009) will be achieved using seedlings from the on-site nursery, which currently contains 21,000 plants at different stages of development.

Alternate bearing is a major constraint in planning a future business strategy. From data collected over the years, it is clear that Barnea cultivars show the highest fruit yield (38 ton/ha) with a 6 percent fresh oil content in a high-yield year, but bear almost no fruit in the following year (Table 13.2). In a normal-yield season, Barnea cultivars produce less fruit (25 ton/ha) with a higher oil content (18 percent). The amount of oil produced is regulated mainly by the amount of mesocarp available for oil biosynthesis (Lavee and Wodner, 2004). The lower oil content in a high-yield season is probably due to the mesocarp/endocarp ratio, which is strongly related to fruit size and thus affected by high and low yields. The fruit size range is significantly higher in low-yielding trees than in high-yielding trees, in accordance with Lavee and Wodner

Table 13.2: Fluctuation in yield and oil content of the three main cultivars grown at Re'em Farm

Cultivar	Commercial purpose	High yield			Normal yield		
		Average oil content (%)	Fruit weight (T/ha)	Following season's fruit weight (T/ha)	Average oil content (%)	Fruit weight (T/ha)	Following season's fruit weight (T/ha)
Barnea	Oil	16	38	< 0.5	18	25	5
Picual	Oil & table	14	20	< 0.5	14	7	–
Manzanillo	Oil & table	10	18	–	10	10	–

(2004). According to current observations, Manzanillo is the most stable cultivar, showing less fruit-bearing fluctuations than the others.

13.4 Major management practices

13.4.1 Water management

A constant and reliable water supply is one of the financial constraints in crop husbandry. Re'em Farm uses approximately twice the quantity of irrigation water used by northern Israeli olive farms, because of the high evapo-transpiration and low precipitation in the desert environment and the additional soil leaching demands. The price of saline water in the southern region of Israel is relatively high (0.692 NIS/m^3), although partly subsidized by the government. Irrigation water used in the Re'em Farm plots consists of wastewater from the fish ponds after bio-filtering, or comes directly from local drilling sites.

All olive plots in the Re'em Farm are drip irrigated with saline water, except for the nursery seedlings and small plants, which are irrigated with fresh water using sprinklers. Leaching is an important aspect of the water regimen. Each hectare of adult olive trees consumes ~9000 m^3 of water annually. In the plot for mature trees, leaching is only applied following 3–5 mm of precipitation. In the plots for younger trees, the soil is leached after every rainfall event. The leaching volume is 150–250 m^3/ha of saline water, depending on the intensity of the rain. Immediately after planting, no leaching is applied.

A common problem is cessation of the water supply during rain or immediately afterwards; this is because during heavy rain all the crop owners in the area increase their water supply to leach the soil and, since water demand increases sharply, the water pressure decreases and the supply sometimes stops completely. Wastewater from aquaculture can then be used as a supplement, temporarily lowering the water levels in the fish raceways by a few centimeters.

13.4.2 Soil management

Olive trees can grow in various soils, including saline desert soil. Re'em Farm has two main soil types; alluvial soils, with a variable pebble content, dominate the adult plot, while the plot for young plants is characterized by aeolian soils containing calcite (20 percent), loess soil and a higher pebble content. Soil depth varies from 30–60 cm in the plot for young plants to 2 m in the adult plot. The soil characteristics can affect the leaching efficiency significantly. Homogenized soil increases the leaching effect because the flow is laminar. The high content of pebbles in the plot for young plants requires a different dripper layout to increase the leaching efficiency – here, instead of one dripper every 70 cm, there are three drippers for each tree (35 cm apart) clustered together with a fixed zigzag water pipe.

13.4.3 Pest management

A pest management strategy consistent with economic, ecological and toxicological requirements can be employed to maintain pests at an economically acceptable level. One of the major
pests at Re'em Farm is the olive fruit fly (*Bactrocera oleae*); another, to a lesser extent, is the
Mediterranean fruit fly (*Ceratitis capitata*).

Olive fruit fly

The olive fruit fly is among the most serious insect pests of the olive agro-ecosystem (Navrozidis
et al., 2000). It can survive and develop in any area of the world where olive trees are grown.
The olive fruit fly is strictly monophagous (Ochando and Reyes, 2000), and the female lays
eggs in olive fruits. The resulting larvae feed and perforate galleries in the mesocarp, which
are quickly infected with fungus. The affected olive fruit is unsuitable for processing and the
olive oil quality is sharply reduced, having a higher oil acidity. It is estimated that the olive fruit
fly accounts for 30–40 percent of the losses in olive production in the Mediterranean region
(Sivinski and Burk, 1989), in spite of the fact that pesticide treatments are applied on a regular
basis to control the population density.

Spinosad is a protein-based insecticide derived by fermentation of the Actinomycete bacterium
Saccharopolyspora spinosa (Milles, 2003), and is a natural mixture of spinosyn A and spinosyn
D (85 : 15). Spinosad kills susceptible species, including fruit flies, by causing rapid excitation
of the insect's nervous system (Mangan *et al.*, 2006). Insects must feed on the bait mixture and
ingest the insecticide in order for it to be effective.

Two approaches to the problem are applied in the adult plot. The first is to use traps containing
spinosad bait percolating from a capsule. The trap is used either to monitor the presence of the
olive fruit fly or to eradicate it completely. Monitoring of the presence of the fruit fly (there
are 23 monitoring stations in 63 ha) is achieved either by placing a trap every 3–5 ha or by
examining 100 randomly picked fruits at each monitoring station. Each fruit is searched for
the number of bites, presence of any eggs, and the dead or alive larva content. Placing a trap
every four trees can eradicate the fly efficiently, and prevent future damage. In years when the
fly causes no substantial damage, the economic cost of this method is not justified.

The second approach is to spray ground bait with the same active ingredient, spinosad
(Success®). Foliar spot spray is applied in the middle of each tree row using an all-terrain
vehicle. Additional fruit examination after spraying can lead to ground foliar cover spray of
a broad-spectrum organophosphate (Rogor® containing Dimethoate) if necessary. One disadvantage of organophosphate spray is its non-specific targeting; as well as eradicating the olive
fruit fly it also eradicates its natural enemies, such as the parasite wasp and some acari, thus
reducing biodiversity. Another shortcoming of using this type of organophosphate spray is the
high residue content in the fruit, which can be transferred to the oil produced. As a general

Table 13.3: Common pests at Re'em Farm, and the most susceptible cultivars on-site

Common name	Scientific name	Susceptible cultivars
Olive fruit fly	*Bactrocera oleae*	Manzanillo
Mediterranean fruit fly	*Ceratitis capitata*	Manzanillo
Olive mite	*Aceria oleae nalepa*	Souri
Black scale	*Saissetia oleae*	No observed difference
Olive psyla	*Euphyllura olivina Costa*	Manzanillo and Picual
Olive scale	*Parlatoria oleae*	No observed difference
Verticillium wilt (disease)	*Verticillium dahliae*	Picual and Souri
Peacock leaf spot (disease)	*Cycloconium oleaginum*	Souri

rule, Rogor® should be applied at a rate of up to one spray per season; however, it is generally not required. Since the Manzanillo variety is especially sensitive to olive fruit fly damage, this specific plot receives one or two sprays per season in addition to bait spraying. One important factor in insecticide treatment relates to the environmental temperature. At Re'em Farm, experience has clearly shown that olive fruit fly activity decreases substantially if the air temperature is above 35 °C; therefore, there is no need to apply insecticide at such temperatures.

Olive mite

Another common pest at Re'em Farm is the olive mite (*Aceria oleae nalepa*), which attacks only olive trees. The degree of sensitivity to the olive mite varies by cultivar (Table 13.3). This mite is a native pest of the Mediterranean and is found in most countries that grow olives (Elhadi and Birger, 1999). The damage is expressed as deformation of the immature terminal olive leaves, and bulges or knobs on the fruits. There is no effect on the quality or quantity of the oil produced. To date, no treatment has been applied against the olive mite at Re'em Farm.

Other pests

Black scale (*Saissetia oleae*), a widespread pest in olive plots (Gullan, 1997), is also found at Re'em Farm. The main damage caused by *S. oleae* is caused by the large amount of honeydew excreted by the scale and the subsequent development of sooty mold fungi.

Verticillium wilt, caused by *Verticillium dahliae* vascular infecting fungi, is also common at Re'em Farm. Infected olive trees seriously defoliate. This problem occurs mainly in Picual and Souri cultivars. The farm has recorded incidents of lethal infection. The spread of the disease is probably due to the establishment of new plots in infested soil and/or the use of *V. dahliae*-infected planting material (Thanassolopoulos, 1993).

Olive psyla (*Euphyllura olivina Costa*) and olive scale (*Parlatoria oleae*) and are also present at Re'em Farm, but no treatment is applied because the damage capability of the pests is relatively low. Only once, in 2006, was substantial damage by olive psyla observed, when inflorescence production was decreased and table-olive fruit quality reduced. The pest can be controlled by using the commercial pesticide foliar spray Marshall®. Some evidence of peacock leaf spot disease caused by *Cycloconium oleaginum* has also been observed, mainly affecting the Souri cultivar. Initial signs become evident in high humidity and rainy conditions, and are small sooty blotches on the leaves, which then drop prematurely. Copper ground cover spray and intensive pruning can reduce the damage.

13.4.4 Weed management

Weeds pose the risk of developmental reduction of many fruit tree crops, particularly damaging young trees during their first two years of life. These non-crop plants compete for moisture, light, nutrients and space. The presence of weeds in an olive plot can also reduce the maximal efficient use of land, by increasing the costs of production and harvest. Reforestation, reduced yield and weed control, as well as root damage resulting from cultivation, are further expenses. Effective weed control is therefore an important part of crop husbandry. In addition to mechanical methods, chemical control strategies are used intensively; however, these have side-effects on the environment, users and non-target organisms (Hurle, 1996).

Manual and mechanical techniques such as pulling and cutting are commonly used at Re'em Farm to control invasive plants such as tamarisk, particularly if the population is relatively small. Tamarisk is a small tree that is native to Europe and Asia (Peter and Scott, 2004), and invades Ramat Negev olive tree plots. Tamarisk can adapt to barren soil, and especially saline soils (Carman and Brotherson, 1982). It can produce seeds throughout the year, and germination occurs in wet soil, with the weeds rapidly sending down deep roots (Everitt, 1980). It is drought-tolerant, capable of surviving reductions in groundwater levels detrimental to native plants, and of obtaining moisture from unsaturated soil as well as groundwater (Cleverly *et al.*, 1997).

Continuous monitoring of the weeds and manual control are generally labor- and time-intensive. Treatments must typically be administered several times in one season to prevent re-establishment of weeds. Mowing is the best method in a large-scale vegetation management strategy. In addition to standard orchard mowers, trim mowers with a trip-mount mechanism are used; these can mow inter-rows.

Routine chemical treatments for limiting weed infestations include ground spraying of Simazine (pre-emergence) and Roundup® (a contact herbicide). Simazine (6-chloro-N,N1-diethyl-1,3,5-triazine-2,4-diamine) is commonly used during the fall and winter to control germinating and broad-leaved weeds at Re'em Farm. Simazine is applied to the soil surface but requires further incorporation into the soil matrix to be effective (Ashton *et al.*, 1989), usually by rainfall. Simazine has moderate persistence in the soil, with a half-life encompassing a few days to several months, depending on soil characteristics and the number of previous herbicide applications

(Kristensen *et al.*, 2001). Simazine can be used in a mixture with Goal® (oxyfluorfen) for improved results.

Roundup® contains the active ingredient glyphosate, a broad-spectrum contact herbicide. Glyphosate is strongly immobilized by soil, and is biodegraded readily (Kennedy and Jordan, 1985), so is essentially non-toxic to animals and birds. Any toxicity of Roundup® is attributed to the added surfactant (Freedman, 1991). Glyphosate is a systematic herbicide that is taken up by the leaves and then translocated through the plant from the point of contact towards the root system. If properly applied it can kill the entire plant, including the roots: thereby preventing later sprouts (Kennedy and Jordan, 1985).

13.4.5 Harvesting and oil extraction

This subsection describes some aspects of olive harvest timing and oil extraction from the intensively farmed commercial desert olive plot at Re'em Farm. Long experience and continuous study over the years has led to knowledge that is helping to achieve high oil content and promote unique marketing strategies. The main guideline in Re'em Farm management is to aspire to as high a level of oil extraction as is possible, even at the expense of a slight reduction in oil quality. Israeli consumer preferences are relatively undeveloped, so there is not enough economic justification for producing less olive oil of higher quality or with different organoleptic properties.

Green Souri, Manzanillo and Picual cultivars designated for table consumption are harvested strictly by hand, beginning in September during the "off" year and October during the "on" year. Black olives from the same cultivars designated for table consumption are harvested beginning in late October during the "off" year and early November during the "on" year. Hand-harvesting of table olives is time-consuming and costly but mandatory, since mechanical harvest damages the fruit, rendering them unsuitable for marketing.

Olive cultivars assigned for oil production are mechanically harvested in a systematic order. The first cultivar to be harvested is Manzanillo, followed by Souri, Barnea, and finally Picual, which matures late in the season (at the beginning of January). Some overlap occurs between the ripening stages of sequential cultivars. Delaying the harvest date until the olives are fully ripe (with a maturity index of at least 2–3, supported by laboratory tests) does not affect economic feasibility; a high amount of oil with a lower polyphenol content is produced. The Barnea cultivar shows the highest variability in the timing of ripening stages. Harvesting of Barnea can take a whole month; the early oil extracted has a higher polyphenol content and is more bitter.

Olive fruit harvesting for oil production is achieved mainly by trunk shakers. The shaker operates by vibrating the trunk (Figure 13.3), while workers complement the action by beating the tree

Figure 13.3: Harvesting olive fruit for oil production using a trunk shaker, complemented by pole beating of branches. Superficial trunk injury can be seen in the smaller picture (low-right), but there is no damage to productivity in the long run.

branches with a pole. Canvasses or nets are placed under the tree to collect the fallen fruits. The shaker must be used at the correct frequency and amplitude in order to avoid tree damage.

The harvest season of 2007 was considered to be a strong "off" year for Barnea and Souri cultivars, and, as "on" and "off" Barnea trees are present within the same area, trunk shakers are not employed constantly. Instead, manual harvesting is used to avoid logistical problems regarding trunk shaker operation in some areas of the plot.

The efficiency of fruit removal by mechanical shaker is inversely related to the fruit removal force (FRF). To reduce the detachment force and facilitate the harvesting operation, abscission chemicals have been used. Some abscission chemicals have been identified that enhance mechanical harvesting efficiency by loosening the pedicle strength of the fruit (see Chapter 8 for further details).

The fruit oil content is determined by the results provided by external olive mills. Laboratory analysis shows that the olive pomace usually contains up to 12 percent oil content on a dry weight basis, and the wastewater contains up to 0.2 percent oil (on a dry weight basis). Since the wastewater volume is relatively high, an oil content of less than 0.5 percent means that further oil extraction is not financially viable.

Re'em Farm markets its olive oil exclusively in bulk (tons of oil weight) instead of bottles. The concept of selling bulk olive oil has developed because the marketing of the final product requires additional expertise. Bottling of the final product, analyzing it and promoting the brand carry high costs, and require logistics and dedicated salespersons – which Re'em Farm does not have.

13.5 A successful approach to intensive olive cultivation in the desert environment

Re'em Farm management is adapted for the commercial cultivation of olive trees in the desert environment. The main advantages of desert agriculture at Re'em Farm are high solar radiation, low land costs and a very low pest density. The main disadvantages are scarcity of water resources and highly saline soil which requires additional irrigation volumes.

Water reuse from aquaculture gives Re'em Farm an additional advantage over similar fruit tree farms in the area. Although geothermal groundwater is not a unique desert environment characteristic, diverting this wastewater source for irrigation lowers irrigation expenses. In addition, it allows the farm to easily overcome problems in the water delivery infrastructure during the winter, caused by leaching requirements and subsequent low water pressure. Saline soils are common in dryland areas and can affect plant growth rates and the root-zone area, thus influencing the canopy structure and productivity of trees. Leaching soluble salts away from the root zone is important. Re'em Farm has no saline-related problems in productivity provided a good soil leaching methodology is applied (see Chapter 6).

Photosynthetically active radiation intercepted by the crop canopy determines the plant growth rate, and is a main factor in dry matter production. The amount of solar radiation energy available sets a limit on potential production (Mariscal *et al.*, 2000). The semi-arid Negev region of Israel is characterized by relatively high average daily irradiation rates and extended daylight periods. The annual average daily radiation intensity at Sde Boker (near Ramat Negev) is 20.5 MJ/m^2 (Ianetz *et al.*, 2000).

The photoperiod has no significant effect on inflorescence production, but low temperatures during winter are required (Hackett and Hartmann, 1964), and the low winter temperature (4.7–16.3 °C) in Ramat Negev enables fruit setting (IMS, 2007). Incomplete chilling delays the release of floral bud dormancy and first flowering (Barranco *et al.*, 1994).

The dryland climate reduces the density of pests drastically because of the low humidity. However, the humidity in the Ramat Negev area is sufficiently high to sustain pest biodiversity. Relative humidity levels as high as 60 percent during early summer mornings create an adequate habitat for olive fruit fly, acari and mites. These pests affect the economic feasibility of olive oil and table-olive production. However, the quantity of pesticides required for control is significantly lower than in non-desert olive plots. Furthermore, when the air temperature increases above 35 °C, the flies become dormant and no fly-related pesticides are required.

To grow olives in the desert, a relatively large amount of water ($\sim$9000 m^3/ha) is needed due to the high evaporation rate (1800 mm). Although water prices in the Negev area are partially subsidized, the large volume necessary has an impact on the cost of oil production. In the case of Re'em Farm, though, irrigation expenses have been reduced by half by reusing fish-culture wastewater.

In spite of the high costs of production, the radical difference in temperature between day and night in the desert environment of the Ramat Negev area substantially affects the positive attributes of olive oil's organoleptic properties, enabling unique brand marketing.

References

Ashton, F. M., Crafts, A. S., & Agamalian, H. S. (1989). Chemical control methods. In: E. A. Kurtz (Ed.), *Principles of Weed Control in California. California Weed Conference* (pp. 115–170). Thompson Publications, Fresno, CA.

Barranco, D., de Toro, C., & Rallo, L. (1998). Epocas de maduracion de cultivares de olivo en Cordoba. *Invest agrar Prod prot vegetales*, 13, 359–368.

Carman, J. G., & Brotherson, J. D. (1982). Comparisons of sites infested and not infested with saltcedar (*Tamarix pentantra*) and Russian olive (*Elaeagnus augustifolia*). *Weed Sci*, 30, 360–364.

Cleverly, J. R., Smith, S. D., Sala, A., & Devitt, D. A. (1997). Invasive capacity of *Tamarix ramosissima* in a Mojave Desert floodplain: the role of drought. *Oecologia*, 111, 12–18.

Elhadi, F., & Birger, R. (1999). A new approach to the control of the olive mite *Aceria* (eriophyes) *Olea Nalepa* in olive trees. *Acta Hort*, 474, 555–558.

Everitt, B. L. (1980). Ecology of saltcedar – a plea for research. *Environ Geol*, 3, 77–84.

Freedman, B. (1991). Controversy over the use of herbicides in forestry, with particular references to glyphosate usage. *J Environ Sci Health*, C8, 277–286.

Gullan, P. J. (1997). Relationships with ants. In: Y. Ben-Dov, C. J. Hodgson (Eds.), *Soft Scale Insects – Their Biology, Natural Enemies and Control* (pp. 351–377). Elsevier Science, Amsterdam.

Hackett, W. P., & Hartmann, H. T. (1964). Inflorescence formation in olive as influenced by low temperature, photoperiod, and leaf area. *Botanical Gazette*, 125, 65–72.

Hurle K. (1996). Weed management impact on the abiotic environment in particular on water and air quality. In: *Proceedings of the Second International Weed Control Congress, Copenhagen, Denmark*, pp. 1153–1158.

Ianetz, A., Lyubansky, V., Setter, I., Evseev, E. G., & Kudish, A. I. (2000). A method for characterization and inter-comparison of sites with regard to solar energy utilization by statistical analysis of their solar radiation data as performed for three sites in the Israel Negev region. *Solar Energy*, 69, 283–293.

IMS (2007). Israel Meteorological Service, http://www.ims.gov.il.

Kennedy, E. R., & Jordan, P. A. (1985). Glyphosate and 2,4-D: the impact of two herbicides on moose browsing in forest plantations. *Alces*, 21, 1–11.

Kristensen, G. B., Johannesen, H., & Aamand, J. (2001). Mineralization of aged atrazine and mecoprop in soil and aquifer chalk. *Chemosphere*, 45, 927–934.

Lavee, S., & Wodner, M. (2004). The effect of yield, harvest time and fruit size on the oil content in fruits of irrigated olive trees (*Olea europaea*), cvs. Barnea and Manzanillo. *Sci Hort*, 99, 267–277.

Mangan, R. L., Moreno, D. S., & Thompson, G. (2006). Bait dilution, spinosad concentration, and efficacy of GF-120 based fruit fly sprays. *Crop Prot J*, 25, 125–133.

Mariscal, M. J., Orgaz, F., & Villalobos, F. J. (2000). Modelling and measurements of radiation interception by olive canopies. *Agric Forest Meteorol*, 100, 183–197.

Milles, M. (2003). The effects of spinosad, a naturally derived insect control agent to the honeybee. *Bull Insectol*, 56, 119–124.

Navrozidis, E. I., Vasara, E., Karamanlidou, G., Salpipiggidis, G. K., & Koliais, S. K. (2000). Biological control of *Bactocera oleae* (Diptera: Tephritidae) using a Greek *Bacillus thuringiensis* isolate. *J Econ Entomol*, 93, 1657–1661.

Ochando, M. D., & Reyes, A. (2000). Genetic population structure in olive fly *Bactrocera oleae Gmelin*: gene flow and patterns of geographic differentiation. *J Appl Entomol*, 124, 177–183.

Peter, L., & Scott, M. (2004). Ecological strategies for managing tamarisk on the CM Russell National Wildlife Refuge, Montana, USA. *Biol Conservation*, 119, 535–543.

Reynolds, W. W., & Casterlin, M. E. (1979). Behavioral thermoregulation and the 'final preferendum' paradigm. *Am Zoologist*, 19, 211–224.

Rothbard O. S., & Peretz Y. (2002). Tilapia culture in Negev, the Israeli desert. *Proceedings of the International Forum on Tilapia Farming in the 21st Century, Tilapia Forum*.

Sivinski, J., & Burk, T. (1989). Pest status: Mediterranean region. In: A. Robinson, G. H. Hooper (Eds.), *World Crop Pests, Fruit Flies: Their Biology, Natural Enemies and Control, 3A* (pp. 39–50). Elsevier, Amsterdam.

Thanassolopoulos, C. C. (1993). Spread of verticillium wilt by nursery plants in olive groves in the Chalkidiki area (Greece). *Bull OEPP*, 23, 517–520.

Further reading

Liu, F., & O'Connell, N. (2003). Movement of Simazine in runoff water and weed control from citrus orchard as affected by reduced rate of herbicide application. *Biores Technol*, 86, 253–258.

Mercado-Blanco, J., Rodríguez-Jurado, D., Pérez-Artés, E., & Jiménez-Díaz, R. (2002). Detection of the defoliating pathotype of *Verticillium dahliae* in infected olive plants by nested PCR. *Eur J Plant Pathol*, 108, 1–13.

CHAPTER 14

Bio-organic desert olive oil cultivation: case study

14.1 Neot Smadar: a model bio-organic farm on the most arid land

Neot Smadar is a kibbutz (collective settlement) situated in the Arava region of southern Negev, 60 km north of Eilat (Figure 14.1). Established in 1989, Neot Smadar Kibbutz is the model for bio-organic olive cultivation in the most arid land in Israel. The kibbutz community comprises 200 members who believe in conserving the fragile balance of the ecosystem by using only organic farming management for sustainable agriculture. Their numbers have been growing, and the vibrant community implements self-sufficiency in a sustainable manner. They feed themselves, planting for the satisfaction of their own needs, and extra produce goes to market.

14.2 Organic farming

The United States Department of Agriculture (USDA) estimated retail sales of organic foods in the USA of approximately $6 billion in 1999, and the number of organic farmers as increasing at a rate of 12 percent per year (USDA, 2000). This clearly indicates the growth in consumer demand for environmentally friendly, chemical-free food, which has led to an increase in demand for organic produce. Consumer interest has also grown in response to repeated food safety scares (Gregory, 2000), animal welfare concerns (Harper and Makatouni, 2002) and general reservations regarding the impact of industrial agriculture on the environment (Grunert and Juhl, 1995).

Figure 14.1: A view of Neot Smadar Kibbutz, situated in the Arava region of Israel. Source: Kibbutz website, published with permission.

The International Federation of Organic Agriculture Movements (IFOAM) has defined the basic principles of organic production and processing (IFOAM, 1998), and these are listed below. It is clear from these statements that the scope of the principles extends beyond simple biophysical aspects to abstract matters such as responsibility.

14.2.1 The principles of organic production and processing as defined by IFOAM (1998)

- To produce food of high quality in sufficient quantity.

- To interact in a constructive and life-enhancing way with natural systems and cycles.

- To consider the wider social and ecological impact of the organic production and processing system.

- To encourage and enhance biological cycles within the farming system, involving micro-organisms, soil flora and fauna, plants and animals.

- To develop a valuable and sustainable aquatic ecosystem.

- To maintain and increase long-term fertility of soils.

- To maintain the genetic diversity of the production system and its surroundings, including the protection of plant and wildlife habitats.

- To promote the healthy use and proper care of water, water resources and all life therein.

- To use, as far as possible, renewable resources in locally organized production systems.

- To create a harmonious balance between crop production and animal husbandry.

- To give all livestock conditions of life with due consideration for basic aspects of their innate behavior.

- To minimize all forms of pollution.

- To process organic products using renewable resources.

- To produce fully biodegradable organic products.

- To produce textiles which are long-lasting and of good quality.

- To allow everyone involved in organic production and processing a quality of life which meets their basic needs and allows an adequate return and satisfaction from their work, including a safe working environment.

- To progress toward an entire production, processing and distribution chain which is both socially just and ecologically responsible.

14.3 Requirements for organic agriculture in Israel

In 1972, the International Federation of Organic Agriculture Movements (IFOAM) was founded in France. IFOAM sets common regulations and guarantees to safeguard organically produced foodstuffs and protect consumers. The Israel Bio-Organic Agricultural Association (IBOAA, http://www.organic-israel.org.il/) is the member of IFOAM for Israel. It was established in the 1980s, and trains new organic producers and ensures inspection and certification in Israel.

A national regulatory framework for organic farming is notable not only for export certification, but also for the reinforcement of the consumers' culture and trust, and consequently for the development of domestic markets (Vincenzo and Damiano, 2002). In the Mediterranean region, only Tunisia, Turkey, Slovenia and Israel have national regulations regarding organic farming.

All organic growers in Israel are members of the IBOAA and of Agro Bio Mediterranean (ABM). The IBOAA promotes, disseminates and develops local organic knowledge through courses, field trips, extension activities and marketing. Israel's organic standard follows standards instituted by its main target markets, namely EU countries; therefore, organic production complies with EU 2091/92 regulations and IFOAM standards. Exporters to the US market are obliged to follow the regulations of the National Organic Program (NOP) of the USDA.

The Plant Protection and Inspection Services (PPIS) of Israel's Ministry of Agriculture and Rural Development are responsible for inspection of fresh organic produce. PPIS regulations lay down the minimum requirements with which all operators dealing with organic agricultural fresh produce, or processed products, must comply.

14.4 The olive plantation at Neot Smadar

The Neot Smadar Kibbutz consists of 40 ha of different organic crops (Table 14.1) producing 200 tons of produce. All plots are relatively small to ensure the blending of crops into the surrounding landscape as part of community practice.

The olive plantation consists of Nabali (20 percent) and Picual (10 percent), planted at a tree density of 360 trees/ha. There is strict prohibition of the use of synthetic chemicals, as in other organic farming systems. Since the Arava soil is typical arid soil, deprived of nutrients and micro-elements, improvement of soil fertility is achieved solely by adding compost (locally made in the kibbutz) once a year, and guano. Guano is a natural fertilizer used in organic farming for nitrogen supplementation, since most of the nitrogen content is lost during preparation of compost. Neot Smadar's compost consists of goat manure and milking waste, organic kitchen waste (mixed with sawdust), grape and olive pomace, and landscaping clippings – $1000\,m^3$ year of shredded date prunings. Compost is applied mechanically by spreading it above ground level (no soil disking).

The productivity of the cultivated olive varieties has averaged out as follows: Barnea about 7 ton/ha; Souri about 5 ton/ha; Nabali about 5 ton/ha; and Picual 4 ton/ha.

Another aspect of desert farming in the Arava region is early ripening of deciduous trees, which makes marketing easier (except in the case of olive oil, which is a final product and not a fresh one) because fruits are available before the beginning of the season in the central and northern parts of Israel. Some fruits, such as apricots, ripen early all over Israel.

Table 14.1: Organic produce grown in Neot Smadar Kibbutz; dates (*Phoenix dactylifera*) and olives provide the main income from agriculture products

Crop	Plot size (ha)	Remarks
Olive	14	Souri, Barnea, Picual, Nabali
Date	20	
Almond	5	
Apricot	3	
Fruit trees	6	Apple, peach, pear, plum
Grapevines	3	

Figure 14.2: Wetlands constructed by the Neot Smadar Kibbutz, for ecologically friendly municipal and agricultural sewage treatment.
Photo source: photohttp://www.sviva.gov.il/bin/en.jsp.

14.5 Major management practices

14.5.1 Water resource management

The kibbutz's only water supply is from the Pharan stream drilling site (300 m deep), since there is no connection to the National Carrier (Israel's main water delivery plant). High-quality fresh water is used for irrigation (250 mg/l of chloride) and domestic use after desalination.

The community's decision to improve its conventional wastewater treatment system as part of its ecological concept led to an innovative pilot project for treating sewage by constructed wetlands. Technical treatment plants in small systems are usually unreliable, especially with a fluctuating population, and consume a great deal of electricity. Constructed wetlands, however, are usually tolerant of load fluctuations and have very low maintenance and energy requirements. In October 2007, the project was inaugurated at the kibbutz. The constructed wetlands site consists of five sequential pools, planted with a diverse assortment of water plants. The municipal and agricultural sewage from the Neot Smadar Kibbutz and Neve Harif flows through the pools, where micro-organisms surround the plants' roots. Purified water is then reused for irrigation (Figure 14.2).

To reduce the expense of energy for pumping, the water company agreed to help create a lake. Neot Smadar therefore has its own 50,000 m³ water reservoir (Figure 14.3), along with an island and a few canoes. Water is pumped to a pool sitting on a hilltop, and irrigates via gravity.

Figure 14.3: Neot Smadar Lake, which provides a water reservoir for low energy irrigation.

Figure 14.4: Above-ground sprinkler irrigation of olive trees, 30 cm above ground level.

Irrigation is applied using water from the lake, and water directly from the Pharan drilling site. The water consumption of the olive plantation is 10,000 m³/ha per year.

Olive trees are irrigated by sprinklers located 30 cm above ground level. Overhead sprinkler irrigation enables effective manual weeding, as no herbicide is applied, and protects the system from damage caused animals such as porcupines (Figure 14.4). Another reason for choosing sprinkler over drip irrigation is the utilization of organic matter from the compost. In order to better extract nutrients from the compost and effectively transfer them to the soil, it needs to be wetted constantly.

Olive trees grown in Neot Smadar are relatively small, and therefore the canopy and trunk sizes do not represent their true age. Constant overhead sprinkler irrigation simulates rain, leading to a soil leaching effect. In contrast to drip irrigation, soluble salts are only partly washed from the root zone. The relatively constant high soil salt concentration (soil salinity) limits root zone development, and consequently productivity. Routine leaching is not applied, since precipitation is extremely low (30 mm annually). A minimum of 10 mm of precipitation is required for a salinity effect on the olive trees in the Arava region. However, overground sprinkler systems are not an effective method of irrigation because of the high evaporation rate, so a sub-surface irrigation method is being planned for the near future.

14.5.2 Pest management

Maintaining an ecological balance while preserving biodiversity and controlling the population density of natural pests is a major aspect of organic farming management. The limitations on the use of conventional pesticides and the lack of sufficient data regarding organic substitutes make this subject even more complicated. The arid climate in the Arava can be an advantage regarding pest control.

The olive fruit fly (Figure 14.5), which is generally eradicated by GF-120, is present in Neot Smadar, although Neot Smadar's remoteness from the pest's origin isolates the crop from potential attack. Early ripening of the fruit in the southern Negev enables growers to disregard the damage inflicted by the fly, since olives are harvested before the flies reach the olive plots. Usually olive fly damage is inflicted during the harvest's last two weeks, affecting mainly Barnea

Figure 14.5: The olive fly.
Source: http://www.cnr.berkeley.edu.

cultivars, so there is no substantial fruit damage or oil yield reduction. In years when fly attacks occur earlier than usual, traps are used.

As with the Ramat Negev climate, high levels of relative humidity (60 percent) during the mornings create a favorable environment for pests to thrive in. Increased agricultural lands have over the years increased relative humidity levels, creating a micro-climate preferred by pests. On the other hand, radical temperature differences (ranging from $-3\,°C$ in the winter to $45\,°C$ in the summer) eliminate the need to cope with conventional pest population densities. One disadvantage of low pest diversity is the lack of natural enemies for new pests, which can lead to increased population and future crop damage.

Olive perlatoria is also present in Neot Smadar, and is biologically controlled by the parasite wasp. The wasp's larva eats the olive fly's larva, reducing fruit damage. Jasmine moth is also a common pest, but because its impact is negligible, no treatment is applied.

14.5.3 Determination of harvest timing

Olives for table purposes are harvested manually early in the season, according to the olive skin color. Green fruits of Souri, Nabali and Picual cultivars designated for table purpose are harvested from September in "off" years and October in "on" years. Black olives from the same cultivars designated for table consumption are harvested from late October in "off" years and early November in "on" years. As in Ramat Negev, hand harvesting of table olives requires a large amount of human and time resources; these are provided by the Neot Smadar community itself.

Determination of optimal harvest timing for oil is more complicated. Date palms provide the main agriculture income for the kibbutz (which has more than 2300 date trees), and harvesting of dates clashes with the olive harvest. From past experience, Barnea harvesting can be slightly delayed because fallout is not intensive in late season, unlike Souri. For this reason, Souri olives are harvested immediately after completion of the date harvest; soon after this, the Barnea harvest begins. All olives designated for oil production are mechanically harvested by a trunk shaker. Another significant aspect of olive harvesting is the remoteness of the olive mill that extracts the oil. This is 370 km away, and the distance restricts the harvest period in order to reduce transportation costs.

14.6 Marketing of organic olive products

Despite limited arable land in Israel, organic production is well developed. Some 400 farmers are organized under the IBOAA. Member growers of the IBOAA grow fresh produce on more than 6000 ha, based on organic principles; this is about 1.3 percent of the arable land in the country. Of this land area, 310 ha is dedicated to organic olive agriculture. The total organic sector in Israel is worth US$180 million; 30 percent of the produce is marketed locally and the remainder is exported (IFOAM, 2006). In Israel there are 17 authorized organic shops, 50 sale points in supermarkets, and also a few hundred health food shops that sell organic products. The total value of organic fresh products exported was around US$35 million in 2003.

Although attributes associated with organic foods may be difficult to identify by visual inspection alone, most consumers purchase organic products because of the perception that these products have unique and superior attributes compared to conventionally grown alternatives (Vindigni *et al.*, 2002). Annual average prices for organic produce are generally about double the non-organic price. As the market matures, these premiums can be expected to decline; however, a rapid increase in consumer demand for organic products in the near term will likely support these high organic premiums. The high price of organic products is due to the growers' use of organic seed, organic fertilizer and organic pesticides, among others. In addition, inputs may be difficult to find or require considerable transportation and labor costs, which can reduce profits. To date, all the organic olive oil produced by the Neot Smadar Kibbutz is sold in stores in Israel selling organic porducts (Figure 14.6).

Although until now the olive oil produced by the Neot Smadar Kibbutz has only been available to domestic markets, the high international demand for organic olive oil has led to plans to export to the international market in the near future. As in Ramat Negev, the extreme temperature differences between day and night create a uniquely flavored olive oil that can be distinguished from others. This unique taste helps in promoting a particular brand in a saturated market.

Figure 14.6: Neot Smadar organic olive oil is marketed through stores in Israel selling organic products.

References

Gregory, N. G. (2000). Consumer concerns about food. *Outlook Agric*, 29(4), 251–257.

Grunert, S. C., & Juhl, H. J. (1995). Values, environmental attitudes and buying of organic foods. *J Econ Psychol*, 16(1), 63–72.

Harper, G. C., & Makatouni, A. (2002). Consumer perception of organic food production and farm animal welfare. *Br Food J*, 104(3–5), 287–299, http://www.cnr.berkeley.edu/biocon/W1185%20Weed%20biological%20control.htm.

IBOAA (2007). Israel Bio-Organic Agriculture Association, http://www.organic-israel.org.il/.

IFOAM (2006). The world of organic agriculture, statistics & emerging trends. (available at http://www.ecoweb.dk/ifoam/).

IFOAM (The International Federation of Organic Agriculture Movements) (1998). *Basic Standards for Organic Production and Processing*. Tholey-Theley: IFOAM.

USDA (2000). Glickman announces new proposal for national organic standards. USDA News Release No 0074.00.

Vincenzo, F., & Damiano, P. (2002). Organic agriculture in the Mediterranean area: state of the art. In: L. Al-Bitar (ed.), *Report on Organic Agriculture in the Mediterranean Area*. Centre International de Hautes Études Agronomiques Mediterranéennes, Série B 40: 9–51.

Vindigni, G., Janssen, M. A., & Jager, W. (2002). Organic food consumption: a multi-theoretical framework of consumer decision making. *Br Food J*, 104(8), 624–642.

CHAPTER 15

Economic aspects of desert olive oil cultivation

15.1 Background

This chapter demonstrates sample costs for the establishment of a high-density olive plantation in the desert, and oil production in the Negev region of Israel. This study includes estimations and the real costs of establishing an olive orchard, and of producing olive oil after establishment. It is intended as a guide only, and can be used to make production decisions, to determine potential returns and to prepare budgets. Practices described in this chapter are based on the production procedures currently being used in high-density olive oil orchards in the Ramat Negev area of the Negev Desert in Israel.

Large-scale (90 ha) conventional desert olive plots have been cultivated in the conventional manner in Ramat Negev for the past 12 years, and a smaller plot at Neot Smadar (14 ha) in the Arava region has been cultivated organically since the 1980s. Both farms are irrigated with saline water. The main olive varieties planted in the Negev are Barnea, a unique olive variety designated for oil production and originating in Israel, followed by Picual (which originated from southern Spain), Souri (which originated from northern Israel and/or southern Lebanon) and Manzanillo (which originated from central Spain).

Detailed costs and revenues are presented, followed by a comparison of organic and conventional farming, and outlines of cultivation and harvesting costs. Sample costs for labor, materials, and equipment and customs services are based on current figures. Some costs and practices presented in this study may have to be updated according to various regional and national parameters. No financing or government subsidy has been taken into consideration.

15.2 Economic aspects of conventional desert olive plot establishment and production

15.2.1 General assumptions in this study

The following assumptions have been made regarding establishment of high-density olive orchards and olive oil production in the Israeli Negev Desert. Some costs, practices and materials may not be applicable to a specific situation, or used every year. Additional practices that are not indicated may be needed. Establishment and cultivation practices can vary according to growers and regions, and variations can be economically significant. Costs are calculated in US dollars on an annual per hectare basis.

Land

A favorable choice for land use in Israel is to lease directly from the government. One disadvantage of governmental land leasing is the prolonged procedure, caused by obtaining a water quota and by bureaucratic negotiation. The fundamental advantage is land proprietary rights, since land value increases over time. Other advantages include a distinctly low land leasing cost, solvency of the property, and almost full security for contract renewal after expiration. The cost of land lease in the Negev is $465/ha per year. Olive plots are established on 95 percent of the land, with the remainder being dedicated to roads, irrigation systems, unused land and a farmstead.

Trees

The olive cultivar used in this study is Barnea, which is the best adapted to high-density olive plantations and mechanical harvesting. It has a high yield (17 ton/ha, 16 percent oil content), and a medium intensity of alternate bearing. It is a vigorous tree of an erect shape that facilitates mechanical harvesting, but it needs more pruning than other varieties. The Barnea seedlings in this study cost $2.00 each. The costs will vary depending on the variety and age of the plant, the quantity purchased and the payment plan. In this study, tree spacing is maintained at 7×5 m, totaling 280 trees per hectare.

Irrigation

The irrigation cost consists of the water consumption volume (m^3) alone. Israeli water prices are unaffected by energy requirements for water delivery, such as pumping costs. The cost of irrigation will vary by region, depending on water prices and other irrigation-related factors. Desert olive plot irrigation is expensive because the evaporation rate is high. In this study, the water price is calculated at $0.16/m^3$ of saline water – not significantly lower than the price of fresh water ($0.17/m^3$). The irrigation rate increases each year as the plot matures until the fifth year, as shown in Table 15.1.

Table 15.1: Irrigation volumes (saline water) and costs for the development of olive trees (calculated irrigation volume is reduced by half)

Year	Irrigation volume (CM)	Irrigation cost ($/ha)
1	700	112
2	2000	320
3	3000	480
4	4000	640
5+	4500	720

In Ramat Negev, the unique feature of reusing fish-culture wastewater for olive plot irrigation decreases the cost of irrigation. As described in Chapter 13, in practice each hectare of adult olive trees consumes $9000\,m^3$ of water. For calculation purposes, the irrigation volume in Table 15.1 is reduced by half.

Labor

Labor rates are $8 per hour for machine operators and $6 for non-machine labor, assuming 240 labor hours in one month. The overhead includes the employer's share of state payroll taxes. Labor costs for operations involving machinery are higher by a third, to account for the extra labor involved in equipment maintenance, field repairs and the professional manpower rate.

15.2.2 Establishment costs
Pre-planting site preparation

Before ground preparation for planting, the desert landscape is typical bare ground with sparse resident vegetation. Site preparation begins in the summer by disking once with a stubble disk to break up the ground. The land is then subsoiled to a depth of 35 cm in a crossing pattern to break up compaction, followed by plowing and tilling of the land twice. Details of pre-planting preparations are listed in Table 15.2. Tree sites can be marked by a global positioning system (GPS), and pits are excavated. System and drip-irrigation infrastructure is laid on the ground for pre-irrigation to help soften the soil. Establishment costs, as described in Table 15.5, are included in the first year.

Planting

In the autumn, trees are planted with spacing of 7 m between rows and 5 m within the rows, totaling 280 trees per hectare. Initially, seedlings are planted to create 560 plants per hectare; every second tree is then thinned out in the eighth year. The estimated planting cost is $26 per hetare. Wilted trees need to be replanted in the second year (2 percent replenishment rate).

Table 15.2: Detailed pre-planting preparation costs ($)

Specific preparation	Cost per unit	Unit	Cost per ha
Irrigation infra-structure construction	2790	1	2790
Soil survey	14	1	14
Disking	47	1	47
Subsoiling	58	1	58
Plowing × 2	29	2	58
Tilling × 2	23	2	46
Measuring and marking plant locations	14	1	14
Miscellaneous	10%		303
Total cost			3330

Training and pruning

Training of new trees starts by tying the seedling to a bamboo pole at intervals as the leader branch grows. Trees are trained to be upright with an open canopy. Pruning is carried out as follows, and prunings are placed in the middle of the row and collected.

- *Year 1.* Two cuts are made, to choose the leader for each tree. One cut is made during a tying pass.

- *Year 2.* The only pruning carried out is to remove the side branches below 60 cm and any suckers arising from the roots.

- *Year 3.* Tying and training continues, and leaders are pruned. Some larger side branches can be left in the first three years to help fill the space and produce more fruits earlier, but these are pruned by hand in the fourth year to avoid excessive shading.

- *Year 4.* Light pruning only is performed.

Pest control

Control of diseases while the olive orchard is being established is normally minimal. It begins in the first year with two sprays, in February and November. Olive mite is a pest that affects seedlings, and is treated with the miticide Vertimec® ($93/l, 0.1 l/ha). Jasmine moth is another pest to which young trees are susceptible, so they are treated twice with Monokron® ($23/l, 0.2 l/ha) ground spray. Vertimec and Monokron are only applied during the first two years.

Although it is less widespread in the Negev than in other areas, the olive fruit fly is also present in the region and regular control is necessary once fruiting begins. In the second year, Simazine

Table 15.3: Weed control in olive plot establishment

Year	Mowing	Herbicide application
1	×	Goal × 1
2	×	Goal × 1
3	×	No herbicide applied
4+	×	Roundup × 2

is applied only once ($12/l, 1 l/ha). Starting from the third year, Simazine is applied five times a year.

Weed control

Floor management is different within and between tree rows. Resident vegetation is allowed to grow between rows, and this vegetation is tilled twice during the growing season, starting in the first year. Immediately after planting, the pre-emergence herbicide Goal® ($56/l, 0.2 l/ha) is applied once to prevent weed germination in a 1.5 m wide strip (Table 15.3). This will effectively prevent most weeds from growing, but will not cause phytotoxicity in case of tree contact. Goal® is only used during the first two years. Roundup® ($4.2/l, 1 l/ha) is applied twice starting from the fourth year.

Fertilization

Typical arid soil has a low nutrient content, and must be replenished to meet the demands of vegetation growth. Nitrogen is the major nutrient required for proper tree growth and optimum yields. Young trees receive dissolved nitrogen (N) and potassium (K) fertilizers through the irrigation system in the form of ammonium sulfate and potassium chloride, respectively. The nitrogen and potassium dose should be increased during plot maturation (Table 15.4) until the fourth year. Liquid phosphate (P) is supplied as phosphoric acid in small quantities in the first year, and in constant amounts from the fourth year onwards. The quantity of materials applied depends on the percentage of active ingredient in each product.

Table 15.4: Nitrogen, potassium and phosphorus fertilizers applied in olive plot establishment

Year	Nitrogen (kg/ha)	Potassium (kg/ha)	Phosphorus (l/ha)
1	120	70	10
2	330	180	0
3	800	350	0
4+	1000	500	100

Wholesale prices: N, $0.2/kg; K, $0.2/kg; P, $0.9/l.

Table 15.5: Costs per hectare for intensive desert olive cultivation (280 trees/ha)[*]

	Year 1	Year 2	Year 3	Year 4
Planting costs				
Land leasing	465			
Land preparation – plowing × 2	58			
Land preparation – tilling × 2	46			
Survey & mark plot	28			
Seedlings	2240			
Dig & plant seedlings	91			
Wrap trees	6			
Trellis system installation	56			
Replant (2%)	0	12		
Total planting costs	2990	12		
Cultivation costs				
Pruning	168	252	420	6
Irrigation	112	320	480	640
Fertilizer – N	24	66	160	200
Fertilizer – P	9	0	0	90
Fertilizer – K	14	36	70	100
Weed control – middle row tilling × 2	30	30	30	30
Weed control – Goal × 1	11	11		
Weed control – Roundup × 2				8
Pest control – olive mite & jasmine moth × 2	28	28		
Pest control – olive fruit fly × 5 (1 in Year 1)		12	60	60
Tractor use	147	147	147	147
Total cultivation costs	543	902	1367	1341
Harvest costs				
Mechanical harvest				2240
Total harvest costs				2240
Total establishment costs	3533	914	1367	3581

Labor rate: non-machine labor, $6/h; machine labor, $8/h.

Total establishment cost

The total establishment cost is the sum of the costs of land preparation, seedlings, planting, and the production expenses of growing olive trees until oil is produced, minus any returns. In this study, production begins in the fourth year. The total establishment cost per hectare is $3533 in the first year and $3581 in the fourth year (Table 15.5).

15.2.3 Production (operation) costs and incomes
Pruning

Pruning manipulates the tree growth to improve mechanical harvesting efficiency and maintain fruit production. Any prunings are transferred to the middle of the row and later given to the local Bedouin community (for stock feed and burning) to reduce costs. Pruning expenses vary by variety, tree vigor, etc. In this study, manual pruning is carried out, and the mechanical cost of clearing the prunings is incorporated into the tractor costs. Pruning is as follows.

- *Year 5*. An open formation canopy is facilitated by pruning leader trees.

- *Year 6*. Non-leader trees are pruned to allow the canopy leader to expand and increase fruiting.

- *Year 7*. Central leader trees are not pruned; non-central leaders are uprooted after the harvest.

- *Year 8*. Pruning practices are similar to those in the fifth year.

Fertilization

Nitrogen and phosphate are applied annually at a rate of 1000 kg/ha, and 100 l/ha, respectively, through the irrigation system. Potassium is dissolved and sprayed at a rate of 500 kg/ha every other year.

Weed control

Starting from the fifth year, the soil is tilled twice and Roundup® spray is applied twice; this is sufficient for weed control as the trees mature.

Harvesting

Harvesting starts in the fourth year. The fruits are delivered to an external mill contractor for extraction. Olives for oil extraction are mechanically picked at an IOC Maturity Index of at least 3, from October to December. A trunk shaker is employed to shake the fruits off the trees. Mechanical harvesting by trunk shaker is estimated to cost $4 per tree, including labor. Since non-leader trees are uprooted after the seventh year (post-harvest), harvesting expenses are calculated for 560 trees per hectare until the eighth year. The fruit weight per hectare also increases over time (Table 15.6). Olive fruit delivery and oil extraction by an external mill contractor is evaluated at 13 percent of oil income (Table 15.7).

Table 15.6: Olive oil prices, oil content and yields as influenced by typical "on" and "off" years

Year	Fruit weight (ton/ha)		Oil content (%)	Olive oil price ($/kg)
	Leader	Non-leader		
5	10	10	16	4
6	7	3	17	4.2
7	17	5.5	16	4
8	5.5	Uprooted	17	4.2

Yields and gross income

High-density desert-planted olives begin bearing an economic crop in the fourth year after planting, and the maximum yield is reached in the seventh year. The amount of extracted oil can vary considerably, depending on the age of the tree, the fruit moisture content (influenced by irrigation and rainfall), the crop load and fruit maturity. Olive oil prices, fruit oil content and yields as influenced by typical "off" and "on" years (the sixth and seventh years) are listed in Table 15.6. Gross income from olive oil sales is presented in Table 15.7.

Table 15.7: Costs and income per hectare per year for intensive oil olive production (280 trees/ha); non-leader trees are uprooted after the seventh year (post-harvest)*

	Year 5	Year 6	Year 7	Year 8
Cultivation costs				
Irrigation	720	720	720	720
Pruning	560	420	420	560
Fertilization (N, K, P)	390	390	390	390
Weed control – Roundup × 2 & tilling × 2	38	38	38	38
Pest control – olive fruit fly × 5	60	60	60	60
Tractor use	113	113	248	168
Total cultivation costs	1881	1741	1876	1936
Harvest costs				
Mechanical harvesting	2240	2240	2240	1120
Olive fruit delivery and oil extraction (13%)	1664	1030	1864	557
Total harvest costs	3405	4075	3270	4104
Gross income				
Oil production	12800	7650	14400	4230
Total gross income	12800	7140	14400	3948

*Labor rate: non-machine labor, $6/h; machine labor, $8/h.

Total production cost

Production begins in the fourth year, and as the plot matures the cultivation and material expenses become more stable. The production cost is the sum of the costs for plot fertilization, irrigation, pruning, weeding and harvesting for oil production.

15.3 Economic comparison and outline of organic and conventional intensive desert olive cultivation

Organic agriculture in general aspires to sustainable productivity with minimal damage to the environment. The organic principle is reflected in the economic aspects of olive plot establishment and production. For example, tree density is less intense in organic plantations, to avoid pests and diseases by better aeration and more sunlight. The final spacing can reach 7×8 m (170 trees per ha), as opposed to the intensive spacing of 7×5 m (280 trees per hectare).

Productivity and yield are also affected by organic cultivation. Trees bear their first economic yield in the fourth year in the conventional system, but as vegetative growth is slower in organically cultivated trees, they begin fruiting later and fruit yield is also lower (Table 15.8).

Table 15.8: Overall outgoings and income from desert conventional and organic farming (US$/y per hectare).

	Intensive farming		Organic farming	
	On year	Off year	On year	Off year
Materials	810	902	1466	1740
Labor	611	933	1397	1324
Tractor	142	180	216	195
Miscellaneous	70	70	93	93
Irrigation	1360	1209	1209	1058
Total production costs	2994	3294	4382	4410
Harvesting and oil extraction costs	3101	1860	2261	1330
Total expenses	6095	5153	6644	5740
Fruit weight (kg)	4566	1660	2993	998
Oil weight (kg)	731	282	509	1800
Oil price ($/kg)	4	4.2	5	5
Total income	12420	5081	11194	3951
Net profit	6325	−72	4550	−1789

Tree density: organic, 286 trees/ha; conventional, 357 trees/ha.
Irrigation costs are calculated for full price (no reuse from fish culture).
Saline water price, $0.65/CM.

Organic fertilization is based on large quantities of solid components, mainly compost and manure. Supplements for intensive fertilization include guano, feathers, flowers and industrial pellets. Transporting large volumes of compost and manure and spreading it in the field are expensive operations, while transporting a low volume of chemical fertilizers is relatively cheap. Delivery of soluble fertilizers through irrigation ("fertigation") is relatively low cost compared with other methods of fertilization, but the cost of compost is significantly higher than other methods of fertilization.

The quantity of compost needed for organic farming is relatively higher in southern than in northern parts of Israel owing to the low soil nutrient content and rapid decomposition of organic matter in the south.

Conventional weed control is carried out both mechanically and chemically. Organic agriculture uses plowing combined with mechanical trimming, which is more costly and less efficient. Pest control is carried out using natural pesticides and/or insects that serve as natural enemies (mainly special species of wasps and acari).

As in conventional farms, olive fly, the most problematic pest in olive cultivation, needs to be controlled in organic orchards. In conventional farming, control is based on bait sprays or mass trapping accompanied by monitoring; organic agriculture, however, uses only bait traps. Infected fruit produces more acidic oil (Mavragani-Tsipidou, 2002), which cannot be classified as extra-virgin oil, thus resulting in significantly lower prices. Successful control of olive fly is therefore essential for economic olive oil production. While conventional farming can correct the damage by applying phosphoric pesticides, organic farming has no margin for errors. As a result, monitoring and trapping is more intensive and expensive, and the risk of infestation of fruits still exists.

Implementing organic olive growth at the same level of intensity as conventional farming is not possible. Following the organic regulations for certification also means more expensive cultivation. The economy of organic olive plantations is therefore conditional upon achieving a much higher price for oil of the same quality. Marketing under a prestigious label with designation of the origin and organic certification can help in achieving the target price. Organic farming's distinct gain is an oil price unaffected by alternate bearing (Table 15.8).

The overall expenses and revenue of conventional and organic olive farming in desert conditions, as practiced in Israel are listed in Table 15.8. Organic agriculture requires intensive cultivation costs, reflected in the labor, mechanical and cultivation expenses, and uses more expensive materials. Harvest operations are more expensive in conventional farming, because the tree density is higher. The income from olive oil sales is significantly lower with organic cultivation, resulting from low productivity (compensated for by higher oil prices).

Reference

Mavragani-Tsipidou, P. (2002). Genetic and cytogenetic analysis of the olive fruit fly *Bactrocera oleae* (Diptera: Tephritidae). *Genetica*, 116, 45–47.

CHAPTER 16

Marketing of desert olive oil

16.1 Olive oil and global marketing

Marketing is a method of business management that places the customer at the epicenter, and at the same time is an action system that forms an inherent part of a firm's business strategy. Marketing identifies customer needs and wants, determines which target markets the organization can serve best, and designs appropriate products, services and programs to serve these markets. Market identification is fundamental to the growth of any business. Product marketing involves consumer segmentation in order to target different types of products to the right type of consumers (Roininen *et al.*, 1999).

The olive oil market is part of the agrifood industry, and as such needs specific marketing strategies. Agricultural marketing, as stated by Siskos *et al.* (2001), is differentiated from general marketing with respect to the product's attributes and natural characteristics, price determination, promotion, advertising and, finally, distribution procedures. Shepherd (1999) stated that the factors influencing food choice are categorized as those related to the food, to the individual making the choice, and to the external economic and social environment within which the choice is made. Shepherd and Farleigh (1986) concluded that psychological differences between individuals, such as personality, may also influence food choice. External factors such as weather and susceptibility to diseases and pests have a significant effect on the output and quality of agrifood products, whose attributes of taste, aroma, color, age, shape, moisture, nutritional value and chemical composition define their quality. The top grades are processed directly to the consumer, whereas lower ones are treated as raw materials. Kohls and Uhl (1990) reported two marketing strategies that food firms frequently follow:

1. Product differentiation by emphasizing a product's unique features, which set it apart from its competitors.

2.   Market segmentation by developing unique product variations that will be better perceived by different consumer classes and market segments.

Siskos *et al.* (2001) determined that agricultural product design and promotion are based on distinct factors such as attributes of the particular variety, possible health effects, and the region of origin. Ward *et al.* (2003) pointed out the importance of product attributes and the necessity for providing information about the products. Senauer (1989) suggested that the main attributes consumers look for when considering food products are quality, taste, convenience, nutrition, wholesomeness and value. There is a constantly increasing demand from consumers for higher-quality food, as a result of increased knowledge about links between diet and health, awareness of quality characteristics and availability of information (Fotopoulos and Krystallis, 2003). Numerous studies (Madu and Kuei, 1994; Rajagopal *et al.*, 1995; Gundogan *et al.*, 1996) have demonstrated the strategic benefits of using quality as a means of gaining a marketing competitive edge. Recently, demand has also risen for agricultural products produced either organically or without the use of synthetic chemicals (Lampkin, 1990; Ophuis, 1991; Schifferstein and Oude Ophuist, 1998; Tzouvelekas *et al.*, 2001; Cicia *et al.*, 2002).

Not every society has the same conventions and values regarding a product's features. Nielsen and colleagues (1998) suggested that the importance of certain attributes of a food product or the most preferred value of an attribute differ between cultures and nationalities. A study of the EU fresh fruit market revealed that European markets differ significantly in *per capita* consumption, taste and preference for varieties, sizes, quality and color of fruit (Martinez and Quelch, 1991). There is a broader distinction between consumers in Mediterranean countries, where excellent-quality fruit is produced and where there are strong preferences for size and quality, and those in Northern European countries, where there are less demanding tastes. A significant difference has been distinguished by Frohn (1991) in the EU countries regarding media in food advertising: in Portugal, Greece and Italy, electronic media prevail, while in Denmark, the Netherlands, Germany, France and Great Britain, printed media are more common.

Olive oil fits well with modern consumers' expectations, with its healthy image, proven gastronomic properties and health benefits. Anania and Pupo D'Andrea (2007) state that the olive oil market is very complex for a number of reasons: production is spread over developed and developing countries; the oil is produced regionally but traded globally; and olive oil consumption is growing, but consumption patterns vary widely (both in quantity and quality). Moreover, in some countries and for some consumers, quality product attributes have come to assume an increasingly important role in consumption decisions. It is widely recognized by consumers that there is an association between certain quality traits of olive oil and the geographic area of production (Scarpa and Del Giudice, 2004). Caporale *et al.* (2006) also confirm that extra-virgin olive oil plays an important role as a typical example of Mediterranean production whose characteristics are strongly affected by the origin of its raw material and the manufacturing technology used. In addition, olive oil production fluctuates from one crop year to another, mainly due to alternate bearing, thus causing an impact on prices, producer income and supply. One way to cope with this lies in the development of high-quality products and markets through increased consumer awareness of the superior elements of olive oil and the differences among olive oil types (Patsis, 1988). For this strategy to be successful, it is very important that the product attributes, packaging design, price and labeling closely meet the specific needs of the end customer. Parameters that should be taken into consideration while planning a marketing strategy for olive oil may include:

- the segment of the market that is being targeted;

- characteristics that define the target customers;

- unique features and benefits of the product that meet the customer's needs;

- unique features that differentiate the product from that of competitors;

- verification that the product is actually in demand from customers;

- a pricing strategy.

In the future, the olive oil market will probably be more segmented on the basis of product quality differentiation, with price playing a relatively less important role than in the past (Mili and Zúñiga, 2001). Moreover, Cicia and colleagues (2002) and Scarpa and Del Giudice (2004) have suggested that the typical product price is often perceived to be an indicator of quality, and thus a high-cost product may be of even greater appeal to consumers. Recently, Anania and Pupo D'Andrea (2007) have suggested that increases in consumption are driven by differentiated consumption patterns, with an increasing share of consumers moving from bulk purchase directly from producers to purchase of bottled and branded olive oil from large retailers; from conventional to organic olive oil; and from olive oils of non-specific origin to oils with a certified geographic indication, as is the case for protected denomination (PDO and PGI) olive oils in the EU.

16.2 Marketing strategies in major EU countries

In this section, marketing strategies for olive oil are discussed as perceived through intensive research of segments and demographics mainly within European countries. In order to penetrate this competitive market, it is vital to know the consumers' wants and needs. Each sector values different olive oil attributes, and hence the marketing strategy must be planned accordingly.

Greece

A survey conducted by Krystallis (2005) attempted to develop the profile of a quality-conscious Greek consumer cluster in terms of olive oil-related purchasing motives and personal values by performing both qualitative and quantitative research.

Table 16.1: Levels of the conjoint analysis (CA) factors selected and their relationships in extra-virgin olive oil (Krystallis, 2005)

Factor	Level no.	Level of description
Organic label	2	1: Yes
		0: No
PDO label	2	1: Yes
		0: No
ISO certification	2	1: Yes
		0: No
HACCP certification	2	1: Yes
		0: No
Health information	4	1: Best before date
		2: Keep until instructions
		3: Additives/preservatives free
		4: Cholesterol free
Glass bottle	2	1: Yes
		0: Other[*]
Country of origin	2	1: Written on the label
		0: Not written on the label
Price levels[**]	4	1: 3.25 €/L
		2: 4.41 €/L
		3: 5.88 €/L
		4: 6.76 €/L

[*] *"Other" usually implies plastic bottle.*
[**] *Price levels were identified from average retail prices in Athens, April–May 2000.*

The survey was in the form of questionnaires given to 40 participants who were chosen for being responsible for food purchasing in their households, and who had purchased at least 1 l of bottled olive oil in the month prior to the survey period (May–September 2000). The participants were selected as being well educated, young, and of above-average income.

The replies to the questionnaires were analyzed using advanced statistical tools and methods. By applying conjoint analysis (CA), the researchers were able to identify the attribute combinations preferred by respondents and the relative importance of each attribute. Eight olive oil attributes relating to the perceived quality and healthiness of olive oil were selected based on the results of a means–end chain (MEC) analysis, as shown in Table 16.1.

Sixteen olive oil profiles were estimated from the combination of the levels of each of the eight factors, as shown in Table 16.2.

Table 16.2: Predicted preference for 16 olive oil profiles according to their total utility, Krystallis (2005)

Rank	Description	Predicted preference
1	Organic and PDO olive oil, with ISO and HACCP, "best before" date and the origin on the label, bottled in glass and with the highest price (6.76€/L)	MOST PREFERRED 9.1460
2	Organic olive oil, with ISO and HACCP, "keep until" instructions and the origin on the label, bottled in glass and with average for organic olive oil price (5.88€/L)	8.2703
3	Organic and PDO olive oil, with ISO, "cholesterol free" sign and the origin written on the label, bottled in glass, with average for organic olive oil price (5.88€/L)	8.1235
4	PDO olive oil with HACCP, "keep until" instructions and the origin written on the label, bottled in glass, with average for conventional olive oil price (4.41€/L)	6.5461
5	Organic and PDO olive oil, with HACCP, "cholesterol free" sign written on the label, bottled in glass, with the highest price (6.76€/L)	6.3486
6	Olive oil with HACCP, "best before" date and the origin on the label, bottled in other than glass, with average for conventional olive oil price (4.41€/L)	INDIFFERENT 5.6956
7	Organic olive oil with ISO, "additives/preservatives free" sign written on the label, bottled in glass, with average for organic olive oil price (6.76€/L)	5.6716
8	Organic and PDO olive oil, with "best before" date written on the label, bottled in glass, with the highest price (6.76€/L)	5.3990
9	PDO olive oil with the "additives/preservatives free" sign and the origin on the label, bottled in glass, priced cheaply (3.25€/L)	LEAST PREFERRED 4.9830
10	Organic olive oil, with HACCP, "additives/preservatives free" sign, bottled in glass, with average for organic olive oil price (5.88€/L)	4.9333
11	Olive oil with ISO and HACCP, with the "cholesterol free" sign written on the label, priced cheaply (3.25€/L)	4.8148
12	Olive oil with "cholesterol free" sign and the origin written on the label, bottled in other than glass and priced cheaply (3.25€/L)	4.6733
13	Organic olive oil with "keep until" instructions on the label, bottled in glass, with average for organic olive oil price (5.88€/L)	4.5245

Table 16.2: (*Continued*)

Rank	Description	Predicted preference
14	PDO olive oil with ISO and HACCP, with "additives/ preservatives free" sign written on the label, bottled in other than glass, with average for conventional olive oil price (4.41€/L)	4.3292
15	Olive oil with ISO and "best before" date written on the label, bottled in glass, with average for conventional olive oil price (4.41€/L)	3.8969
16	PDO olive oil, with ISO, "keep until" instructions written on the label, bottled in other than glass, priced cheaply (3.25€/L)	3.8887

Results indicated that the most preferred brand was one that had a "best before date" on the label, organic and PDO signs, ISO and HACCP certification, was presented in a glass bottle, with country of origin information, and had a price of 6.76€/l.

The most important olive oil attributes were found to be country of origin, organic labeling, and health-related information. An olive oil brand that had no quality label, country-of-origin sign or glass bottle was perceived indifferently by the participants. It represented the most common olive oil brands with which consumers are familiar. The "additives/preservatives free" health information was, perhaps surprisingly, the only factor level with negative utility. However, Greeks are generally aware of the fact that olive oil does not contain any kind of chemical additives or preservatives, and hence the "additives/preservatives free" claim may be considered to be irrelevant or even misleading for a food product such as olive oil. All the remaining factor levels had positive worth, representing preferred product attributes. Price in particular, at all the specified levels, resulted in positive scores for consumers. It seems that the notion of a value-for-money olive oil price is satisfied by the actual price levels of Greek olive oil brands.

Spain

Mili and Zúñiga (2001) conducted a survey designed to forecast the main trends and likely developments affecting the Spanish olive oil business and exports over the next decade. In this survey the Delphi method was applied on a panel of eight experts from the olive oil sector, all having a high degree of responsibility and experience, and coming from the business, academic, and administration arenas in Spain. The survey was carried out in 1999 in two rounds, using two different questionnaires. It was constructed on the basis of an initial review and critical analysis of existing information on recent trends in the world olive oil market as a whole, and of Spanish olive oil exports in particular. In addition, the profile and preferences of the potential consumer were identified, making the distinction between traditional and non-traditional consumer markets.

The experts were requested to define the main features of consumption and potential uses of olive oil in the markets where demand is expected to increase over the coming decade, making distinctions in their assessments between Producer Countries (PCs) and Non-Producer Countries (NPCs). The experts pointed to the increased importance in the future of the quality factor as against price, and agreed that there is a wide margin for growth in consumption if there is a change in the price ratio favorable for olive oil. At the same time, they expected growth and greater diversification over the coming years in both food and non-food uses for olive oil. Moreover, they stressed the existence of a large margin for growth in olive oil consumption amongst young, urban populations, and in consumption outside the home.

In view of the strong international competition, both between olive oil and other fats and oils and actually within the olive oil market, the criteria for differentiating Spanish olive oils in foreign markets were explored. In comparison with substitution products, Spanish olive oils would need to be differentiated in foreign markets based mainly on the attributes of a healthy, natural, and good-tasting product, with special attention being paid to presentation and the environmental benefits. For oils from competing countries, the main variables for differentiation would likely be the brand and the country image, without forgetting the use of denominations of origin and varietal differentiation according to the location of cultivation.

With respect to the attributes of olive oil that are considered most appropriate for efficient marketing in potential markets, both in PCs and in NPCs, the experts identified four relevant attributes that have been classified according to their relative importance, as shown in Table 16.3. These attributes are fruitiness, low acidity, "personality," and mildness.

Table 16.3: Attributes of olive oil for effective marketing in potential markets: relative importance of factors (Mili and Zúñiga, 2001)

Particulars	Producer countries (PCs)		NPCs (Non-producing countries)	
	Z	CV	Z	CV
Fruity oils	4.18	18.18	4.05	24.2
Oils with low acidity	4.15	16.39	4.25	16.23
Oils with "personality"	3.95	22.02	3.47	29.97
Mild oils	3.57	22.13	4.33	15.93

The values of Z are on a scale of 1 to 5, where 1 = not significant, 2 = very limited, 3 = limited, 4 = significant and 5 = very significant.

Table 16.4: International marketing mix: relative importance of factors (Mili and Zúñiga, 2001)

Particular	Z	CV
Product		
Quality assurance	4.64	11.64
Oil type	4.43	16.93
Packaging	4.20	16.90
Labeling	4.15	14.70
Branding	4.05	21.97
After sale warranty	3.82	25.39
Price		
Price discrimination within the range of olive oils	3.87	20.67
Price discrimination with respect to substitute oils	3.62	22.93
Price discrimination with respect to domestic market	3.23	31.58
Distribution		
Regularity of supply	4.61	10.63
Permanence on the shelf	4.49	11.14
Delivery terms	4.30	14.65
Price stability	4.21	16.39
Terms of payment	3.59	22.28
Communication media		
Information at points of sale	4.42	13.80
Information in mass media	3.98	16.58
Presence at trade fairs and shows	3.95	18.30
Contents		
Information on dietary and nutritional benefits	4.66	11.80
Information on differences from other oils	4.42	16.97
Information on culinary uses	4.36	13.99
Information on the natural and environmental value of oil	4.31	18.10
Information on the geographic origin of the oil	3.79	24.54

The values of Z are on a scale of 1 to 5, where 1 = not significant, 2 = very limited, 3 = limited, 4 = significant and 5 = very significant.

Various dimensions of the four basic variables of the marketing mix that the experts considered to be the most relevant for international trade over the coming decade were ranked according to the degree of their relative importance (Table 16.4).

It is clear from the results that, taken together, decisions on product characteristics, pricing strategies, links with distribution channels, and communication policies will be increasingly important factors for success in foreign markets. It can be observed that the importance assigned to pricing aspects was lower than that assigned to those related to the other three variables,

Table 16.5: Attributes and levels used in the ranking survey (Scarpa and Del Giudice, 2004)

Attributes	Level	Level of description
Appearance	2	Turbid
		Limpid
Certification	3	PDO/PGI
		Organic
		No certification
Geographic origin	3	North–central Italy
		Southern Italy
		Unknown origin
Price	3	4.00 €/l
		6.00 €/l
		7.50 €/l

meaning that, according to the experts, price will rarely be the main variable for competing in foreign markets.

Italy

Choice behavior of Italian olive oil consumers may be perceived through the study of Scarpa and Del Giudice (2004), where urban consumers of extra-virgin olive oil from three Italian cities (Naples in the south, Rome in the center and Milan in the north) were compared. These cities were chosen in order to measure and characterize the product perception and preferences of consumers living at a distance from the regions where extra-virgin olive oil is produced.

The study involved 300 consumers of extra-virgin olive oil who regularly buy the product from large retailers. The survey instrument consisted of a questionnaire that included several sections gathering information about the consumers' purchase and consumption behavior regarding extra-virgin olive oil. Respondents were asked to rank, according to their preference, 12 alternative labels of extra-virgin olive oil. Each label was described according to four main attributes (Table 16.5).

Results showed that people in the northern city of Milan tend to favor oil produced in the center/north of Italy, but exactly the opposite is true for residents of the southern city of Naples, with residents of Rome rating the two geographical origins of product very closely. Moreover, consumers in Naples showed a dislike for oil from the south. This suggests that products from the south, which are very welcome elsewhere, are still subject to suspicion, possibly based on die-hard prejudice. Certification for organic oil and PDO/PGI status was found to vary across the country. The intensity of preference for organic oil was found to be lower on moving from the north to the south of the country, with the preference intensity for PDO/PGI always being

Table 16.6: Consumer perceptions and attitudes for the olive oil products (Siskos *et al.*, 2001)

S. no.	Hypothesis	Consumers (%)		
		Agree	Unaware	Disagree
1	Olive oil is very healthy	66.2	22.5	11.3
2	Olive oil makes food tasty	91.7	5.4	2.9
3	Olive oil can be used for frying more than once	4.9	40.7	54.5
4	Olive oil is not fattening	25.5	34.3	42.2
5	Olive oil does not increase cholesterol	30.9	46.6	22.5
6	Olive oil is good for the heart	40.7	49.0	10.3
7	Olive oil does not contain other additional substances	51.0	21.6	27.4
8	Olive oil is relatively expensive	92.6	0.0	7.4

higher than that for the organic nature of the product in all towns. Price, on the other hand, was found to have an effect in Rome and Naples, but not in the Milan district. This may indicate that consumers in south and central Italy perceive the product price as a cue for quality.

France

A consumer study conducted by Siskos *et al.* (2001) tried to identify and determine French consumer behavior with respect to the product's attributes and special features. The consumer profile and the factors contributing to the selection of a product were determined by data analysis. The survey was conducted in Paris, and included 205 olive oil consumers. The results showed that nearly all consumers utilized olive oil in general, while the frequency of their use allowed them to be classified into two broad categories: regular consumers (75.1 percent) and occasional consumers (24.9 percent). These olive oil consumers mainly use olive oil for salads, hot dishes and cooking. The results of specific attitude questions are given in Table 16.6. It is important to note that, for instance, French olive oil consumers believe that olive oil is very healthy, makes food taste better, but is significantly expensive.

The perception of criteria among consumers is summarized in Table 16.7; 77 percent considered the color of the olive oil to be important at a weighted significance level with respect to the other criteria of company image (80 percent), price (64 percent), taste (61 percent) and odor (51 percent). Based on the criteria ranking according to the level of significance, it is also feasible to segment the olive oil market through criteria or complex combinations of them.

A conclusion from this analysis is that olive oil color and company image are the most significant criteria for the choice of a specific olive oil brand, while price and packaging are of medium importance to French consumers. Finally, the taste and odor of the oil are the least significant determinant criteria for consumer behavior.

Table 16.7: Market segmentation through criteria combinations (Siskos *et al.*, 2001)

Particulars	Color (%)	Image (%)	Price (%)	Packaging (%)	Taste (%)	Odor (%)
Color	77	61	49	58	48	47
Image		80	53	54	50	46
Price			64	46	37	35
Packaging				70	43	44
Taste					61	41
Odor						57

16.3 Marketing strategies in Australia

The Australian case is presented by McEvoy and Gomez (1999), who compiled a quantitative and qualitative marketing research study that was conducted through a field survey divided into seven sections. The target population included the main grocery buyer of households, both users and non-users of olive products, in the metropolitan areas of Brisbane, Sydney and Melbourne.

The survey findings indicated that awareness of olive oil in the marketplace is close to 100 percent – the highest of all edible oils. However, awareness of specific types of olive oil is considerably lower. The findings also confirmed the increasing trend in household consumption in recent years, as 77.1 percent of respondents said they used olive oil. Of the oil users, 41.8 percent stated that their preferred oil in the household is virgin olive oil, although according to retail figures, the sales by volume of pure olive oil are higher than for extra virgin. A possible explanation for this apparent discrepancy is that extra-virgin olive oil is used by consumers in much smaller quantities than pure olive oil. Olive oil is not commonly bought once a week or once a fortnight, as with many other grocery products, but is purchased once a month or once every three months.

Respondents who had purchased olive oil were asked to indicate, from a list of reasons, which appealed to them when buying olive oil, and then to nominate the main reason for buying olive oil. The survey's results are shown in Table 16.8.

Respondents perceived health benefits as the major influence on their purchase decision. Other significant factors were taste, and ideas obtained from recipes. More than 57 percent of consumers had a basic understanding of the health benefits of olive oil, even though they could not always explain why this was the case, and more than 37 percent had become accustomed to its distinctive taste. This suggests that olive oil can be easily differentiated from competing oils, as consumers are aware of its unique characteristics. The qualitative research found that consumers who buy olive oil because of its health benefits later become accustomed to its flavor. The fact that consumers perceived olive oil as "healthy" provides the base for a promotional campaign emphasizing such benefits and possibly explaining the reasons for them.

Table 16.8: Reasons for buying olive oil (McEvoy and Gomez, 1999)

Reason	All reasons for buying	Main reasons for buying
Taste	372	164
Health benefits	574	428
Trendy image	19	2
Price	64	15
Attractive packaging	18	2
Promotion	31	6
Recipe ideas	290	122
Other	31	25
Don't know	1	7

It is interesting to note that most consumers rated other attributes above price in their purchasing decisions. It could be inferred from these responses that consumers are in general comfortable with the current price of olive oil.

Although olive oil consumption is common to all demographic groups in Australia, the analysis of the survey has provided a profile of a specific market segment that could be targeted by the Australian industry. According to the survey findings, a typical olive oil consumer is aged 25–55, highly educated, has a skilled profession and a household income of more than $45,000, and is possibly (but not necessarily) of European origin. This profile could be the basis for the promotional and market distribution strategies that need to be developed by the Australian olive industry. It should be noted that since consumers meeting the target profile identified in this research have a strict notion of quality, special consideration should be given to the design of labeling and packaging.

16.4 The future of olive oil marketing

After reviewing a significant number of studies carried out in the field of olive oil marketing, the potential strategies that might be implemented when choosing a marketing campaign started to become clear. The main attributes investigated by each researcher, suggesting the main criteria that affect consumers' opinion regarding olive oil are summarized in Table 16.9. It is notable, as suggested in earlier research, that the cultural differences are linked to different notions of desired attributes, and sometimes lead to diverse valuation of the same property.

16.5 Factors affecting consumer preferences
Quality

Quality is the main attribute consumers ascribe to olive oil. Today's increased knowledge of the healthy components that comprise the oil is the main reason customers purchase it. In

Table 16.9: Overall attributes previously investigated in research into olive oil marketing

	Greece	Spain	Italy	France	Australia
Label	X	X			X
Certification	X	X	X		
Health information	X	X		X	X
Origin	X	X	X		X
Price	X	X	X	X	X
Taste and/or odor		X		X	X
Appearance of oil			X	X	
Package	X	X		X	X
Image				X	X

countries newly introduced to olive oil, as in the case of Australia, information regarding its health benefits should be mentioned on the label, or be a leading message in promotional campaigns. On the other hand, quality certification at this stage would have less impact than in more traditional countries, where a sign of quality is sometimes obligatory. In traditional olive oil countries, for example Greece, stating the obvious in regard to health benefits may lead to the opposite effect. International certifications of quality, such as ISO, HACCP, PDO, PGI, have a huge role in traditional countries. Consumers and experts in Greece, Spain and Italy stated that quality assurance was a highly desired attribute of the oil. Organic oil was found to have a positive influence on customer's decisions, although, as in the case of Italy, preferences changed throughout a country. In addition, a distinct preference was shown for PDO/PGI certification over organic produce.

Country of origin

In traditional countries, the origin of the oil was a major factor affecting consumers' decisions. Many consumers thought the origin of the oil should be mentioned on the label. This suggests a degree of awareness of the effect of origin. The research carried out in Spain suggested the origin of the oil as being the main attribute that differentiates known high-quality oils from one another. However, in some cases the origin of the oil may have a negative effect, as in Italy, owing to die-hard prejudice. In Australia, which is considered to be a non-traditional olive country, European oils are considered to be more desirable, partly due to a perception of the better quality of imported products.

Price

All the studies dealt with the issue of olive oil and its price. Olive oil is known to be more expensive than other vegetable oils. Surprisingly, even though consumers found the price of olive oil to be high, they felt it did not affect their choice of whether to purchase the oil, suggesting a certain degree of comfort regarding the cost. This may be due to the fact that consumers sometimes connect high costs with quality. A portion of the market believes that by paying more money they achieve better quality. From this point of view, selling olive oil for a reasonably higher price may have a positive influence on sales. However, the price of olive oil may discourage unfamiliar buyers or lead them to buying a lesser-quality olive oil, as in the case of Australia or France.

Organoleptic properties

Taste, odor and appearance are typical characteristics of olive oil, though not always determinant criteria for buying it. French consumers, for instance, think olive oil improves the food's taste, but when asked to rank how important the taste is to them the results indicated that taste was the least significant factor. On the other hand, the most important factor to French consumers was the color of the oil. Australian consumers, unlike the French, thought the oil's taste was the main reason for purchasing it. In more traditional olive countries these characteristics are well known to consumers, and are thus not always relevant in marketing.

Package

Greek and Spanish consumers and experts gave high significance to the oil's packaging, with clear tendencies towards glass bottles. French and Australian consumers implied that packaging was not highly important to them.

Image

This criterion was researched only in French and Australian consumers, who exhibited opposite results. While French consumers felt that the image of the company selling the oil was extremely important, Australian consumers thought that a trendy image was one of the least important factors. According to these findings, it is feasible to market a specific olive oil to French consumers by focusing on the company's image.

References

Anania, G., & Pupo D'Andrea, M. R. (2007). The global market for olive oil: actors, trends, policies, prospects and research needs. 103rd EAAE Seminar on Adding Value to the Agro-Food Supply Chain in the Future EuroMediterranean Space, Spain, 23–25 April 2007.

Caporale, G., Policastro, S., Carlucci, A., & Monteleone, E. (2006). Consumer expectations of sensory properties in virgin olive oils. *Food Qual Pref*, 17, 116–125.

Cicia, G., Del Giudice, T., & Scarpa, R. (2002). Consumers' perception of quality in organic food: a random utility model under preference heterogeneity and choice correlation from rank-orderings. *Br Food J*, 104(3/4/5), 200–213.

Fotopoulos, C., & Krystallis, A. (2003). Quality labels as a marketing advantage: the case of the "PDO Zagora" apples in the Greek market. *Eur J Marketing*, 37(10), 1350–1374.

Frohn, H. (1991). Food promotion and advertising in the single European market. In: *Proceedings of the 25th Seminar of the EAAE, 24–26 June*, pp. 343–356. Columbus, OH: EAAE.

Gundogan, M., Groves, G., & Key, J. M. (1996). Total quality management: a way towards total integration. *Total Qual Manag*, 7(4), 127–144.

Kohls, L. R., & Uhl, N. J. (1990). *Marketing of Agricultural Products* (7th edn.). Macmillan, New York, NY.

Krystallis, A. (2005). Emergence of quality conscious consumer segments: insights from the Greek olive oil market. In: P. Soldatos & S. Rozakis (eds.), *The Food Industry in Europe*. Athens: Erasmus Intensive Programme in Agri-business Management, pp. 61–74.

Lampkin, N. (1990). *Organic Farming*. Farming Press, Ipswich.

Madu, C. N., & Kuei, C. (1994). Strategic total quality management – transformation process overview. *Total Qual Manag*, 5, 255–266.

Martinez, I. J., & Quelch, A. J. (1991). David Del Curto S.A. In: A. J. Quelch, D. R. Buzzell, R. E. Salama (Eds.), *The Marketing Challenge of Europe 1992*. Addison-Wesley, Reading, MA.

McEvoy, D. G., & Gomez, E. E. (1999). *The Olive Industry: A Marketing Study*. A report for the Rural Industries Research and Development Corporation, Publication No. 99/86.

Mili, S., & Rodrigez-Zúñiga, M. (2001). Exploring future developments in international olive oil trade and marketing: a Spanish perspective. *Agribusiness*, 17(3), 397–415.

Nielsen, N. A., Bech-Larsen, T., & Grunert, K. G. (1998). Consumer purchase motives and product perceptions: a laddering study on vegetable oil in three countries. *Food Qual Pref*, 9(6), 455–466.

Ophuis, O. (1991). The importance of health and environment as product attributes for foods. In: *Proceedings of the 25th Seminar of the EAAE, 24–26 June*, pp. 295–306. Columbus, OH: EAAE.

Patsis, G. P. (1988). Publicity and prices of olive oil and its substitutes. *Olivae*, 21, 22–24.

Rajagopal, S., Balan, S., & Scheuing, E. E. (1995). Total quality management strategy: quick fix or sound sense? *Total Qual Manag*, 6, 335–344.

Roininen, K., Lahteenmaki, L., & Tuorila, H. (1999). Quantification of consumer attitudes to health and hedonic characteristics of foods. *Appetite*, 33, 71–88.

Scarpa, R., & Del Giudice, T. (2004). Market segmentation via mixed logit: extra-virgin olive oil in urban Italy. *J Agric Food Ind Org*, 2, 7.

Schifferstein, H. N. J., & Oude Ophuist, P. A. M. (1998). Health-related determinants of organic food consumption in the Netherlands. *Food Qual Pref*, 9(3), 119–133.

Senauer, B. (1989). *Major Consumer Trends Affecting the US Food System*. Staff paper, Department of Agricultural and Applied Economics, University of Minnesota, 89(16), 1–23.

Shepherd, R. (1999). Social determinants of food choice. *Proc Nutr Soc*, 58, 807–812.

Shepherd, R., & Farleigh, C. A. (1986). Preferences, attitudes and personality as determinants of salt intake. *Human Nutr Appl Nutr*, 40A, 195–208.

Siskos, Y., Matsatsinis, N. F., & Baourakis, G. (2001). Multicriteria analysis in agricultural marketing: the case of French olive oil market. *Eur J Oper Res*, 130, 315–331.

Tzouvelekas, V., Pantzios, C. J., & Fotopoulos, C. (2001). Technical efficiency of alternative farming systems: the case of Greek organic and conventional olive-growing farms. *Food Policy*, 26, 549–569.

Ward, R. W., Briz, J., & de Felipe, I. (2003). Competing supplies of olive oil in the German market: an application of multinomial logit models. *Agribusiness*, 19, 393–406.

CHAPTER 17

Developing the Negev Desert olive oil brand name

Having looked at the global olive oil economic and marketing data presented in previous chapters, there is no doubt that the chance of a profitable desert olive oil industry lies in branding of the product.

Based on the common olive oil trends noted by the various researchers, consumers mainly want healthy, well-flavored olive oils. Owing to the fact that competition in the olive oil market is becoming tougher each year, production of a unique, high-quality olive oil is very important in order to succeed.

In order to address this topic, we used the Negev Desert olive oil cultivation model, assuming that this model may later be adopted and modified for many other desert areas in which an advanced olive oil industry is now starting to develop.

Initially, we screened the main principles specifically apt to the development of a new Negev Desert olive oil brand name. Every brand name is based on a strong consumer cover story backed by solid core data and information.

The following approaches seem to be most suitable for development of brand names:

a. *The historical approach.* Olive oil is well known for having been cultivated in the Negev area since biblical times, when it was a highly important and well-recognized natural material. It was used in the coronation ceremonies of King David and King Solomon. It is believed that at least some genetic progeny of olive trees from ancient times are still living in the Middle Eastern region. Using advanced genetic markers to complete the genetic identification and fingerprinting of ancient olive trees remaining in this area, with an emphasis on the Judea desert area, may significantly contribute to a solid cover story that could be used for a new unique and attractive brand name. As described in Part I, the Nabatean people used to live in the Negev Desert area, where they developed a unique and very advanced method of agriculture. Archeological findings clearly show that the Nabateans used to cultivate grapes and olives, and established sophisticated industrial facilities in the Negev area. These findings are of great benefit in developing a new brand name, bearing in mind that the olive genetic material used by the Nabateans is indeed available and used today for olive oil cultivation in the area. Furthermore, a religious brand name, with an emphasis on the beginning of Christianity in Judea, is another excellent option for branding Negev Desert olive oil. Even now, there are some well-protected trees in areas near Jerusalem, mainly on Olive Mountain, that are considered to originate from the time of early Christianity. Other such trees are distributed throughout various dryland areas in the Middle East, and can also be used in developing a new local brand name.

b. *The environmental approach.* Generally, people today are becoming more aware of the contribution of agricultural products to global environmental issues. Desert olive oil production helps to combat global desertification, and may also contribute to reduction of the greenhouse effect caused by desertification in recent decades. Using drylands to increase agriculture production sounds beneficial to people. Of course, the desert aspect of olive oil production cannot stand alone, and will not convince consumers to choose this oil unless it is of high quality and has healthy properties as well (see below). However, as clearly demonstrated in various parts of the present book, isolated areas that are easily available in the Negev Desert provide good opportunities for the cultivation of olive oil either under fully organic farming systems or at least with significantly reduced synthetic pesticide use. These approaches are currently a good fit with the global trend of supplying of safer agricultural products, of significantly higher prices, throughout developed countries' markets.

c. *The health approach.* Owing to the unique desert environmental conditions of the Negev Desert area, with high temperatures and increased irradiation, low humidity, and irrigation with moderately saline water, the Negev Desert olive oils contain relatively high levels of antioxidant compounds. These compounds consist of polyphenols, tocopherols, phytosterols, etc. The production of these antioxidants is stimulated for protection of the triglyceride composition and integrity of the oil. Following many medicinal studies reported in recent years, the health properties of these naturally produced compounds are well known among most consumers all over the world. Moreover, the new findings presented in this book with regard to the unique TAG composition of oils from various olive varieties cultivated in the Negev Desert area clearly demonstrate the uniqueness of these oils. Providing clear evidence regarding the health properties of Negev Desert olive oil, combined with flavor and taste properties (see below), may increase attention to these oils.

d. *The flavor and taste approach.* Both genetic and environmental factors have an effect on aroma, as they control the volatiles and other taste-related compounds in olive oils. The high adaptability of a wide range of olive oil varieties to saline irrigation in Negev Desert farming enables the development of new blends of extra-virgin olive oils. The blending may yield new flavors and tastes that can be tailored to fit well with the consumers targeted by new Negev Desert olive oil brands. Industrial tailoring of the most favored desert olive oil flavor may take advantage of headspace technologies developed in recent years, and may carefully regulate and control the new blended olive oils produced in the Negev Desert area, so that they will be unique and differentiated from other olive oils and blends. For example, a unique strongly flavored Negev Desert olive oil blend may include Souri, Barnea, Picholin and Picual, whereas another unique but more moderately flavored oil could be obtained from a blend of Leccino, Koroneiki, Barnea and Arbequina.

e. *The peace approach.* The Negev Desert area, as with most drylands throughout the world, is characterized as a difficult region for agriculture and farming, and is therefore populated with relatively poor people who lack good sources of income. This causes many conflicts among people – the Middle East provides a good example. There is the opportunity to reduce tension between people by improving their economic status and directing their energy toward olive oil production and additional industrial by-products. Branding of high-quality Negev Desert olive oil may attract the attention of consumers who are open to the political situation in the Middle East. If such a branded olive oil, with healthy properties combined with environmentally friendly production, is released to some targeted markets, there is a good chance of positive response from consumers. Indeed, some initiatives in developing Middle Eastern locally branded olive oils are already in progress, with the involvement of some Bedouin, Palestinian, Israeli and Jordanian producers of olive oil. This local "Peace Olive Oil" is mostly based on Nabali, Souri and some additional milder olive oils.

In summary, it is very important to obtain a full and clear ID for each brand of Negev Desert olive oil, including the following information:

- the chemical profile, based on IOC standards;

- the organoleptic profile, based on IOC standards;

- a specific cover story, based on solid data and information sources;

- definition of the segment of the market to which the brand is targeted;

- definition of the features that differentiate each new brand from those of competitors.

Part 4
Summary

CHAPTER 18

The future of desert olive oil, and concluding remarks

Owing to its unique health-promoting activity and its nature as a major component of the Mediterranean diet, the consumption of olive oil has increased tremendously in recent years. The awareness of health and its connection to olive oil is not confined to the West; it has also slowly reached other parts of the globe, including some of the most populated countries in Asia, such as China and India. Furthermore, as the economy of these countries has grown rapidly, there has been a corresponding increasing international demand for olive oil and other health foods. In the case of olive oil, this trend is clear from data showing higher figures for consumption than for production in the past three years. At the global level, the increase in production of olive oil is much lower than the increase in consumption; this means that there is a higher probability of increased prices in the near future. The current trends in the global olive oil industry can be summarized as follows:

- there is an increased global demand;

- there is greater appreciation in non-traditional countries;

- domestic consumption is remaining high in all major producing countries;

- small growers are becoming organized into cooperative organizations to enhance production and marketing;

- high proportions of the products are being marketed directly, and are more closely related to rural tourism;

- both investment and production in Mediterranean countries are increasing;

- advances are being made in production and processing technologies, including cropping patterns, harvesting, and oil extraction.

This scenario leads to the potential for reduction of consumption in developed countries, mainly the USA and EU, and a move to other vegetable oils, if the production of olive oil does not expand to meet the demand. It is therefore urgent that the global production of olive oils is increased.

In the meantime, it is clear that in the past few decades expansion of the land under olive culti-vation has increased beyond the non-traditional areas – i.e., to areas outside the Mediterranean basin. These new areas are mainly located in the USA, South America and Australia, as well as in India, China, South Africa and the Middle East. These new olive-growing regions mainly lie in comparatively arid deserts and drylands. Olives have been growing quite successfully in these new areas, and there are plenty of opportunities for future olive cultivation in the form of both natural and human resources. This indicates that the cultivation trend in these areas will continue more rapidly in the future.

There is no doubt that, in these circumstances, one of the best ways to increase production is to grow olives in the marginal lands or environmentally challenging areas where other crops – mainly cereals – cannot be grown. This indicates the strong prospects for olive oil cultivation in deserts. Such production will have some very interesting consequences; in fact, an increase in income can be linked with an increase in food consumption and an increase in exports to the global olive market.

Furthermore, the increasingly global and demand-driven economic scenario now offers new challenges and opportunities for all the agricultural and food sectors, including the olive oil sector. To succeed in achieving a sustainable competitive advantage in production, organizational and commercial strategies play a crucial role. In this sense, olive oil cultivation in desert land is clearly appropriate. Successful production of olive oil in desert and/or new dryland areas will certainly fulfill the consumers' demand for olive oil without any reduction in quality. To achieve this, greater focus should be paid not only to selecting the best site and growing the plants, but also to the extraction, storage and delivery of the oil, while maintaining a quality standard.

Though olives have been growing in dryland or desert areas for thousands of years, the cultivation of olives in a systematic manner, with intensive practices, on a commercial scale using modern technological know-how, is a new strategy. Like other industries that need long-term investment, judicious decisions are very important for the olive industry as well. Cultivation of agricultural plants, especially fruit trees, in the desert or an economically challenged area may mean many constraints for the investors, especially in the early stages of the project. Although traditionally olive plants are able to bear fruits for hundreds of years, in modern intensive cultivation cost-effective management decisions should be followed. Generally, an olive orchard can be expected to last for 40–50 years; after that, it is suggested that, even though the trees may continue to bear fruit, renovation is necessary in order for the project to remain economically viable.

The decade-long experiment carried out by the team at Ben-Gurion University in the Negev Desert of Israel, and research performed elsewhere by other scientists as mentioned in the earlier chapters of this book, have led to the conclusion that, by using a proper environmentally friendly agro-biotechnologies, most of the prevailing olive cultivars can be successfully grown in desert conditions, even with moderately saline irrigation (i.e., EC 4.2 dS/m). Some varieties, such as Koreineki, Arbequina and Picual, even show better performance with moderately saline irrigation than with fresh water irrigation (i.e., EC 1.2 dS/m). However, based on experience and close observation of the flowering, maturity phenomenon and uniqueness of the individual cultivars for their flavor and taste, a well-balanced mix of olive varieties, including selected local cultivars, traditional Mediterranean varieties and newly introduced varieties from Central Asia, is recommended. This is because past experience has shown that although most cultivars will set some fruit in mono-cultivar cultivation, they benefit greatly from cross-pollination. When selecting the varieties, criteria such as the productivity per unit area, chemical and organoleptic tastes, maturation and ripening time, fitness for mechanical harvesting and storability quality, susceptibility to pests and diseases, and tendency to alternate bearing should be taken into consideration. For this, the present book will be a valuable asset.

One of the limiting factors for growing olives in desert conditions is the lack of fresh water for irrigation. Most drylands have a scarcity of fresh water and in the majority of cases the main available source of water is underground saline or brackish water. As discussed in Chapter 5, a special design and planning are necessary for saline irrigation, because the use of such irrigation is the main difference between traditional and desert cultivation. Saline water is generally found underground, so pumping and mixing systems, reservoirs and delivery infrastructure are needed. Furthermore, pumping from generally very deep aquifers costs a huge sum of money, so government or public institution support is very important unless the plantation is run by a large commercial firm. Another significant point regarding saline water irrigation is the practice of leaching to avoid excessive salt in the root zone. Evidence has clearly shown that even though the olive is considered to be a moderately salt-tolerant plant, long-term irrigation with saline water of more than 3 dS/m leads to the need for leaching. Furthermore, the irrigation schedule requires a modern irrigation system: surface or sub-surface drip irrigation is mandatory for saline water irrigation in desert conditions, and conventional irrigation cannot be utilized.

Another peculiarity of desert cultivation is its relative isolation. This means that much of the necessary infrastructure is frequently not available on site, so additional government and/or public support may be needed to develop the fundamental construction. From a cost-effective point of view, desert olive farms should be at least 100 ha in area.

As quality and/or purity are the prime concerns for the olive oil industry, desert olive oil producers should not have to compromise on anything that will affect the oil quality. As mentioned in Chapter 11, the quality of olive oil is dependent not only on genetic factors and agro-ecological conditions, but also on other factors such as harvesting, extraction, delivery, and storage conditions. In order to maintain production of high-quality oil, improved technical facilities should be established. A good di-phase olive mill is recommended. An environmentally sound facility for oil extraction, decanting, bottling and storage should also be built, and a small quality-analysis laboratory established so that basic quality parameters can be checked in each batch of oil.

The basic oil quality parameters can be checked by growers with a small amount of know-how. For detailed quality analysis, growers can send an oil sample to the nearest IOC-approved laboratory. In order to sell olive oil in the international market, and in some cases even in the domestic market, the oil must pass IOC tests according to the grade. Besides the quality of the oil, proper management of waste produced from the olive mill should be a high priority as an environmentally friendly production principle.

Regarding the quality, desert-cultivated oil has shown some peculiar characteristics – high levels of antioxidants (mainly polyphenols, tocopherols and phytosterols) and these qualitative attributes have put desert-cultivated olive oil in a good position for marketing. Interestingly, in desert-cultivated olive oil even high levels of polyphenols and other antioxidants did not affect the organoleptic qualities of the oil, including taste and aroma. Generally, olive oils produced in traditional areas taste bitter if they have a high polyphenol content. Other quality parameters, such as free fatty acids and peroxide values, of most desert-produced olive oils are within the range for extra-virgin grading. This further emphasizes that there is no doubt regarding the quality aspects of desert-cultivated oil.

As in many other different agricultural products, high-quality olive oil produced to organic standards is very desirable in the international market and among Western consumers. This is the reason why the number of organic olive growers is increasing. Although it is difficult to grow plants according to the principles of organic culture, it is easier, especially under desert conditions, for olives than for many other agricultural products. The Neot Smadar Kibbutz in Israel is a good example. One of the reasons it is comparatively easy to maintain an organic farm in the desert is the isolated area; this isolation leads to a comparatively low incidence of pests and diseases.

This book describes two case studies of olive farming in the desert: Re'em Farm provides a model of a conventional intensive farm (see Chapter 13), while Neot Smadar provides a good example of an organic farm (see Chapter 14). Re'em Farm not only produces different varieties of olive oils and table olives in an intensive manner; it also incorporates fish-culture into the olive plantation system. The incorporation of fish-culture into olive farming not only provides a supplemental income that increases economic feasibility; it also helps to establish geothermal groundwater by passing it through the fish farm runways and into a storage water reservoir (Figure 13.2). This system also helps to maintain the irrigation water at a desirable temperature. Neot Smadar utilizes the wastewater by treating it in a special wetland construction site consisting of five sequential pools planted with a diverse assortment of water plants (Figure 14.3). This also provides the opportunity to treat the municipal wastewater and utilize it in agriculture, while also helping to protect the environment.

Olive oil produced in desert conditions, using special methods including saline water irrigation and other cultural practices, may enable the establishment of a specialist olive oil brand that can also be blended if required. Today's consumers are looking for diverse specialties, but do not want to compromise on quality. Here, there is a good opportunity to produce a specialist organic desert brand, to IOC standards, that may ultimately attract a large number of international consumers – especially those who are well aware of the quality and taste implications of the oil.

As well as describing the special techniques used for olive oil cultivation in desert conditions, this book identifies specific biotechnologies for olive oil production that can be effectively employed not only in desert areas but also in other regions, including the main traditional areas for olive production. Among these technologies, the following are novel:

- reduction of alternate bearing and increase of olive oil yield by supplemental foliar nutrition;

- monitoring of root zone soil buffering using the minirhizotron system;

- prediction of the optimal olive harvesting date by a chemo-optic system;

- increasing the efficiency of mechanical harvesting by using effective control release formulation loosening agents;

- increasing olive oil extraction by using proper extraction techniques, including lipase enzymatic production;

- utilization of solid waste for growth stimulation of field crops, and for fermentation of PGRs.

To date, most olive plantations in dryland areas have been established in comparatively developed countries. However, most desert areas are found in the poorer, developing countries of the world. Technologies for desert oil cultivation have not yet reached these dryland areas, so there is a strong need to deliver them. Olive plantations on these poor dryland areas may not only provide an essential and healthy source of edible fat for the local people, but also allow them to earn currency by exporting to developed countries via international markets. Moreover, increasing income and employment for local people may contribute significantly to a reduction in violence in such regions. Desert olive oil cultivation may also contribute to the urgent global initiative of the UN to combat rapid climate change, with a focus on limiting desertification in developing countries. Olive oil cultivation cannot, of course, solve this problem by itself, but it will make a positive contribution towards the global environmental challenge. Desertification is a problem for human beings, too; more and more arable lands are becoming degraded and useless for the production of the common and/or traditional agricultural crops that have grown there for hundreds of years, and in this scenario crops that can be produced in comparatively drier or desert areas will play a vital role in the future of agriculture. Cultivation of desert olive oil will gradually increase the supply to the market, which is important because the worldwide demand for olive oil is expanding rapidly; this is mainly because of its health implications and the numbers of new consumers in both developed and developing countries where olive oil has not traditionally been consumed. The consumption in traditional countries will remain at current levels or may even grow further. The prospects for the cultivation of desert olive oil are therefore very good, and this book may have a role in encouraging such cultivation.

This book also highlights the comprehensive protocol for the determination of triacylglycerol and/or fatty acid profiles by a novel approach using MALDI-TOF mass spectrometry. MALDI-TOF/MS methodology is a rapid, easy and reliable method for determining not only intact TAG composition but also the fatty acid profiles of oil, without requiring any derivatization. There has been a long-standing need for such a technique, because conventional methods of determining these parameters using various chromatographic tools such as thin layer chromatography (TLC), high-pressure liquid chromatography (HPLC), and gas chromatography (GC) are time-consuming and tedious.

Index

Printed and bound by CPI Group (UK) Ltd, Croydon, CR0 4YY

03/10/2024

01040313-0002